Statistical Image Processing Techniques for Noisy Images

An Application-Oriented Approach

Statistical Image Processing Techniques for Noisy Images

An Application-Oriented Approach

François Goudail and Philippe Réfrégier
Fresnel Institute, ENSPM
Marseille, France

Springer Science+Business Media, LLC

Library of Congress Cataloging-in-Publication Data

Goudail, François.
    Statistical image processing for noisy images/an application-oriented approach/
François Goudail and Philippe Réfrégier.
      p. cm.
    Includes bibliographical references and index.
    ISBN 978-1-4613-4692-0   ISBN 978-1-4419-8855-3 (eBook)
    DOI 10.1007/978-1-4419-8855-3
      1. Image processing—Statistical methods. I. Réfrégier, Philippe. II. Title.

TA1637.G69 2003
621.36'7—dc22

                    2003058900

ISBN 978-1-4613-4692-0

©2004 Springer Science+Business Media New York
Originally published by Kluwer Academic/Plenum Publishers, New York 2004

http://www.wkap.nl

10 9 8 7 6 5 4 3 2 1

A C.I.P. record for this book is available from the Library of Congress

Contents

Preface

We are pleased to have the opportunity to thank colleagues from different organizations whose contributions have been essential for this book.

First of all we would like to thank our colleagues who have been Ph.D. students: Stéphanie Cabanillas, Christophe Chesnaud, Jean Figue, Frédéric Galland, Olivier Germain, Frédéric Guérault, Vincent Laude, Vincent Pagé, and Olivier Ruch. We also thank Rafal Kotynski and Henrik Sjöberg, research visitors in our laboratory, for their contributions. Many shorter-term students have also collaborated with us. They are too numerous to be cited, but their contribution is acknowledged. This book has been written in the Physics and Image Processing research group; we would like to acknowledge Nicolas Bertaux and Muriel Roche for their support and the very pleasant ambiance within the team which was very helpful in writing this book.

The material presented in this book has been inspired by the research work of the authors. One of the authors has been an engineer in the Central Research Laboratories of Thomson-CSF, then a professor in the Signal and Image Laboratory of the École Nationale Supérieure de Physique de Marseille and then in the Physics and Image Processing Group of the Fresnel Institute. We would like to thank all our colleagues whose discussions on the topics of this book have been fruitful and who have created a productive environment for research.

This research has been conducted in collaboration with many institutional partners in France. We acknowledge our colleagues of the Fresnel Institute and of the Ecole Nationale Supérieure de Physique de Marseille, the GDR ISIS, organized by CNRS and the French Optical Society. We also acknowledge the CNES, the DGA, and colleagues from different companies and especially from Thalès.

FRANÇOIS GOUDAIL AND PHILIPPE RÉFRÉGIER

Statistical Image Processing Techniques for Noisy Images

An Application-Oriented Approach

Chapter 1

INTRODUCTION

1.1. General introduction

Digital images are now present in many human activities. One can cite for example medical diagnostic, artificial vision for automatic inspection, remote sensing for earth monitoring, robotic applications, telecommunications, scientific instrumentation for astronomy or biological research, military optoelectronic, and radar surveillance systems. The tremendous development of these fields of application requires the design of appropriate digital image processing techniques since the amount and the rate of data are often incompatible with human intervention. Many different reasons can be put forward in order to explain this situation. Two of them are of prime importance for the subject of this book: the progress, the diversity, and the diffusion of different imaging sensors and the increase of the computational power of digital processors.

Due to the importance of the digital image processing applications, many different approaches have been intensively investigated in the last thirty years. Image processing is thus a very wide and rapidly evolving research field which becomes more and more difficult to delimit. As a few examples of the different approaches, one can cite geometry-based techniques, multi-resolution and wavelet methods, functional analysis approaches based on partial differential equations, and neural network-based methods.

Among these different image processing techniques, statistical methods have been thoroughly investigated and have led to innovative and powerful techniques. The main advantage of these methods lies in their ability to efficiently take into account the nature of the different fluctuations and noises present in the images. Furthermore, they often make it possible to define a rigorous methodology for addressing a given signal or image processing problem. The

1

first goal of this book is to introduce statistical image processing techniques, emphasizing two characteristics: noise robustness and rigorous methodology.

Another important challenge in the design of statistical processing techniques is to determine a good trade-off between realistic image models and simple and fast processing methods. Since the goal is to determine efficient processing techniques, data modeling does not have to be exact but has to include the main characteristics which allow one to get satisfactory results. This is also a goal of this book to illustrate the methodology which can lead to such trade-offs.

The objective of the authors has not been to survey all the recently developed statistical algorithms and techniques but rather to analyze simple methods in deeper detail and to illustrate their capacities with application examples. They have thus chosen to describe a methodology and to discuss techniques which are directly related to their recent research works. This approach gave them the opportunity to present in-depth description of the processing algorithms and to describe many concrete applications. We believe and hope that this approach will allow the reader to follow more clearly the coherence of the proposed methodology.

1.2.   Image processing tasks

There exist many different types of information that can be extracted from an image depending on the application or, in other words, on the "task" which has to be performed with this image. In the course of this book, several image processing tasks will be addressed (see Figure 1.1). In order to be concrete in this introduction, we propose to briefly describe some of them, which will be the subject of more precise analysis in the main part of the text.

Object detection.   This task consists in deciding if an object is present or absent in an image. The information that is extracted from the image during this task is thus binary, since it consists in choosing between two hypotheses: target present (Hypothesis H_1) or target absent (Hypothesis H_0).

Parameter estimation.   One of the most classical examples of such task is object location. In this case, the information extracted from the image is the position of the object in the scene, which is defined by two-dimensional coordinates (τ_x, τ_y).

The problem of target location can be generalized to the estimation of other geometrical parameters of the object. For example, attitude estimation consists in estimating the orientation of the object in the 2D space (it is then defined by a single angle θ) or in the 3D space (it is then defined by two angles). The information that is extracted from the image during this task is in general of continuous nature, since it consists in estimating real-valued parameters.

Object segmentation. In this task, the information extracted from the image is the shape of the object of interest. This shape can be defined by the set of image pixels which are spanned by the object. However, if the object is large, it is more efficient to describe its shape by its contour, that is, the border of the object region. Different curve models can be used to represent this contour. For example, it can be represented by a polygon, which is then defined by the position of its nodes.

Image partition. In some applications, it is necessary to partition the whole image in regions which share similar properties. These properties can be the average gray level, the texture, or other features. Here, the information which is extracted from the image is its complete partition in homogeneous regions.

Object recognition. The primary objective of many image processing systems is to recognize the nature of the object(s) present in an image, i.e., is it a car or a truck, or what type of airplane is it ? In order to be exploitable by signal and image processing techniques, such applications are usually defined as the problem of assigning the object of interest to one among K possible classes $C_k, k \in [1, K]$. The information extracted from the image in this case is thus the index of the class to which the object belongs.

Of course, in a global image processing system, several of these tasks may have to be performed jointly. For example, when detecting an object in an image, one usually does not know its position. The task of detection must thus be performed at each possible location in the image, which results in a compound detection/location problem. As another example, prior to being recognized, the object may have to be detected, localized, and often segmented in order to estimate its shape.

Although these tasks are interdependent in most practical image processing problems, it appears fruitful to first study them separately. This approach will allow us to precisely assess their difficulties as well as the properties and potential performances of the algorithms designed to solve them. The statistical theory of detection and estimation provides a powerful framework to define the interrelations between these tasks in a rigorous way with the concept of nuisance parameters. Of course, if the optimal solution obtained with this approach is too time consuming, heuristic solutions can be designed by using as building blocks the algorithms defined for each subtask. It is clear that such heuristic solutions will be all the more efficient so that these building blocks have been separately optimized and carefully characterized. This is the methodology we will follow in this book.

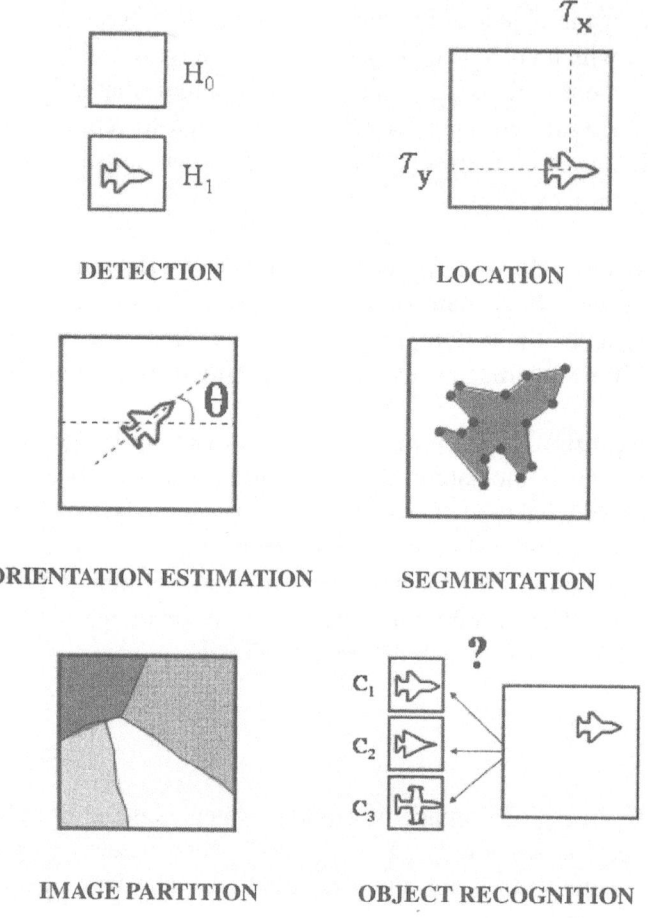

Figure 1.1. Examples of image processing tasks.

1.3. Statistical decision and estimation theory

Statistical decision and estimation theory is a theoretical framework which permits one to handle rigorously the problem of information extraction from a signal or an image. Within the framework of this theory, the problem to be solved is generally composed of the three following elements:

- The *image model* describes the observed image in terms of the parameters of interest – which represent the information to be extracted from the image – and in terms of the nuisance parameters – which represent information present in the image but which is of no interest. For example, in a localization application, the image model may consist of the shape of the object to be located, the parameter of interest is the position of the object, and the orientation of the object may be a nuisance parameter.

- The *noise model* is a probabilistic representation of the perturbations of the observed image with respect to the ideal image model. The nature of this noise depends on the imaging system. For example, in a Synthetic Aperture Radar (SAR) image, it follows from physical considerations that Gamma laws are among the simplest models for the graylevel fluctuations. On the other hand, in images formed with a very low flux of photons, the fluctuations will rather be distributed with a Poisson law. Since in general, the precise values of the parameters of the probability laws which describe these fluctuations are unknown, they can be considered as nuisance parameters.

- The *performance criterion* quantifies the quality of the information extracted from an image by a given processing algorithm. For example, in detection applications, the performance criterion can be the probability of detection for a given false alarm rate. In a location application, it can be the mean square error between the true position and the estimated one.

A goal of statistical decision and estimation theory is to define reasonable performance criteria and to provide tools for designing algorithms that optimize them. Indeed, one of the main advantages of this theory is to define precisely an image processing problem in terms of information extraction, that is, what is the image model, what is the information to be extracted, what are the perturbations that affect the measures, what are the nuisance parameters, and what is the performance criterion that the processing algorithm optimizes.

1.4. An application-oriented approach

The main focus of the present book is on the applications of decision and estimation theory to practical image processing problems. We will address the problems of target detection and location, which will be illustrated on practical tracking problems in video images. We will also discuss edge detection in SAR images, shape segmentation and image partition in video and SAR images. In general, we have interleaved in each chapter the definitions of theoretical concepts and their illustration with real-world applications.

Furthermore, in Chapter 7, we will illustrate the previously described techniques and methodologies on a specific example (i.e., polarimetric images). This extended example will allow us to show how the field of application of these techniques can be extended and how the obtained performance can be precisely characterized.

Some criticisms are traditionally made when it comes to applying statistical approaches to practical image processing problems. The first one is often to claim that the image models that can be handled by this approach are too simplistic, and do not fully represent the real image, especially in terms of noise modeling. On the other hand, if one tries to make the models more complex in order to be closer to reality, the optimal solutions become rapidly too computationally intensive to be implemented with the limited resources that are often available in an image processing system. This is indeed the main problem encountered when designing image processing algorithms based on statistical decision and estimation theory: to consider models sufficiently complex to accurately represent real perturbations, but sufficiently simple to lead to computationally efficient implementations.

One of the main goals of this book will be to show that simple models can lead to algorithms which are rapid and efficient on real-world images.

1.5.   Outline of the book

Chapter 2 is devoted to the design of correlation filters that perform an optimal trade-off between several heuristic performance criteria, and that can solve such problems as parameter estimation and object recognition. In this approach, the structure of the processing algorithm, i.e., correlation, is imposed *a priori*, but it will be shown in the following chapters that they are optimal in the presence of additive Gaussian noise. We also address in this chapter the important notion of stability of filtering techniques. Indeed, images are complex signals which are not easy to model accurately, and it is often necessary to use filters that are robust to variations of the object or of the noise parameters with respect to the model. This concept of stability is very general and will have to be kept in mind when analyzing the techniques discussed in the following chapters. Finally, the chapter is concluded by an application of the correlation methods to the estimation of airplane attitude in an image.

In Chapter 3, we analyze the noise conditions for which the heuristic filters derived in Chapter 2 may not be efficient. The basics of statistical decision and estimation theory, which are privileged candidates to solve these problems, are then presented. In the sequel of the chapter, examples of application of this theory to the design of object location algorithms in the presence of different types of noises are presented. The classical case of additive Gaussian noise with known spectral density is first analyzed. The case of additive Gaussian noise with unknown power spectral density is then addressed. Finally, the important case of nonoverlapping Gaussian noise is studied in detail. It is shown that statistical decision and estimation theory leads to different algorithms in function of the *a priori* knowledge which is assumed on the target graylevel fluctuations.

In Chapter 4, the case of nonoverlapping noise is presented in a more general context. We define the Statistically Independent Region (SIR) image model which accounts for nonoverlapping noise and random target texture in the presence of different types of noise. By varying its parameters, this model can be used to solve the image processing tasks listed in Section 1.2. It is also shown that if the noise probability density function (pdf) belongs to the exponential family, the resulting processing algorithms can be made computationally efficient and quite fast. Three examples of application of this approach to practical object location and edge detection/location problems are then discussed for correlated Gaussian noise, for binary noise, and for SAR images perturbed with Gamma noise.

Chapter 5 is devoted to object segmentation. It is shown that a statistical active contour – or snake – algorithm can be obtained from the SIR image model. A polygonal model for contour description is then used and the method is developed when the noise pdf belongs to the exponential family. This case is of theoretical as well as practical importance since a fast snake optimization algorithm can be determined rigorously in simple cases. These techniques are illustrated on practical applications to object tracking in video image sequences and to edge location accuracy improvement in SAR images.

Chapter 6 presents some improvements and generalizations of statistical active contours based on the SIR model. This method is first generalized to vectorial images and is illustrated with the segmentation of multidate SAR images. The second improvement concerns the estimation of the number of nodes of the polygonal snake. It is shown that this approach permits one to simplify the description of the shape and to improve the segmentation accuracy without introducing *ad hoc* parameters as is the case with other regularization or stabilization techniques. Finally, the snake technique is generalized to a statistical grid which permits one to perform a partition of the whole image into homogeneous regions. The efficiency of this algorithm is demonstrated on applications to target segmentation over nonhomogeneous backgrounds and to the segmentation of agricultural fields in SAR images.

We show in Chapter 7 how the techniques presented in the previous chapters can be applied in a nonstandard physical imagery system. More precisely, the statistical image processing algorithms designed throughout the book are illustrated on coherent active polarimetric images. Such images represent the polarization state of the light coming from each point of a scene. Since the polarization pdf does not belong in general to the exponential family, the previously described techniques have to be adapted. It is thus shown how the data can be transformed so that the algorithms developed in the previous chapters can be used with near-optimal performance. Second, we show how a general method for algorithm performance evaluation can be implemented in this complex situation by the precise definition of contrast parameters. These concepts are illustrated on target detection and object segmentation problems.

Chapter 2

LINEAR FILTERS: HEURISTIC THEORY AND STABILITY

Linear filters are among the most widely used tools in image processing for such applications as target detection, localization and classification (or recognition). They are based on correlations between the searched objects (or a linear combination of them) and the analyzed scene. The correlation operation has several advantages in the context of image processing. First, it is an intrinsically position invariant recognition method which makes it possible to recognize an object whatever its location in the image. Second, correlation is quite robust to noise, and thus often constitutes an efficient way of processing very noisy images. Finally, it can be computed at a relatively low computational cost if fast methods are available, based for example on Fast Fourier Transform (FFT).

The main drawback of linear filters in image processing is that they are not robust to such variations of the object shape as rotation or scale changes. This is why many different solutions have been proposed for filter design [1-14] in the last thirty years. For example, when invariant or tolerant pattern recognition is required, specific methods for improving recognition capabilities have been introduced. When distortions cannot be described analytically, a supervised learning method can be used in which the filter is determined by using training images that are sufficiently descriptive of the expected distortions. A well known example of such filters are the Synthetic Discriminant Function (SDF) filters and their different variations [15-17].

Efficient linear filters are obtained through the optimization of several criteria, such as noise robustness, discrimination ability, and robustness to shape variations. These criteria are most of the time antagonist, which means that increasing one of them will decrease the others. For example, increasing the discrimination ability of a filter to a nontarget object will generally decrease the robustness to shape variations of the objects of interest. In order to synthesize efficient filters, it is thus of prime importance to determine a good trade-off between these different criteria, using for example an Optimal Trade-off (OT) approach [17].

The criteria which are optimized during filter design usually take into account a limited number of expected distortions. For example, the optimization of noise robustness (which corresponds to the maximization of the signal-to-noise ratio) is dependent on a noise model. A filter optimal in this context can be greatly suboptimal for other types of distortions or noise models [18]. Filters whose performance sharply decreases with a slight variation from the conditions in which they are optimal are termed *unstable* [19, 20]. Images are very complex signals, which are very difficult to model accurately, and stability is thus a highly desirable feature for image processing filters. This is why we will analyze precisely in this chapter different solutions to stabilize – or *regularize* – filtering methods.

We will begin this chapter in Section 2.1 with a general discussion of the different approaches to the design of image processing filters: the empirical, heuristic, and decision theoretic methods. We will then expose in Section 2.2 different methods for designing linear filters from heuristic criteria. Section 2.3 will be devoted to the evaluation of the stability of these filters and Section 2.4 to regularization methods aimed at improving this stability. We will conclude this chapter in Section 2.5 with an application of SDF filters to the estimation of in-plane rotation of objects.

2.1.   The different approaches to filter design

We briefly discuss in this section the different approaches that can be used to design image processing filters.

With the **empirical approach**, the mathematical expression of the filter is chosen *a priori* without optimality consideration. This approach can provide powerful techniques capable of solving problems unsolved with the known optimal methods. This has been the case for example with Nonlinear Joint-Transform Correlators (NLJTCs) proposed in Refs. 21 and 22. These processors were not first defined by the optimization of some criteria, but they have been shown to be attractive in many difficult instances because of their high discriminating performance [22-24]. However, the empirical approach is far from being satisfactory since it does not elucidate in which context the considered method is superior to others. Systematic simulations and experiments are necessary to answer this question. However, in the field of image processing, such an answer is hard to obtain due to the huge variety of input scenes that can occur and to the large number of existing processing techniques which all have their own parameters. This approach may thus not be of particular interest for efficient filter design.

With the **heuristic approach**, the filtering method is obtained by optimizing heuristic criteria over an *a priori* chosen class of algorithms (in general, linear filters). This is a powerful and versatile technique which has been the subject of deep investigations [3, 16, 17, 25, 26]. Furthermore, with this approach, it is quite easy to determine filters that are invariant to some expected distortions of the observed scene compared to the reference objects [4,15-17,27,28]. It is also possible to determine filters which perform a trade-off between several antagonist criteria [17]. The present chapter will be devoted to the heuristic approach: We will explain its main results and illustrate them with applicative examples.

The main limitation of the heuristic approach is that the criteria are in general chosen *a priori* by the designer, as well as the structure of the algorithms. Of course, there have been many discussions in the scientific community about the different criteria [28], but none of them has been shown to be adapted to all the different situations which can occur in image processing. In particular, as far as noise robustness is concerned, all classical criteria are adapted to an additive noise with known power spectral density and to a target with known gray levels. The third approach to the design of image processing algorithms is based on **statistical decision and estimation theory** and makes it possible to go beyond these limitations. With this approach, the structure of the filter is no longer *a priori* imposed, but results from the optimization of a statistical criterion. The filter structure is thus adapted to the nature of the noise actually present in the processed images. This approach will constitute the main topic of the next chapters of this book.

2.2.   Heuristic criteria and optimal filters

Our goal in this section will be to determine linear filters that perform a prescribed task in an optimal way. To give a mathematical sense to this assertion, we have to define numerical *criteria*, which are mathematical functions of the filter impulse response that correspond to a quantitative evaluation of a given property of the filter. This property can be for example the robustness to additive noise or the capacity to discriminate between several objects with different shapes. To each of these properties will thus correspond a numerical criterion.

In this section, we present the criteria which are classically considered in filter design for image processing. In order to determine the filters that optimize these criteria, we will make use of the heuristic approach in which the structure of the filtering technique is imposed. This structure is in general linear filtering, but we will also describe an optimal nonlinear filter which optimizes discrimination capability. Finally, at the end of this section, SDF filters, which simultaneously optimize several antagonist criteria, will be introduced and discussed.

2.2.1   Noise robustness characterization and matched filter

Let us first define the image model that we will consider. For the sake of simplicity, we will consider one-dimensional images, but the obtained results are straightforwardly generalizable to two-dimensional images (see Appendix 2.A). Let s be the image to be processed. It consists of a vector with N real-valued components $\{s_i \mid i \in [0, N-1]\}$ which represent the values of the N pixels of the image. We consider that this image is corrupted by an additive noise, so that:

$$\mathbf{s} = \mathbf{r} + \mathbf{n} \qquad (2.1)$$

where $\mathbf{r}$ is the object of interest (i.e., the "signal") and $\mathbf{n}$ the noise. This noise is assumed to have zero mean and to be cyclostationary. This means that its covariance matrix $S = <\mathbf{n}\mathbf{n}^\dagger>$ is symmetric and circulant ($< . >$ represents ensemble averaging, the symbol $\dagger$ denotes transconjugation[1]). Consequently, its Fourier transform $\tilde{S}$, which represents the power spectral density (PSD) of the noise, is diagonal (see Appendix 2.A).

The operation which will be applied to process the image is a correlation with a linear filter of impulse response $\mathbf{h}$, so as to obtain the *correlation plane* $\mathbf{c}$:

$$\mathbf{c} = \mathbf{h} \star \mathbf{s} = \mathbf{h} \star \mathbf{r} + \mathbf{h} \star \mathbf{n} \qquad (2.2)$$

where the symbol $\star$ denotes the correlation operation, which is defined as follows (see Appendix 2.A):

$$c_j = [\mathbf{h} \star \mathbf{s}]_j = \sum_{i=0}^{N-1} h_{i-j}\, s_i \qquad (2.3)$$

The term $\mathbf{h} \star \mathbf{r}$ in Eq. 2.2 constitutes the useful signal in the correlation plane, and $\mathbf{h} \star \mathbf{n}$ the noise term. The correlation process is illustrated in Figure 2.1. In general, the filter $\mathbf{h}$ is designed to provide a peak at the location of the object of interest. In target location applications, the location of the center of the so-called *correlation peak* yields an estimate of the position of the target. In target recognition/discrimination applications, the height of the peak is compared to a threshold, or compared to the height of peaks obtained with other reference objects, and a decision is taken. In the following, with no loss of generality, we will assume that the object is centered in the analyzed image so that the correlation peak is located at the origin. Its value will thus be denoted by c_0. We note that the correlation peak c_0 can be written as:

$$c_0 = \mathbf{h}^\dagger \mathbf{r} + \mathbf{h}^\dagger \mathbf{n} \qquad (2.4)$$

---

[1]Transconjugation means the complex conjugate of the transpose: Let M be a matrix, $M^\dagger = (M^T)^*$ where T denotes transposition and $*$ complex conjugation. If the matrix is real-valued, transconjugation is equivalent to transposition.

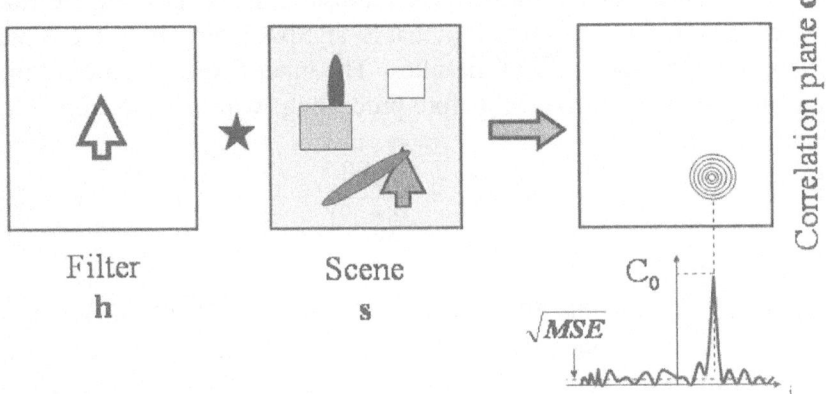

Figure 2.1. Principle of processing an image with a linear filter.

Noise robustness can be characterized by the output variance (or mean square error, MSE) of the correlation peak:

$$MSE = <(c_0 - <c_0>)^2> \qquad (2.5)$$

From Eqs. 2.2 and 2.4, and taking into account that $<n>=0$, it is clear that $<c_0>=\mathbf{h}^\dagger\mathbf{r}$. Consequently,

$$MSE = \left\langle \mathbf{h}^\dagger\mathbf{nn}^\dagger\mathbf{h} \right\rangle = \mathbf{h}^\dagger \left\langle \mathbf{nn}^\dagger \right\rangle \mathbf{h}$$

and thus

$$MSE = \mathbf{h}^\dagger S\mathbf{h} = \tilde{\mathbf{h}}^\dagger \tilde{S}\tilde{\mathbf{h}} \qquad (2.6)$$

where $\tilde{\mathbf{h}}$ is the Fourier transform of $\mathbf{h}$. The last term of the previous equation is obtained by using the Parseval relation (see Eq. 2.A.5). Since the covariance matrix S is assumed circulant and symmetric, the PSD matrix $\tilde{S}$ is diagonal and one has:

$$\tilde{\mathbf{h}}^\dagger \tilde{S}\tilde{\mathbf{h}} = \sum_{k=0}^{N-1} |\tilde{h}_k|^2 \tilde{S}_k \qquad (2.7)$$

It is easily seen that the MSE depends on the norm of $\mathbf{h}$. The SNR (signal-to-noise ratio) is usually preferred as a measure of noise robustness since it is independent of this norm:

$$SNR = \frac{|<c_0>|^2}{MSE} \quad \text{with} \quad <c_0>=\mathbf{h}^\dagger\mathbf{r}. \qquad (2.8)$$

$| < c_0 > |^2$ represents the intensity of the useful signal in the correlation plane. Note that this is an output SNR, that is, an SNR after processing, which characterizes the performance of the filter. The input SNR, which quantifies the amount of noise in the signal before processing, will be denoted SNR_{in} and defined as follows:

$$SNR_{in} = \frac{||\mathbf{r}||^2}{\langle ||\mathbf{n}||^2 \rangle} \tag{2.9}$$

where $||\mathbf{n}||^2 = \sum_{i=0}^{N-1} |n_i|^2$.

Let us now determine the filter **h** which maximizes the SNR. Several techniques can be used to obtain this result. The one we will choose has the advantage of being easily generalizable to a combination of several criteria. One can remark that maximizing the SNR amounts to minimizing its denominator while maintaining its numerator at a constant value. Formulated in this way, this problem can be solved by using the method of Lagrange multipliers (see Appendix 2.B). In order to do so, one defines the following Lagrange function:

$$L(\mathbf{h}) = \tilde{\mathbf{h}}^\dagger \tilde{S} \tilde{\mathbf{h}} - \lambda \, \tilde{\mathbf{h}}^\dagger . \tilde{\mathbf{r}} \tag{2.10}$$

where λ is the Lagrange multiplier. Annulling the derivative of this function w.r.t. each element $\tilde{h}_k$ of $\tilde{\mathbf{h}}$ and taking into account the fact that $\tilde{S}$ is diagonal leads to:

$$\tilde{h}_k = \lambda \frac{\tilde{r}_k}{\tilde{S}_k} \tag{2.11}$$

Since λ acts simply as a multiplicative factor, it does not need to be identified. The filter defined in Eq. 2.11 is the well-known *matched filter*.

Figure 2.2 illustrates the performance of the matched filter adapted to white noise, that is, $\tilde{S}_k$ is constant. Image 2.2.a represents the reference object **r** and image 2.2.b a noisy scene containing this object and additive white noise so that $SNR_{in} = -20$ dB. Figure 2.2.c represents the correlation plane obtained with the matched filter, that is, the inverse Fourier transform of $\tilde{h}_k^* \tilde{r}_k$ where $\tilde{h}_k$ is defined in Eq. 2.11. Finally, Figure 2.2.d represents the maximum of each line of Figure 2.2.c. This result shows that the matched filter is quite robust to the presence of additive noise in the signal. Indeed, among the class of linear filters, it is the one which is the most robust to noise, in the sense that it maximizes the output SNR.

However, the matched filter is designed to detect or identify a single object **r**. In practice, if several objects are to be recognized or identified, an important limitation of the matched filter is its low discrimination capability and the presence of sidelobes in the correlation plane which can result in false detection. In order to design filters more efficient for discrimination tasks, other criteria have to be considered.

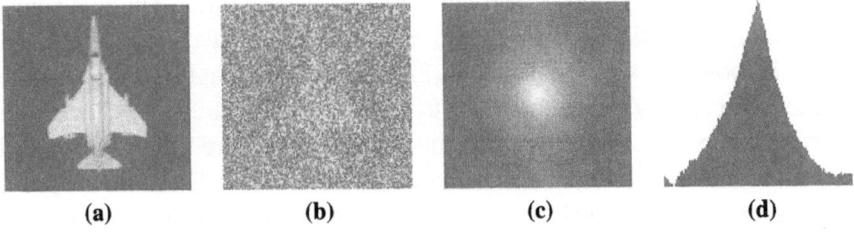

(a)	**(b)**	**(c)**	**(d)**

Figure 2.2. Illustration of the performance of the matched filter adapted to white noise. **(a)** Reference object. **(b)** Noisy scene, $SNR_{in} = -20$ dB. **(c)** Correlation plane obtained with the matched filter (see Eq. 2.11). **(d)** Maximum of each line of (c).

2.2.2 Sharpness of the correlation peak and inverse filter

In order to control the whole correlation plane and to minimize the possibility of large sidelobes which could result in false detection, the Peak to Correlation Energy (PCE) [28] is classically optimized:

$$PCE = \frac{|\langle c_0 \rangle|^2}{CPE} \tag{2.12}$$

where the Correlation Plane Energy (CPE) is defined as: $CPE = \langle \mathbf{c} \rangle^{\dagger} . \langle \mathbf{c} \rangle = \langle \tilde{\mathbf{c}} \rangle^{\dagger} . \langle \tilde{\mathbf{c}} \rangle$. This is the integral of the square modulus of the average correlation plane:

$$CPE = \sum_{i=0}^{N-1} |\langle c_i \rangle|^2 = \sum_{k=0}^{N-1} \langle |\tilde{c}_k \rangle|^2 = \sum_{k=0}^{N-1} |\tilde{h}_k^* \tilde{r}_k|^2$$

This criterion measures the "sharpness" of the correlation peak. It can be shown that the sharper the correlation peak, the more discriminant the filter with respect to objects whose shapes are different from the reference.

Using the Lagrange multipliers technique as above, it is easily shown that the filter which optimizes the PCE is the inverse filter:

$$\tilde{h}_k = \frac{\tilde{r}_k}{|\tilde{r}_k|^2} \tag{2.13}$$

In this equation, the useless multiplicative constant has been omitted. This filter is equivalent to a matched filter designed for noise with PSD matrix equal to $|\tilde{r}_k|^2$, which means that the noise and the target would have the same spectral density. At first sight, this situation may appear to be the most unfavorable one. We will see in the following section that this is not the case. Other criteria have been proposed in order to characterize the sharpness of the correlation peak and to limit the existence of sidelobes (see Ref. 28 for a critical discussion about these criteria). However, they do not lead to solutions having a simple analytical expression.

Figure 2.3 illustrates the performance of the inverse filter. The input image 2.3.a contains two objects, Figure 2.3.b represents the result of the inverse filter built for the object in the right side of 2.3.a. We can see that the searched object is correctly located whereas the other is not. Moreover, it can be checked that the correlation peak is very sharp. On the other hand, Figure 2.3.c represents the result of the matched filter, and it can be seen that it is not discriminant enough since it gives a high correlation for the two types of objects. Figure 2.3.d represents a noisy version of Figure 2.3.a with $SNR_{in} = 20$ dB. It can be seen in Figure 2.3.e that the inverse filter is not robust to noise, since none of the objects is detected, whereas the matched filter response is only slightly modified by noise (compare Figures 2.3.f and 2.3.c).

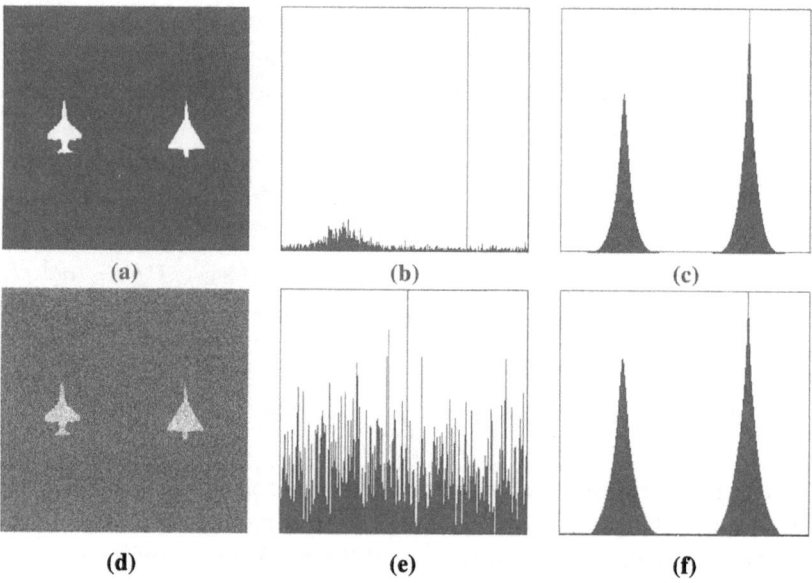

Figure 2.3.   Comparison of the performance of the inverse filter and the matched filter. **(a)** Scene with two objects with different shapes. **(b)** Result of the application of the inverse filter adapted to the object in the right side of (a). **(c)** Result of the application of the matched filter adapted to the object in the right side of (a). **(d)** Noisy version of (a) with $SNR_{in} = 20$ dB. **(e)** Result of the application of the inverse filter adapted to the object in the right side of (a). **(f)** Result of the application of the matched filter adapted to the object in the right side of (a). Figures (b), (c), (e), and (f) represent the maximum of each column of the correlation plane.

2.2.3   Optical efficiency and phase-only filter

During the last thirty years, a large effort has been put on research about optical correlators. These devices can compute the correlation between a scene and a filter, based on a well-known property of light diffraction: In the far field,

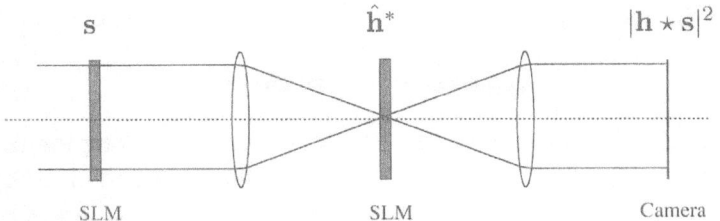

Figure 2.4. Principle of an optical correlation architecture (Vander Lugt correlator).

the angular distribution of light intensity diffracted by an aperture is equal to the Fourier transform of the transparency function of the diffracting aperture. Since correlation amounts to a multiplication in the Fourier space, it is possible to perform a correlation optically thanks to a setup as represented in Figure 2.4. The principle of this setup is the following. A collimated beam of coherent light (from a laser for example) goes through a Spatial Light Modulator (SLM) on which the scene s is encoded. The SLM is an electronically addressed device which modulates the polarization, the phase, or the intensity of the light that goes through it. After being diffracted by the SLM, the beam is focused by a lens. According to the property mentioned above, the spatial distribution of the beam at the focal plane of the lens is equal to the Fourier transform $\tilde{s}$ of the scene. Thus in the focal plane is placed a second SLM on which the complex conjugate of the Fourier transform of the filter $\tilde{h}^*$is inscribed. The spatial distribution of the light after this second SLM is thus the product of the Fourier transforms of the scene and of the filter. The beam finally goes through a second lens, which performs again a Fourier transform, so that the spatial distribution of the amplitude of light in its focal plane is equal to the correlation product of the scene and the filter $c = h \star s$. This distribution is acquired by a camera, which is a quadratic device and thus measures the intensity $|c|^2$ of the correlation plane. The principle of this optical correlator was introduced by Vander Lugt in 1964 [29]. Since then, other architectures, such as the Joint Transform Correlator (JTC) [21, 30], have been proposed and much research has been aimed at solving the technological problems associated with this method and at improving the design of filters.

In optical correlators, it is important to consider the amount of light which will be detected by the camera for the determination of the correlation function. Indeed, if a linear filter is implemented with the Vander Lugt correlator, the transparency of the first SLM cannot be higher than 1, and thus one has, for each spatial frequency, $|\tilde{h}_k| \leq 1$. On the other hand, the closer $|\tilde{h}_k|$ is to 1, the better the *optical efficiency*. This optical efficiency is characterized by the following criterion   [3]:

$$\eta_0 = \| <c_0>^2 \| = \|\mathbf{h}^\dagger \mathbf{r}\|^2 \tag{2.14}$$

which is simply the intensity of the correlation peak. Among the optically realizable filters (those for which $|h_k| \leq 1$, $\forall k$), the one which maximizes the optical efficiency has a phase equal to that of $\mathbf{r}$, and a modulus uniformly equal to 1. This filter is called the Phase-Only Filter (POF) [7] and is written:

$$\tilde{h}_k = \frac{\tilde{r}_k}{|\tilde{r}_k|} \tag{2.15}$$

Figure 2.5 illustrates the performance of the POF on the same image as in Figure 2.2. It can be seen that the correlation is sharper, but the correlation plane is noisier than in the case of the matched filter adapted to white noise.

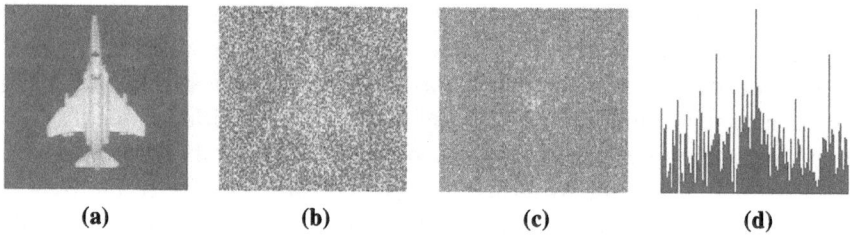

| (a) | (b) | (c) | (d) |

Figure 2.5.   Illustration of the performance of the POF. **(a)** Reference object. **(b)** Noisy scene, $SNR_{in} = -20$ dB. **(c)** Correlation plane obtained with the POF (see Eq. 2.15). **(d)** Maximum of each line of (c).

Although the POF has been derived from a hardware-oriented criterion, which has *a priori* no relation with signal processing performance, it is possible to obtain an interesting interpretation of this result in comparison with the matched filter. As mentioned above, an important question that arises when using the matched filter is to choose the PSD for the noise model from which it will be built. A possible approach may be to consider the worst case, or, in other words, the PSD which minimizes the output SNR for a fixed value of the total noise power, defined as follows:

$$\mathcal{P}_{total} = \sum_k \tilde{S}_k \tag{2.16}$$

This approach thus consists of minimizing the SNR defined by Eq. 2.8, under constraint 2.16. Here again, the Lagrange multipliers technique can be used, and it is shown that the worst-case PSD is: $\tilde{S}_k = \beta |\tilde{r}_k|$. The matched filter for this noise spectral density is then: $\tilde{h}_k = \tilde{r}_k/|\tilde{r}_k|$, which is the POF. The POF can thus be considered as the filter matched to the worst possible noise. It is interesting to note that the PSD of this noise is not the spectral density of the reference ($|\tilde{r}_k|^2$) but its square root, whereas we have seen that the optimization of the CPE leads to a matched filter optimal for a noise with a PSD matrix equal to $|\tilde{r}_k|^2$.

2.2.4   Discrimination capability

Until now, the criteria that we have considered only took into account a single reference object **r** and noise. However, in pattern recognition applications, it is important to be able to improve the discrimination capabilities of the filter with respect to some known objects. This can be the case for example in production control, when the number of possible objects is limited and one wants to identify one of them among the others. Let $\mathbf{y}^\ell$ denote (with $\ell = 1, ..., P$) the objects to be rejected. For improving the discrimination capabilities of the filter, a possible method consists of minimizing the energy of the correlation between the filter and the images $\mathbf{y}^\ell$, with the constraint that the correlation with the object of interest **r** is equal to a given value. In this case, the criterion to minimize is:

$$D_E = \sum_{\ell=1}^{P} \sum_{k} |\tilde{h}_k^* . \tilde{y}_k^\ell|^2 \tag{2.17}$$

Let us define $\tilde{S}_k^{(av)} = \sum_{\ell=1}^{P} |\tilde{y}_k^\ell|^2$ as the average spectral density of patterns to be rejected. The optimal filter for this purpose is determined with the Lagrange multipliers technique and has the following expression:

$$\tilde{h}_k = \frac{\tilde{r}_k}{\tilde{S}_k^{(av)}}$$

The interpretation of this filter in terms of matched filtering is straightforward: Optimizing the discrimination capabilities leads us to consider a matched filter adapted to a noise PSD equal to $\tilde{S}_k^{(av)}$. Optimizing discrimination is a very attractive way of designing filters, since even realizations of noise can be considered as images to be rejected. However, the practical question of inferring the appropriate images $\mathbf{y}^\ell$ which correspond to objects to be rejected is still not obvious since in many applications, these objects are not perfectly known.

An alternative approach can be used to overcome this drawback [31, 32]. Let $\tilde{s}$ denote the Fourier transform of the input scene. A discriminant filter for this image is obtained by minimizing the energy of the correlation function:

$$||\mathbf{c}||^2 = \sum_{k} |\tilde{h}_k|^2 |\tilde{s}_k|^2 \tag{2.18}$$

Of course, the minimization of Eq. 2.18 leads to the null filter. However, if this minimization is performed under the constraint that there is still a correlation peak with the target (i.e., $\sum_k \tilde{h}_k^* \tilde{r}_k = \alpha$), we will show in Section 2.4.2.4 that an optimal nonlinear filtering technique can be obtained. We will also show that this approach can be very attractive if it is correctly regularized. Right now, we can observe that the criterion (2.18) is dependent on the input image. Since a different filter is obtained for each input image, this method can be termed *adaptive*.

2.2.5   Optimal SDF filters

In some practical situations of pattern recognition, the object to recognize may not be unique. For example, one may want to recognize a class of objects, say, cars, which may contain different items (different types of cars). Moreover, each of these types of objects may appear with different view angles and/or at different scales. In the following, we will call the different possible aspects of the objects belonging to a single class their *attitudes*. The problem of identifying a class whatever the observed attitude is known as invariant pattern recognition and is a crucial issue in image processing. In particular, linear filters designed for a single attitude (i.e., for a given type of object, at a given in-plane, out-of-plane rotation angles and scale) are in general unable to recognize the same object in other attitudes.

To solve this problem, two main approaches can be considered. The first one is brute force: A filter is designed for each attitude which can appear in the input image. The input image is then correlated with all the different filters. One can show that this approach is in general the best in terms of probability of good recognition, detection, or location (see Section 3.3). However, if the number of different possible attitudes is large, this optimal approach can lead to the computation of a very large number of correlations and thus may be incompatible with the requirement of fast processing induced by most image processing applications.

In order to overcome this bottleneck, the second approach consists of applying suboptimal but faster strategies based on filters which have some invariance or tolerance properties. A possible solution is to build a single filter from a learning set of reference images representing the different possible attitudes. With this filter, it will be possible to recognize the object at different attitudes with a single correlation. This is the basic concept of the Synthetic Discriminant Function (SDF) filter that we describe now.

Let $\mathbf{r}^\ell$ with $\ell = 1, ..., P$ denote the P reference images. Some of these images correspond to the objects and attitudes to be recognized, and the others to the objects and attitudes that must be rejected. SDF filters are defined in order that the patterns belonging to the training set yield specified values of the correlation peak. In other words, the SDF filter must satisfy the following constraints:

$$\tilde{\mathbf{h}}^\dagger \tilde{\mathbf{r}}^\ell = d_\ell \; ; \; \forall \ell = 1, ..., P \iff \tilde{\mathbf{h}}^\dagger \tilde{R} = \mathbf{d}^\dagger \tag{2.19}$$

where the d_ℓ are in general equal to one if the pattern $\mathbf{r}^\ell$ has to be recognized and is equal to zero otherwise. $\tilde{R} = [\tilde{\mathbf{r}}^1, \tilde{\mathbf{r}}^2, ..., \tilde{\mathbf{r}}^P]$ is the matrix whose columns are the Fourier transforms of the reference images.

There exists an infinite number of solutions to Eq. 2.19 for the filter $\mathbf{h}$ since in general the number of pixels N of the image is superior to the number of constraints P. Consequently, it is possible to optimize some relevant criteria together with satisfying the constraints in Eq. 2.19, and thus to generalize the previously described correlation criteria to the SDF problem. For the sake of simplicity, we will consider here the MSE and the CPE but not the optical efficiency (an analysis of this point can be found in Ref. 33).

Let us for example consider the minimization of the MSE with the constraints of Eq. 2.19. The Lagrange function associated with this problem is:

$$L(\tilde{\mathbf{h}}) = \tilde{\mathbf{h}}^\dagger \tilde{S} \tilde{\mathbf{h}} - \sum_{\ell=1}^{P} \lambda_\ell \tilde{\mathbf{h}}^\dagger \tilde{\mathbf{r}}_\ell = \tilde{\mathbf{h}}^\dagger \tilde{S} \tilde{\mathbf{h}} - \tilde{\mathbf{h}}^\dagger \tilde{R} \boldsymbol{\lambda}$$

where $\boldsymbol{\lambda}$ is the P-dimensional vector of Lagrange parameters: $\boldsymbol{\lambda} = [\lambda_1, \dots, \lambda_P]$. Annulling the derivative of $L(\tilde{\mathbf{h}})$ with respect to $\tilde{\mathbf{h}}$ leads to:

$$\tilde{\mathbf{h}} = \tilde{S}^{-1} \tilde{R} \boldsymbol{\lambda} \tag{2.20}$$

The Lagrange parameters are identified by enforcing that $\tilde{\mathbf{h}}$ verifies the constraints in Eq. 2.19. One obtains:

$$\boldsymbol{\lambda} = \left[\tilde{R}^\dagger \tilde{S}^{-1} \tilde{R} \right]^{-1} \mathbf{d} \tag{2.21}$$

Injecting this expression of $\boldsymbol{\lambda}$ in Eq. 2.20, one obtains the expression of the SDF filter:

$$\tilde{\mathbf{h}} = \tilde{S}^{-1} \tilde{R} \left[\tilde{R}^\dagger \tilde{S}^{-1} \tilde{R} \right]^{-1} \mathbf{d} \tag{2.22}$$

This filter is called the Minimum Variance SDF (MVSDF) filter [15]. It is equivalent to the matched filter defined in Eq. 2.11, but for multiple reference objects. Consequently, this filter shows the same limitation as the matched filter and leads to broad correlation peaks with large sidelobes and poor discrimination capability (see Figure 2.2).

For this reason, it is thus natural to generalize the CPE criterion to SDF filters. The Average Correlation Peak Energy (ACPE) defined from all the training patterns is:

$$ACPE = \frac{1}{P} \sum_{\ell=1}^{P} \sum_{k=0}^{N-1} |\tilde{r}_k^\ell \tilde{h}_k|^2 = \tilde{\mathbf{h}}^\dagger \tilde{D} \tilde{\mathbf{h}} \tag{2.23}$$

with D_k a diagonal matrix of elements $\tilde{D}_k = 1/P \sum_{\ell=1}^{P} |\tilde{r}_k^\ell|^2$. Using the same Lagrange technique as above, the minimization of the $ACPE$ leads to the Minimum Average Correlation Energy (MACE) filter [16]:

$$\tilde{\mathbf{h}} = \tilde{D}^{-1} \tilde{R} \left[\tilde{R}^\dagger \tilde{D}^{-1} \tilde{R} \right]^{-1} \mathbf{d} \tag{2.24}$$

Here again, the MACE filter is the multireference equivalent of the inverse filter defined in Eq. 2.13, and thus shares its main drawback: It is very sensitive to the presence of noise in the analyzed image (see Figure 2.3).

We thus see that, as in the single reference case, the SDF filters which optimize the MSE or the ACPE are overspecialized. The solution to this problem is to define filters which perform a compromise between the good noise robustness of the matched filter (or the MVSDF) and the good discrimination ability of the inverse filter (or the MACE). This is the principle of the Optimal Trade-off filters [17] that we describe in the next section.

2.2.6   Optimal Trade-off filters

We outline here the general concept of Optimal Trade-off (OT) filters. Some examples will be given in the next section.

As seen above, optimal filters are subject to some constraints. For example, for invariant filtering, the central values of the correlation functions are imposed for a set of training patterns. These constraints impose that the filter h can only lie in a subset $\mathcal{D}$ of $\mathbf{R}^N$. For example, for SDF filters, $\mathcal{D}$ is the hyperplane defined by the constraints of Eq. 2.19.

On the other hand, in order to obtain good processing properties, h must also minimize some criteria, which often are *antagonist*, which means that a decrease of one criterion will lead to the increase of the others. Let $E_1(\mathbf{h})$, $E_2(\mathbf{h})$, ..., $E_M(\mathbf{h})$ denote the values of the M considered criteria for the filter h (for example, MSE and $ACPE$). Each criterion is defined in order to correspond to a minimization problem. OT filters are defined in order to lead to the *best* trade-off between the considered criteria with the constraint that the filter h belongs to $\mathcal{D}$. The notion of *best* trade-off is defined as follows. Consider a filter $\mathbf{h}^{(1)}$ in $\mathcal{D}$. The values of the criteria corresponding to this filter are: $E_1(\mathbf{h}^{(1)})$, $E_2(\mathbf{h}^{(1)})$, ..., $E_M(\mathbf{h}^{(1)})$. If there exists a filter $\mathbf{h}^{(2)}$ in $\mathcal{D}$ such that:

$$E_j(\mathbf{h}^{(2)}) < E_j(\mathbf{h}^{(1)}) \quad \forall j = 1, ..., M$$

then it is clear that $\mathbf{h}^{(2)}$ is a better filter than $\mathbf{h}^{(1)}$. OT filters are the filters $\mathbf{h}^{(OT)}$ for which there is no filter h in $\mathcal{D}$ such that:

$$E_j(\mathbf{h}) \leq E_j(\mathbf{h}^{(OT)}) \quad \forall j = 1, ..., M \tag{2.25}$$

with at least one strict inequality. In other words, a filter is OT if it is better than any other filter for at least one of the criteria. This means that if, for one of the considered criteria, a filter $\mathbf{h}$ leads to a smaller value than $\mathbf{h}^{(OT)}$, then it leads to higher values for at least one of the other criteria. It can be shown that the filters which verify this property are obtained by minimizing the following function in $\mathcal{D}$[34, 35]:

$$E(\mathbf{h}) = \sum_{j=1}^{M} \lambda_j \, E_j(\mathbf{h}) \qquad (2.26)$$

where the λ_j are positive numbers that allow us to balance the weights of the different criteria in the optimization. In other words:

$$\mathbf{h}^{(OT)} = \arg \min_{\mathbf{h} \in \mathcal{D}} E(\mathbf{h}) \qquad (2.27)$$

The OT filters are thus obtained by minimization of a linear combination of the criteria in Eq. 2.26. It is important to note that this quite simple result does not arise from empirical considerations, but constitutes the rigorous way of obtaining filters having the property defined in Eq. 2.25.

2.2.7 Optimal Trade-off SDF filters

As stated above, the MACE filter is very sensitive to input noise on the analyzed image, while the MVSDF filter is not sufficiently discriminant in general. As an example of application of the approach described in the previous section, we will thus determine the Optimal Trade-off SDF (OTSDF) filter for noise robustness and discrimination ability, that is, the filter which performs a trade-off between the MSE and the $ACPE$ [34]. According to Eq. 2.26, the function to minimize is thus:

$$E(\mathbf{h}) \quad = \quad \lambda_1 \tilde{\mathbf{h}}^\dagger \tilde{S} \tilde{\mathbf{h}} + \lambda_2 \tilde{\mathbf{h}}^\dagger \tilde{D} \tilde{\mathbf{h}} = \tilde{\mathbf{h}}^\dagger \tilde{A} \tilde{\mathbf{h}} \qquad (2.28)$$

where

$$\tilde{A} = \lambda_1 \tilde{S} + \lambda_2 \tilde{D}$$

The function to optimize is thus simply a quadratic form. The set $\mathcal{D}$ of possible filters being defined by the constraints in Eq. 2.19, one uses the Lagrange multipliers technique to find the solution. Since the problem consists in optimizing a quadratic functional with respect to the constraints in Eq. 2.19, we obtain a similar result as in Section 2.2.5:

$$\tilde{\mathbf{h}} = \tilde{A}^{-1} \tilde{R} \left[\tilde{R}^\dagger \tilde{A}^{-1} \tilde{R} \right]^{-1} \mathbf{d} \qquad (2.29)$$

A strictly equivalent filter is obtained by setting $\lambda_1 = (1 - \mu)$ and $\lambda_2 = \mu$ in the expression of $\tilde{A}$, where $\mu \in [0, 1]$, so as to obtain:

$$\tilde{\mathbf{h}} = \tilde{B}^{-1} \tilde{R} \left[\tilde{R}^\dagger \tilde{B}^{-1} \tilde{R} \right]^{-1} \mathbf{d} \qquad (2.30)$$

where

$$\tilde{B} = (1 - \mu)\tilde{S} + \mu\tilde{D} \tag{2.31}$$

This filter is called the OTSDF [34]. The parameter μ allows us to balance optimally between the different criteria. If $\mu = 0$, the MVSDF filter is obtained (see Eq. 2.22), while if $\mu = 1$, it is the MACE filter (see Eq. 2.24).

Let us comment on Eq. 2.30. $\tilde{B}$ is a diagonal matrix of size $N \times N$ and R is a rectangular matrix of size $N \times P$, so that $\tilde{R}^\dagger \tilde{B}^{-1} \tilde{R}$ is a matrix of size $P \times P$. The inverse matrices $\tilde{B}^{-1}$ and $\left[\tilde{R}^\dagger \tilde{B}^{-1} \tilde{R}\right]^{-1}$ must be understood as inverse matrices in the space spanned by the eigenvectors associated with nonnull eigenvalues. Let us denote:

$$b_\ell = \sum_{n=1}^{P} \left(\left[\tilde{R}^\dagger \tilde{B}^{-1} \tilde{R}\right]^{-1} \right)_{\ell,n} d_n$$

we can then write:

$$\tilde{h}_k = \sum_{\ell=1}^{P} b_\ell \frac{\tilde{r}_k^\ell}{\tilde{B}_k} \tag{2.32}$$

which means that $\tilde{\mathbf{h}}$ is a weighted sum of filters matched to the objects $\mathbf{r}^\ell$ of the database.

Let us illustrate graphically the compromise performed between the different criteria by OT filters. If μ is close to 0, we obtain the MVSDF filter, and if it is close to 1, we tend to the MACE filter. By construction of the OT filter, the compromise obtained for any value of μ is optimal in the sense that for a given value of the MSE it provides the lowest value of the $ACPE$ (taking into account the constraint that the filter is linear). This concept is illustrated in Figure 2.6, which represents the Optimal Characteristic Curve (OCC), that is, the locus of the OT filter in the plane $(MSE, ACPE)$ when μ varies.

As an example, Figure 2.7 illustrates the influence of μ on the correlation peak between the OT filter $\mathbf{h}^{(OT)}$ and a scene corrupted with a strong additive noise. We have supposed here that there is a single object $\mathbf{r}$ in the database and that the noise is white, so that its PSD $\tilde{S}_k$ is constant. The OT filter can thus be written:

$$\tilde{h}_k^{(OT)} = \frac{\tilde{r}_k}{(1 - \mu) + \mu|\tilde{r}_k|^2} \tag{2.33}$$

When $\mu = 0$, one gets the matched filter. It can be seen in Figure 2.7 that it locates correctly the object, but produces a broad correlation peak. As the value of μ increases, the correlation peak gets sharper (the ACPE decreases). For values of μ greater than 0.5, the correlation plane is more and more noisy (the MSE increases). When $\mu = 1$, one gets the inverse filter, which is very sensitive to noise and does not permit us to locate correctly the object in the

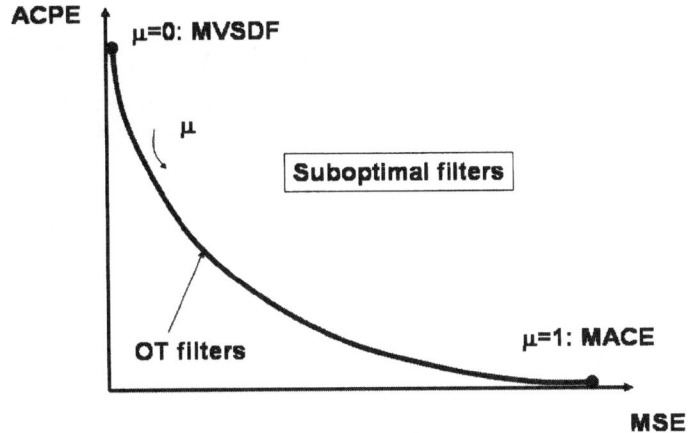

Figure 2.6. Symbolic representation of the Optimal Characteristic Curve (OCC) for an OT filter. Each point of the curve corresponds to a different value of μ (see Eq. 2.33).

noisy scene. It is clearly seen that the OT filter with $\mu = 0.5$ leads to a good compromise in that case.

The way the MSE and the ACPE criteria are balanced in Eqs. 2.31 and 2.33 is quite simple and could have been inferred empirically. However, it should be kept in mind that these expressions are the result of an optimization process. In other terms, no other filter is better than the OT filter in the sense defined above.

Of course, there are other ways of introducing a trade-off between these criteria. For example, the MINACE filter [27] performs a trade-off between $ACPE$ and MSE in the following way:

$$\tilde{\mathbf{h}} = \tilde{N}^{-1}\tilde{R}\left[\tilde{R}^\dagger\tilde{N}^{-1}\tilde{R}\right]^{-1}\mathbf{d} \quad \text{with} \quad \tilde{N}_k = \max[(1-\mu)\tilde{S}_k, \mu\tilde{D}_k] \quad (2.34)$$

where $\max[x, y]$ means maximum value between x and y. However, contrary to the OT approach, the choice of Eq. 2.34 does not lead to an optimal trade-off.

Finally, we can remark that there is another way of considering the SDF. Indeed, its mathematical expression is:

$$\tilde{\mathbf{h}} = \tilde{A}^{-1}\tilde{R}\left[\tilde{R}^\dagger\tilde{A}^{-1}\tilde{R}\right]^{-1}\mathbf{d} \quad (2.35)$$

where $\tilde{A}$ is a diagonal matrix which corresponds to the optimization of the criterion $E = \sum_k |\tilde{h}_k|^2 \tilde{A}_k$. The choice of a matrix $\tilde{A}$, and therefore of the

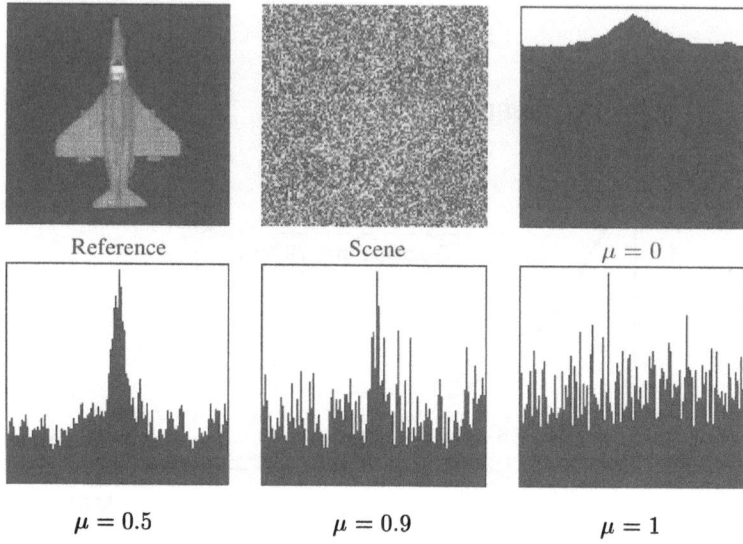

Figure 2.7. Processing of a very noisy scene with an OT filter with different values of μ (see Eq. 2.33). The scene has been constructed by adding a heavy white noise to the reference. The last four images represent the maximum of each line of the correlation planes obtained for four different values of μ.

criterion which is optimized, can be considered as an inference of the spectral density of the noise model (i.e., $\tilde{S}_k = \tilde{A}_k$). In other words, the OTSDF and MINACE filters can be considered as MVSDF filters with noise spectral density equal to $\tilde{B}_k$ or $\tilde{N}_k$, respectively.

2.3.   Analysis of the stability of linear filters

Until now, we have presented methods for designing filters adapted to a set of training patterns while optimizing some criteria. We will now turn our attention to the stability of the obtained solutions [19, 20]. Before explaining the notion of filter stability, let us recall that the filter synthesis problem can be written as:

$$\mathbf{h}^{opt} = \arg\min_{\mathbf{h} \in \mathcal{D}} E(\mathbf{h}) \qquad (2.36)$$

which means that $\mathbf{h}^{opt}$ is the value of $\mathbf{h}$ in $\mathcal{D}$ which minimizes $E(\mathbf{h})$.

Let us first consider the case for which there is a single pattern in the training set. The optimal filter defined in Eq. 2.36 is a function of the reference pattern. Let us emphasize this point by writing: $\mathbf{h}^{opt} = \mathbf{h}^{opt}[\mathbf{r}]$. A filter is stable [19] if a small variation in the reference pattern ($\mathbf{r} \longrightarrow \mathbf{r} + \delta\mathbf{r}$ with $||\delta\mathbf{r}|| << 1$) does not induce a large variation in the impulse response of the filter ($\delta\mathbf{h}^{opt}[\mathbf{r}] = \mathbf{h}^{opt}[\mathbf{r} + \delta\mathbf{r}] - \mathbf{h}^{opt}[\mathbf{r}]$). It is clear that if the filter is not stable, a small deviation $\delta\mathbf{r}$ of the reference will result in a large mismatch between the filter $\mathbf{h}^{opt}[\mathbf{r}]$

and the reference $\mathbf{r} + \delta\mathbf{r}$. In other words, a filter which is not stable against small perturbations in the pattern $\mathbf{r}$ will be very discriminant but will have very poor tolerance capabilities against distortions of the input image. Possible distortions can be the addition of a noise with different statistical properties than those expected [18], but also small deformations due for example to in-plane or out-of-plane rotation.

This concept of stability can be generalized when there are several patterns in the training set [19]. In that case, the filter will be considered stable if it is stable for each possible modification: $\mathbf{r}^\ell \longrightarrow \mathbf{r}^\ell + \delta\mathbf{r}^\ell$ ($\forall \ell = 1, ..., M$). For example, since in general all the possible distortions cannot be included in the learning set, it is important that the SDF filters be robust to small attitude variations which are not included in the training set.

Taking into account stability is thus very important to design filters which are robust to input images with properties slightly different from the ones expected during the filter design. We will study in this section how to characterize the stability of filters [19].

For simplicity's sake, we will first consider that there is a single pattern in the training set. Furthermore, we will directly consider OT filters since they include the matched and the inverse filters as limit cases. The mathematical expression of the OT filter is: $\tilde{h}_k = \tilde{r}_k / \tilde{B}_k$, where $\tilde{B}_k = (1 - \mu)\tilde{S}_k + \mu|\tilde{r}_k|^2$. If there is a frequency k for which $\tilde{B}_k$ is small, then a small perturbation in $\mathbf{r}$ may result in a large variation of $\mathbf{h}$ and of $\mathbf{h} \star \mathbf{r}$. More precisely, if $\delta\tilde{\mathbf{r}}$ is a small perturbation in $\tilde{\mathbf{r}}$, the induced perturbation $\delta\tilde{\mathbf{h}}$ in $\tilde{\mathbf{h}}$ is:

$$\delta\tilde{h}_k \simeq (1 - \mu)\frac{\tilde{S}_k}{[\tilde{B}_k]^2}\delta\tilde{r}_k - \mu\frac{\tilde{r}_k^2}{[\tilde{B}_k]^2}\delta\tilde{r}_k^*$$

The norm of $\delta\tilde{h}_k$ can be large if $\tilde{B}_k$ is small. On the other hand, if $\mathbf{h}$ is fixed ($\tilde{h}_k = \tilde{r}_k / \tilde{B}_k$), the variation of the central value of the correlation function induced by $\delta\mathbf{r}$ is:

$$\delta c_0 = \sum_k \frac{\tilde{r}_k^*}{\tilde{B}_k}\delta\tilde{r}_k$$

Thus, it is clearly seen that when $\tilde{B}_k$ is small, neither the correlation function nor the filter is stable against small perturbations of the pattern $\mathbf{r}$. However, when the noise model is white and μ different from 1, the OT filter is stable. On the other hand, if the noise is low frequency, the PSD vanishes for high frequencies and leads to instabilities. In other words, the condition of stability is that, for all k, $(1 - \mu)\tilde{S}_k$ is not too small. On the other hand, the inverse filter is clearly unstable. An equivalent approach (although the calculations are more tedious) would show that the MINACE filter defined in Eq. 2.34 has the same behavior as the OTSDF with respect to stability.

In the same way, let us analyze the stability of the POF. We have seen in Eq. 2.15 that this filter can be written: $\tilde{h}_k = \tilde{r}_k/|\tilde{r}_k|$. If $\delta\tilde{\mathbf{r}}$ is a small perturbation in $\tilde{\mathbf{r}}$, the induced perturbation $\delta\tilde{\mathbf{h}}$ in $\tilde{\mathbf{h}}$ is:

$$\delta\tilde{h}_k \simeq \frac{1}{2|\tilde{r}_k|}\delta\tilde{r}_k - \frac{[\tilde{r}_k]^2}{2|\tilde{r}_k|^3}\delta\tilde{r}_k^*$$

The norm of $\delta\tilde{h}_k$ can be large if $|\tilde{r}_k|$ is small, which typically happens for usual objects at high spatial frequencies. Thus it is also clear that the POF will not be stable in general. This is the main reason for its high level of discrimination and low level of tolerance to such variations as noise or rotation for example. We proposed in Section 2.2.3 an interpretation of the POF as a matched filter designed with the PSD which corresponds to the worst situation. We see that this approach leads to an unstable filter, i.e., which is too specialized to this particular noise model.

Let us now consider the case of the OTSDF filter (which contains as limit cases the MVSDF and the MACE filter). The expression of the OTSDF filter in Eq. 2.32 ($\tilde{h}_k = \sum_{\ell=1}^{P} b_\ell \tilde{r}_k^\ell / B_k$) leads to a simple analysis. Indeed, the stability conditions of SDF filters can be deduced from the stability conditions of the previously studied single-reference filters.

Letting $\delta\tilde{\mathbf{r}}^\ell$ denote a small perturbation in $\tilde{\mathbf{r}}^\ell$, the induced perturbation $\delta\tilde{\mathbf{h}}$ in $\tilde{\mathbf{h}}$ is:

$$\delta\tilde{h}_k = \sum_{\ell=1}^{P} b_\ell \left[\frac{\tilde{B}_k - \mu|\tilde{r}_k^\ell|^2}{[\tilde{B}_k]^2} \delta\tilde{r}_k^\ell - \mu \frac{(\tilde{r}_k^\ell)^2}{[\tilde{B}_k]^2}(\delta\tilde{r}_k^\ell)^* \right] \tag{2.37}$$

Here again, the norm of $\delta\tilde{h}_k$ can be large if $\tilde{B}_k$ is small. The MACE filter is unstable because the spectral density of images is generally very small at high spatial frequencies. The MVSDF filter is stable for a white noise model, but is unstable for low-frequency noise models. The OTSDF filter, as well as the MINACE filter, shows a similar behavior. They are stable if they have been designed with a white noise model and unstable if they have been designed for a low-frequency noise model.

2.4.   Regularization of filters

It has been shown in Section 2.2.6 that the general problem of filter synthesis can be formulated as the minimization of some criterion over a definition set $\mathcal{D}$. The obtained filters can be unstable in the sense defined in Section 2.3 because this inverse problem is ill-posed. In order to regularize this inverse problem, different solutions can be employed. We will first analyze the simple method of regularization by truncation and then review some well-known heuristic approaches. Then we will consider the regularization of inverse problems with the use of a stabilizing functional [19].

2.4.1   Truncature method for regularization

A classical approach to the regularization of inverse problems is to forbid the inversion of small values. Let us first consider the case of filters designed for a single reference object. Their general expression can be written: $\tilde{h}_k = \tilde{r}_k/\tilde{A}_k$ (which optimizes the criterion $\sum_k |\tilde{h}_k|^2 \tilde{A}_k$). The truncature regularization method replaces this filter with:

$$\tilde{h}_k = \begin{cases} \tilde{r}_k/\tilde{A}_k & \text{if } \tilde{A}_k > \epsilon \\ 0 & \text{otherwise} \end{cases} \tag{2.38}$$

In particular, with a non-white noise model in the previous methods, the filter may be unstable. The truncature-based regularization will guarantee its stability. Of course, it is easily verified that the larger ϵ, the more regularized the solution, but the smaller the correlation peak and the larger the value of the optimized criterion $\sum_k |\tilde{h}_k|^2 \tilde{A}_k$.

This approach is easily generalized to optimal SDF filters. In that case, the truncature regularization leads to:

$$\tilde{\mathbf{h}} = \tilde{F}^{-1} \tilde{R} \left[\tilde{R}^\dagger \tilde{F}^{-1} \tilde{R} \right]^{-1} \mathbf{d} \tag{2.39}$$

where

$$\tilde{F}_k = \begin{cases} \tilde{B}_k & \text{if } \tilde{B}_k > \epsilon \\ 0 & \text{otherwise} \end{cases}$$

remembering that $\tilde{B}$, and then $\tilde{F}$, are diagonal in the Fourier domain.

It is also interesting to remark that truncature regularization applied to the MINACE filter leads to a filter analog to Eq. 2.39 but with kernel:

$$\tilde{F}_k = \begin{cases} \tilde{N}_k & \text{if } \tilde{N}_k > \epsilon \\ 0 & \text{otherwise} \end{cases}$$

where N_k is defined in Eq. 2.34.

2.4.2   Stabilizing functional

A well-known method for regularizing the solution of inverse problems is to introduce a stabilizing functional $\Omega(\mathbf{h})$ in order to limit the sensitivity of the solution to small perturbations in the data (i.e., training patterns). The regularized solution is simply defined as the solution of the following problem:

$$\mathbf{h}^{reg} = \arg\min_{\mathbf{h} \in \mathcal{D}} \left[E(\mathbf{h}) + \alpha\Omega(\mathbf{h}) \right] \tag{2.40}$$

The parameter α can be considered as a Lagrange parameter introduced in order to satisfy the inequality:

$$\Omega(\mathbf{h}^{reg}) \leq \epsilon \qquad (2.41)$$

where $\epsilon > 0$. In general, the value of the Lagrange parameter α is not determined, but it is clear that the larger α, the smaller $\Omega(\mathbf{h}^{reg})$ [19].

Considering quadratic stabilizing functionals is very convenient since an explicit mathematical equation for the filter can still be found (this would not be the case, for example, with maximum entropy methods [36-38]. Furthermore, as mentioned above, an interesting aspect of the stabilizing functional approach is to introduce *a priori* knowledge in the filter synthesis. Indeed, there are some *a priori* properties of the filter which are not easily defined in terms of numerical criteria. This is in particular the case for insensitiveness to distortions of the input pattern since all the possible distortions cannot be simply described nor included in a training set.

In the following, some stabilizing functionals which improve the robustness to input distortions are discussed as examples.

2.4.2.1   The minimum norm stability functional

Let us consider a filter $\mathbf{h}$ and the perturbed version $\mathbf{r} + \delta\mathbf{r}$ of an input pattern $\mathbf{r}$. The induced variation of the correlation function is:

$$\delta\mathbf{c} = \mathbf{h} \star [\mathbf{r} + \delta\mathbf{r}] - \mathbf{h} \star \mathbf{r} = \mathbf{h} \star \delta\mathbf{r} \qquad (2.42)$$

This variation of the correlation function can be bounded using the Cauchy-Schwarz inequality:

$$\| \delta\mathbf{c} \| \leq \| \mathbf{h} \| \| \delta\mathbf{r} \| \qquad (2.43)$$

Thus, imposing a maximum value for $\| \mathbf{h} \|$ in turn imposes a bounding value on the norm $\| \delta\mathbf{c} \|$ of the variation of the correlation function for a given norm $\| \delta\mathbf{r} \|$ of the perturbation $\delta\mathbf{r}$. In order to obtain a quadratic criterion, it is appropriate to choose the following stabilizing functional:

$$\Omega(\tilde{\mathbf{h}}) = \| \tilde{\mathbf{h}} \|^2 \qquad (2.44)$$

which corresponds to a classical regularization method [39]. When the criterion $E(\mathbf{h})$ is also quadratic (i.e., $E(\mathbf{h}) = \tilde{\mathbf{h}}^\dagger \tilde{B} \tilde{\mathbf{h}}$), the regularized filter $\tilde{\mathbf{h}}^{reg}$ is given by:

$$\tilde{\mathbf{h}}^{reg} = \arg\min_{\tilde{\mathbf{h}} \in \mathcal{D}} \left[\tilde{\mathbf{h}}^\dagger \tilde{B} \tilde{\mathbf{h}} + \alpha \tilde{\mathbf{h}}^\dagger . \tilde{\mathbf{h}} \right] \qquad (2.45)$$

It is clear that $\tilde{\mathbf{h}}^\dagger \tilde{B} \tilde{\mathbf{h}} + \alpha \tilde{\mathbf{h}}^\dagger . \tilde{\mathbf{h}} = \tilde{\mathbf{h}}^\dagger \left[\tilde{B} + \alpha\, I_d \right] \tilde{\mathbf{h}}$, where I_d is the identity matrix in C^N (i.e., diagonal with unitary elements). Consequently, the solution of this equation is immediately found to be:

$$\tilde{\mathbf{h}}^{reg} = \left[\tilde{B} + \alpha \, I_d\right]^{-1} \tilde{R} \left[\tilde{R}^{\dagger} \left[\tilde{B} + \alpha \, I_d\right]^{-1} \tilde{R}\right]^{-1} \mathbf{d} \qquad (2.46)$$

When there is only one pattern in the training set, the regularized filter has the following expression:

$$\tilde{h}_k = \frac{\tilde{r}_k}{\tilde{B}_k + \alpha} \qquad (2.47)$$

where $\tilde{B}_k$ is defined by Eq. 2.30. Then, if there exist some values of k for which $\tilde{B}_k$ is small, the filter will be approximately equal to $\tilde{r}_k/\alpha$ at these frequencies, which limits the effect of a small perturbation of $\tilde{\mathbf{r}}$ on the modulus of the correlation function. Indeed, for a small perturbation $\delta\tilde{\mathbf{r}}$, we now obtain:

$$\delta\tilde{h}_k \simeq \frac{\tilde{B}_k + \alpha - \mu \, |\tilde{r}_k|^2}{[\tilde{B}_k + \alpha]^2} \, \delta\tilde{r}_k - \mu \, \frac{\tilde{r}_k^2}{[\tilde{B}_k + \alpha]^2} \delta\tilde{r}_k^*$$

whose modulus is bounded. It is easily seen that for an OT filter designed with a white noise model, this kind of regularization is naturally obtained without regularizing functional.

2.4.2.2 The minimum variation stability functional

In the previous section, no particular form for the perturbation $\delta\mathbf{r}$ was assumed. However, in general, it is needed that the filter be robust to small distortions of the input image, which can be introduced for example by small variations of the attitude of the object (such as scale, view angle). It is then natural to impose that the filter variations be bounded for perturbations $\delta\mathbf{r}$ similar to those induced by small attitude variations. Such small attitude variations emphasize mainly the edges of the object. In order to support this conjecture, we show in Figure 2.8 the difference between an image and a small attitude variation of itself. Figure 2.8.a shows an image of an airplane, and Figure 2.8.b the modulus of the subtraction of this image with its 5-degree rotated version. It is clearly seen that the difference image $\delta\mathbf{r}$ is essentially an edge-enhanced image of the airplane.

Thus, a first approximation of $\delta\mathbf{r}$ might be:

$$\delta\mathbf{r} = \mathbf{g}_{hp} \star \mathbf{r} \qquad (2.48)$$

where $\mathbf{g}_{hp}$ is a high-pass filter, enhancing the edges of $\mathbf{r}$. Using the approach introduced in the previous section, the Cauchy-Schwarz inequality leads to the minimization of the following stabilizing functional:

$$\Omega(\mathbf{h}) = \| \, \mathbf{h} \star \mathbf{g}_{hp} \, \|^2 \qquad (2.49)$$

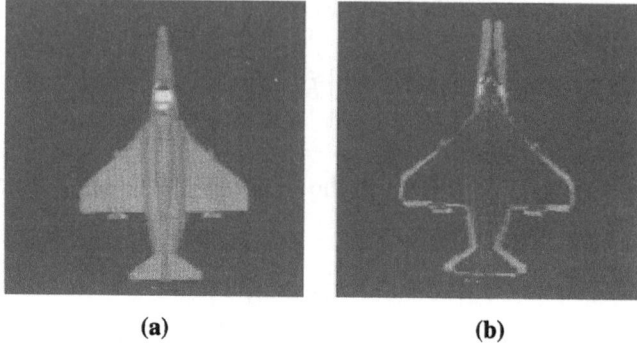

(a)   **(b)**

Figure 2.8. **(a)** Image of an airplane. **(b)** Modulus of the subtraction of the previous image with its rotated version by an angle of 5 degrees. ©1993 SPIE.

When the criterion $E(\mathbf{h})$ is quadratic, the regularized filter $\mathbf{h}^{reg}$ is given by:

$$\tilde{\mathbf{h}}^{reg} = \left[\tilde{B} + \alpha\,\tilde{G}\right]^{-1} \tilde{R} \left[\tilde{R}^{\dagger}\left[\tilde{B} + \alpha\,\tilde{G}\right]^{-1}\tilde{R}\right]^{-1}\mathbf{d} \qquad (2.50)$$

where $\tilde{G}_k = \left|[\tilde{g}_{hp}]_k\right|^2$ is the modulus square of the frequency transfer function of the high-pass filter. Then, if there are some high frequencies k for which $\tilde{B}_k$ is small, the filter will be approximately equal to $\tilde{r}_k/\alpha\tilde{G}_k$ at these frequencies, which bounds the modulus of the effect of a small perturbation of $\mathbf{r}$ due to a small edge shifting.

2.4.2.3   The minimum elastic energy stability functional

The problem of regularization can also be seen in the following alternative way. The filter needs to be robust to small distortions of the input image which can be induced by small variations of the object (such as scale, view angle). This can be achieved if the impulse response $\mathbf{h}$ of the filter does not vary rapidly in the image domain. In other words, it is required that the filter's impulse response be a smooth function of the spatial coordinates (u, v). For the sake of simplicity, let us first consider a nonsampled, continuous image version of the filter $\mathbf{h}$. The filter will be denoted $h(u, v)$. It can be imposed to the filter to be a smooth function by minimization of the elastic energy:

$$\int\int \left|\frac{\partial^2 h(u, v)}{\partial u^2} + \frac{\partial^2 h(u, v)}{\partial v^2}\right|^2 du\,dv \qquad (2.51)$$

Using the Fourier transform of continuous functions, and after sampling, it is easily shown [19] that this approach leads to the following stabilizing functional

(2-D notation is used here):

$$\Omega(\mathbf{h}) = \sum_{k_u} \sum_{k_v} \left[k_u{}^2 + k_v{}^2 \right]^2 \left| \tilde{h}(k_u, k_v) \right|^2 \qquad (2.52)$$

When the criterion $E(\mathbf{h})$ is quadratic, the regularized filter $\mathbf{h}^{reg}$ is given by:

$$\tilde{\mathbf{h}}^{reg} = \left[\tilde{B} + \alpha \, \tilde{L} \right]^{-1} \tilde{R} \left[\tilde{R}^\dagger \left[\tilde{B} + \alpha \, \tilde{L} \right]^{-1} \tilde{R} \right]^{-1} \mathbf{d} \qquad (2.53)$$

where

$$\tilde{L}_{k_u,k_v} = \left[k_u{}^2 + k_v{}^2 \right]^2$$

It is seen again that if the frequencies for which $\tilde{B}_{k_u,k_v}$ is small are high frequencies, $\tilde{L}$ will stabilize the filter.

We can note that the minimum elastic energy stability functional presents a strong analogy with the minimum variation stability functional since both measure the energy of high frequencies.

2.4.2.4   Application to optimal nonlinear discriminant filters

We have seen in Section 2.2.4 that the discrimination capabilities of filters are generally optimized indirectly either by maximizing the sharpness of the correlation function or by minimizing the energy of the correlation with objects to be rejected and background models. We have also discussed the limitations of this approach and introduced a criterion leading to an adaptive filter. With the use of the minimum norm stabilizing functional, we are now able to determine a method for optimizing the discrimination capabilities of the filter which does not require *a priori* knowledge of objects to be rejected nor of the background [32].

Let $\mathbf{r}$ and $\mathbf{s}$ denote respectively the reference and input image. We will consider that the output of the algorithm is still written $c_j = \sum_{i=1}^{N} h_{i-j} s_i$, but $\mathbf{h}$ may now be a function of $\mathbf{s}$, in which case $\mathbf{h}$ no longer represents a linear filter. When the input image is the reference object, it is imposed that the filter produces the correlation peak $c_0 = d$. This leads to the constraint:

$$\sum_{j=1}^{N} h_j r_j = \sum_{k} \tilde{h}_k^* \tilde{r}_k = d \qquad (2.54)$$

As mentioned in Section 2.2.4, in order to optimize the discrimination capabilities of the filter, one considers that the energy of the correlation function with the input image $\mathbf{s}$ must be minimal. This is equivalent to minimizing:

$$E_s[\mathbf{h}] = \sum_{k} |\tilde{h}_k|^2 |\tilde{s}_k|^2 \qquad (2.55)$$

while maintaining Eq. 2.54 as a constraint. Equation 2.55 can also be written: $E_s[\mathbf{h}] = \| \mathbf{c} \|^2 = \| \tilde{\mathbf{c}} \|^2$ with $\tilde{c}_k = \tilde{h}_k^* \tilde{s}_k$.

In order to be robust to modifications or distortions of the reference image $\mathbf{r}$, and since we do not want to introduce any *a priori* knowledge on the perturbations, we consider the minimum norm stabilizing functional: $\Omega(\mathbf{h}) = \| \mathbf{h} \|^2$. The problem is now clearly defined as:

$$\mathbf{h}^{reg} = \arg \min_{\mathbf{h} \in \mathcal{D}} \left[\| \mathbf{c} \|^2 + \alpha \| \mathbf{h} \|^2 \right] \tag{2.56}$$

where $\mathcal{D}$ is defined by the constraint: $\sum_k \tilde{h}_k^* \tilde{r}_k = d$. It can be solved with Lagrange multipliers by optimizing the following functional:

$$F(\mathbf{h}) = \sum_k |\tilde{h}_k|^2 |\tilde{s}_k|^2 + \alpha \sum_k |\tilde{h}_k|^2 - \lambda \sum_k \tilde{h}_k^* \tilde{r}_k \tag{2.57}$$

Annulling the derivative of $F(\mathbf{h})$ with respect to h_k and taking into account that d is arbitrary, one obtains:

$$\tilde{h}_k^{reg} = \frac{\tilde{r}_k}{\sigma^2 + |\tilde{s}_k|^2} \tag{2.58}$$

where $\sigma^2 = \alpha$ is a positive number. It is obvious from Eq. 2.58 that this optimal filter is nonlinear since it requires a nonlinear transformation of the input image. Furthermore, it is adaptive since the filter function is dependent on the energy spectrum of the input image.

It is interesting to note that if $|\tilde{s}_k|^2$ is replaced by $|\tilde{r}_k|^2$ in Eq. 2.58, one obtains the OT filter for peak sharpness and noise robustness with a white noise model. Moreover, this approach can be generalized to any of the previous stabilizing methods or functionals. For example, it is easily shown that the regularized solution with the minimum norm functional of the optimal trade-off filter between discrimination and correlation peak sharpness (PCE) leads to the following filter [32]:

$$\tilde{c}_k = \frac{\tilde{r}_k}{\sigma^2 + \mu |\tilde{r}_k|^2 + (1 - \mu) |\tilde{s}_k|^2} \tag{2.59}$$

with $\mu \in [0; 1]$.

2.4.3 Some processing examples with stabilized filters

We illustrate in this section the effect of stabilization of filtering techniques with numerical simulations. The minimum norm stabilizing functional is first illustrated with the optimal nonlinear discriminant filter. The target used for these different numerical simulations is shown in Figure 2.9.a and the scene in Figure 2.9.b.

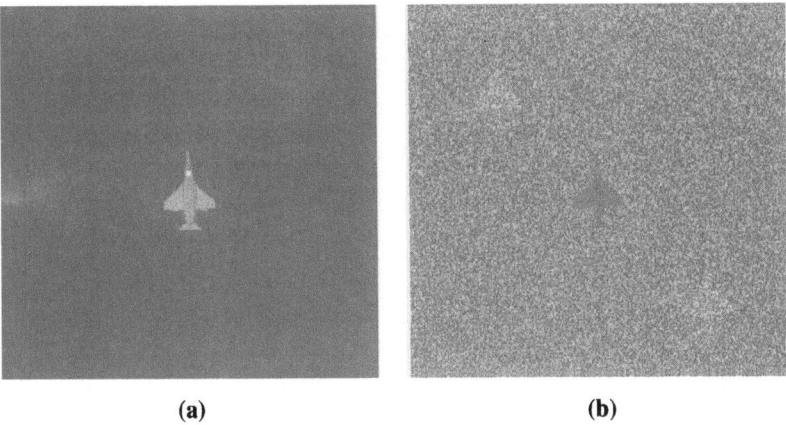

(a)   (b)

Figure 2.9. **(a)** Target used in numerical simulations. **(b)** Input image used for numerical simulations with the optimal discriminant nonlinear filter. ©1993 SPIE.

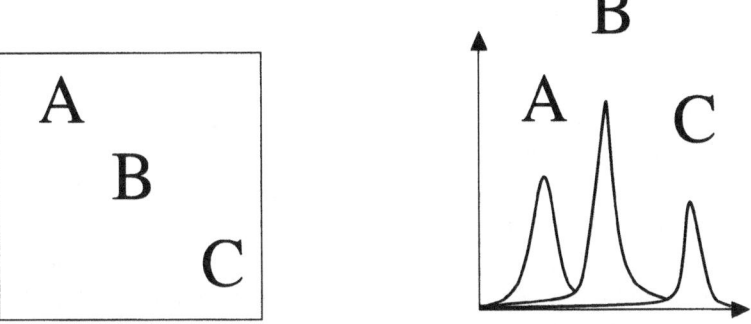

Figure 2.10. Conventions for the different objects present in the input image, and correspondence with the correlation peaks. ©1993 SPIE.

We show in Figure 2.10 our conventions for the different objects in the input image, and the correspondence with the different correlation peaks. The airplanes appear buried in white noise. At the center (position B), the noise is nonoverlapping with the target. However, at positions A and C, the noise is additive (overlapping) with the target. Moreover, the airplane in position A is rotated by 5 degrees with respect to the reference object. In Figure 2.11.a, we show the result of the nonlinear filter of Eq. 2.59 with $\sigma^2 = 0$ and $\mu = 1/2$, i.e., without regularization. It is clearly seen that the correlation peaks are buried in noise, especially the peak corresponding to the rotated airplane. In Figure 2.11.b, a minimum norm regularization is considered (i.e., $\sigma^2 \neq 0$ and $\mu = 1/2$). The efficiency of the minimum norm regularization is clear, since now the correlation peaks can easily be detected.

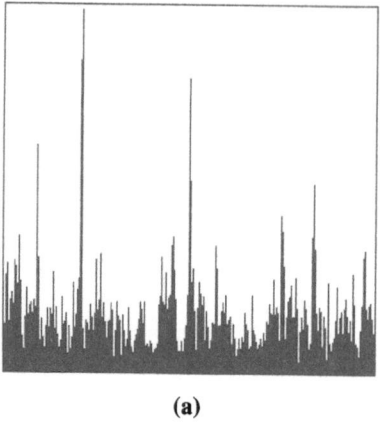 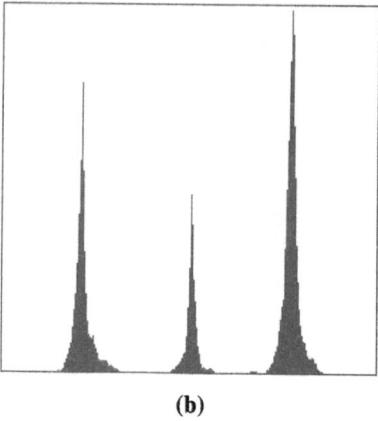

(a)   **(b)**

Figure 2.11. Result of processing the scene in Figure 2.9 with nonlinear filters. **(a)** Correlation with the optimal discriminant nonlinear filter (see Eq. 2.58) without regularization ($\sigma^2 = 0$). **(b)** Correlation with the optimal discriminant nonlinear filter with regularization ($\sigma^2 \neq 0$). ©1993 SPIE.

We now illustrate the effect of the minimum elastic energy stabilization of the OT filter when a single object is present in the learning set. The target is still the airplane of Figure 2.9.a and the input image is now Figure 2.12. The airplanes are embedded in colored noise, with a spectral density equal to $1/[k_x^2 + k_y^2]$ (which will be denoted $1/k^2$ in the following). The noise is overlapping with the target (i.e., additive). At the center (position B), the target has the same orientation as the reference target. At location A, the target is rotated by an angle of 10 degrees and at location C, the target is rotated by an angle of 5 degrees.

In Figure 2.13.a appear the results of the OT filter (see Eq. 2.53) without regularization (that is, with $\alpha = 0$). There is only a thin peak at the location of the nonrotated airplane. The two rotated airplanes are not correctly localized. This is a consequence of the fact that the PSD of the noise decreases in $1/k^2$ and thus tends to zero as the frequency k increases, which makes the OT filter unstable. In Figure 2.13.b, a version of the OT filter regularized with the elastic energy stabilizing functional is considered. The efficiency of this regularization, which is defined in order to stabilize the response of the filter to small variations of the reference object, is clearly seen. Other numerical simulations analyzing the performance of regularized OTSDF filter can be found in Ref. 19. It is important to notice that the MVSDF and MACE filters are limiting cases of the OTSDF filter and they would thus show very unstable behavior in this example.

These two examples clearly illustrate the advantage of using stabilizing functionals when the input image is perturbed. In the next section, we conclude this

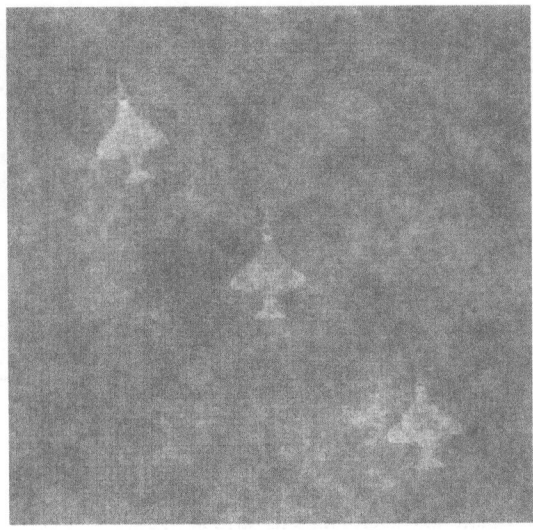

Figure 2.12. Input image used for numerical simulations with OT filters. ©1993 SPIE.

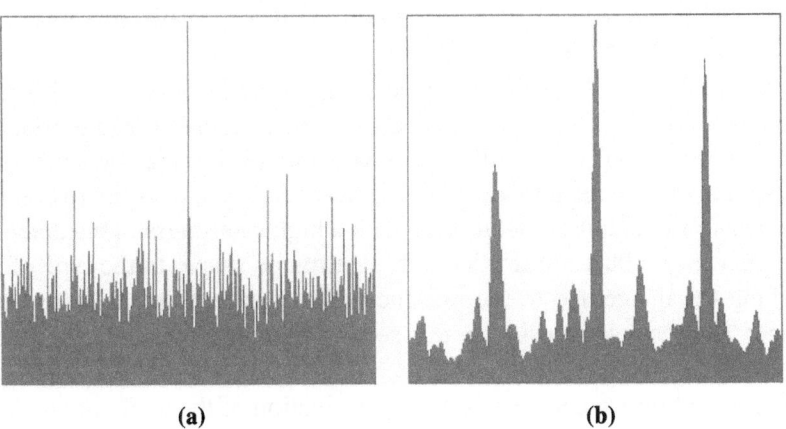

Figure 2.13. Result of processing the scene in Figure 2.12 with OT filters.**(a)** Correlation function with an OT filter, optimized for $1/k^2$ noise and without regularization. **(b)** Correlation function with an OT filter, optimized for $1/k^2$ noise and regularized with the minimum elastic energy stabilizing functional. ©1993 SPIE.

chapter by describing a solution to an image processing problem using OTSDF filters.

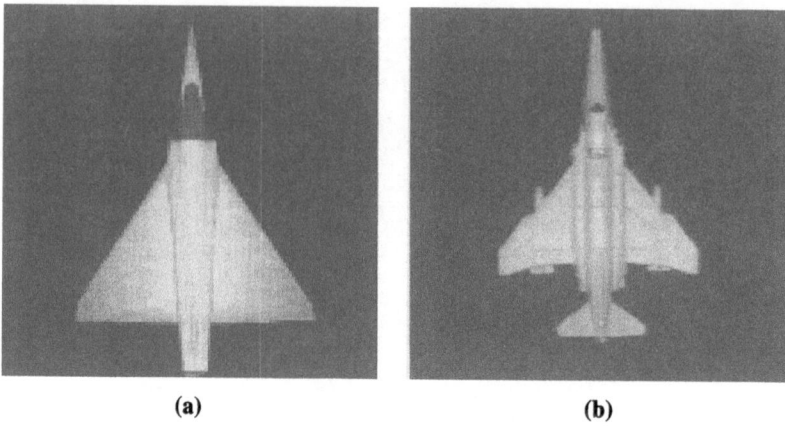

(a)   (b)

Figure 2.14. **(a)** Image used for the determination of the OCCs. **(b)** Airplane model used for synthesizing the filter for attitude determination. ©1994 SPIE.

2.5.   An example of application: Angle estimation of two-dimensional objects

As stated above, OTSDF filters can perform pattern recognition with a certain amount of invariance. This can be used for estimating the in-plane orientation and the position of an object [40]. We show in this section that the multicorrelation technique implemented with OTSDF filters makes it possible to determine the in-plane orientation of an airplane with a high robustness. This determination is also very robust to variations of the airplane shape, to the type of noise perturbing the image and to the presence of hidden parts (occlusions) on the airplane. The results discussed in this section have been presented in Refs. 40 and 41.

The application we consider is the determination of the angle of an airplane model with respect to a given axis (taken as a reference). The filters will be optimized for white noise. Indeed, it can be shown that considering a white noise model and finding optimal trade-off is less sensitive to the noise model than optimizing the filter for colored noise [42]. The multicorrelation technique discussed here consists of a coarse but robust angle estimation (by sectors of 10 degrees). This coarse estimation requires a limited number of 36 different OTSDF filters. It has also been shown in Ref. 40 that another method can be implemented for fine angle determination within the predetermined 10-degree sector. Unless otherwise specified, the image considered for the simulations is the 128×128 pixel image of Figure 2.14.a.

The expression of the OTSDF filter has been given in Eq. 2.30, with vector-matrix notations:

$$\tilde{\mathbf{h}} = \tilde{A}^{-1}\tilde{R}\left[\tilde{R}^{\dagger}\tilde{B}^{-1}\tilde{R}\right]^{-1}\mathbf{d} \quad \text{where} \quad \tilde{B} = (1-\mu)\tilde{S} + \mu\tilde{D} \quad (2.60)$$

We describe here the simple algorithm which allows one to compute an OTSDF filter in practice.

1 For each image $\mathbf{r}^{\ell}$ of the training set, compute its Fourier transform $\tilde{\mathbf{r}}^{\ell}$ and determine $\tilde{D}_k = \sum_{\ell=1}^{P} |\tilde{r}_k^{\ell}|^2$.

2 Having chosen a spectral density model $\tilde{S}_k$ for the noise (for example a constant for a white noise model), determine $\tilde{B}_k = (1-\mu)\tilde{S}_k + \mu\tilde{D}_k$.

3 Compute the matrix $M_{n,m} = \sum_k [\tilde{r}_k^n]^* . \tilde{r}_k^m / \tilde{B}_k$ and determine its inverse $A = M^{-1}$.

4 Let $a_n = \sum_m A_{n,m} d_m$ and compute $\tilde{h}_k = \sum_n a_n \tilde{r}_k^n / \tilde{B}_k$.

5 Repeat these operations with different values of the parameter μ and choose the value such that the OT filter yields a value of the MSE not larger than twice the value obtained with $\mu = 0$. This heuristic method improves significantly the sharpness of the correlation peak without degrading drastically the noise robustness.

The OTSDF filters that we will use will be called *sector filters*. Their role is to characterize the presence or the absence of an airplane with an orientation in a certain angular sector. Their response (i.e., the maximum value of the square modulus of the correlation function) is 1 if the orientation of the airplane lies inside the considered sector and 0 otherwise. The constraints (vector $\mathbf{d}$) imposed on such filters are thus 1 for attitudes of the airplane inside the sector, and 0 for attitudes outside the sector. The whole circle is divided into 36 sectors of 10 degrees, and 36 filters are designed, each being aimed at detecting if the airplane attitude lies in this sector. The sector of 10 degrees containing the orientation of the airplane is thus determined by measuring the maximum value of the separate responses (square modulus of the correlation functions) of each of the 36 sector filters. The filter yielding the maximum response indicates the good sector.

There are many degrees of freedom in designing the sector filters, and the constraints for the filter synthesis are determined experimentally. Indeed, if a large number of constraints is imposed, the noise robustness of the filter is degraded. This is clearly illustrated in Figure 2.15 where the OCCs of three OTSDF filters with different constraint sets are plotted. Filter 1 was constrained only on the attitudes which have to be recognized, every 5 degrees in the sector, that is, for angles $[0°, 5°, 10°]$ (i.e., 3 constraints). Filter 2 corresponds to the following constraints:

$$\text{angle} = [0°, 5°, 10°] \qquad \text{with} \qquad d_{\text{angle}} = 1$$
$$\text{angle} = [-5°, 15°, 115°, 120°, 125°, 130°, \qquad \text{with} \qquad d_{\text{angle}} = 0$$
$$135°, 235°, 240°, 245°, 250°, 255°]$$

(2.61)

Filter 3 was synthesized with constraints every 5 degrees between 0 and 360 degrees (i.e., 72 constraints). Thus, Filter 3 is more constrained than Filter 2, which is more constrained than Filter 1. The OCCs clearly show that for a given value of the ACPE, the MSE increases (i.e., the noise robustness is degraded) when the filter is more constrained.

The second point to take into account is of course that the response of the filter needs to be sufficiently small for angles outside the sector compared to angles inside the sector. Satisfying this point allows one to obtain good discrimination capabilities between the different sectors. In order to make a decision, one has to determine until which noise power each filter is still able to correctly recognize the good sector. Filter 1 cannot be used since it has no discrimination capability. On the other hand, it has been shown in Ref. 40 that Filter 2 is preferable to Filter 3 since it is more robust to noise, and we will thus use Filter 2 in the following simulations.

In order to illustrate the capabilities of OT filters for in-plane angle determination with different kinds of perturbations, an image acquisition system with a digital camera has been implemented [40]. Airplane models are placed under a camera and the analyzed image is sampled with 64 x 64 pixels and 7 bits of graylevels. The filters were determined with the airplane model of Figure 2.14.b.

Examples of images correctly classified by the sector filters are shown in Figure 2.16. These results demonstrate the tolerance of the method to additive noise, out-of-focus and motion blurred input images, changes of airplane model, structured background, and so on. These situations are very different from the additive white noise spectral density model considered for the synthesis of the filters. These results clearly demonstrate the robustness of the correlation operation against different kinds of perturbations provided that adequate filters are used.

Furthermore, the system described above is also tolerant to a change in scale of the object with respect to the reference. The magnitude of the admissible scale variations lies between -20 % and $+30$ % [40].

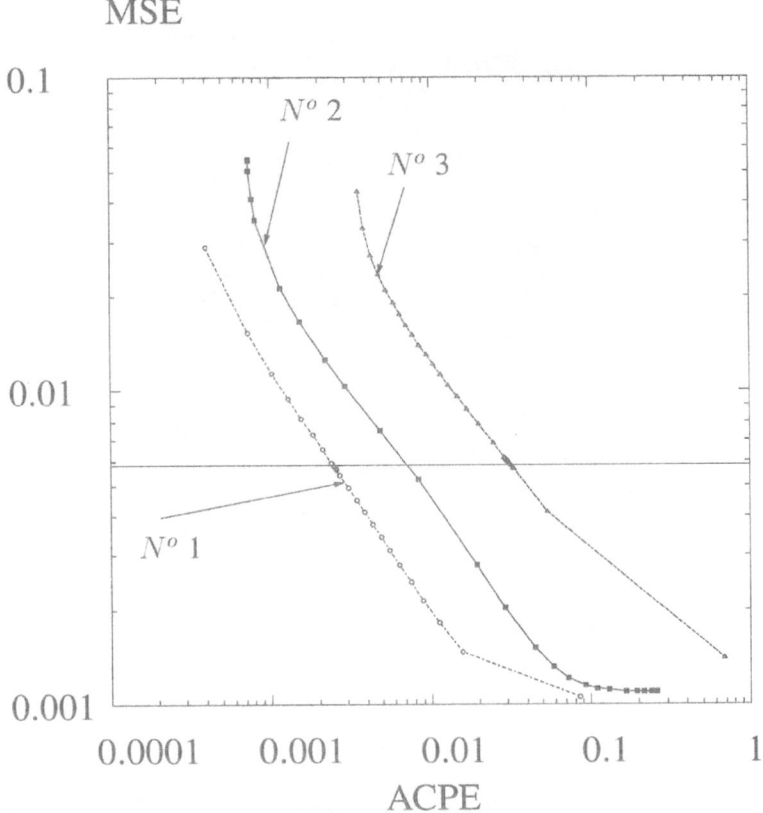

Figure 2.15.   OCCs with three kinds of constraints. See text. ©1994 SPIE.

2.6.   Conclusion

In this chapter, we have presented the main elements of the theory of linear filters for pattern recognition and given some examples of their practical applications. In particular, we have explained how to determine linear filters that rigorously optimize the trade-off between several predefined quality criteria. These Optimal Trade-off filters are thus the best solutions to the optimization of these criteria.

However, the criteria considered in this chapter, such as MSE or ACPE, are heuristic. Although they are likely to lead to filters with good processing qualities, they are not directly linked to the criteria of final interest in pattern recognition, which are the probability of detection, the probability of correct classification, the variance of position estimation or of angle estimation, etc. The filters determined in this chapter are not guaranteed to optimize these fig-

Figure 2.16.   Example of images for which the airplane rotation is correctly estimated. ©1994
SPIE.

ures of merit. The solution to this problem is to use statistical decision and
estimation theory, which makes it possible to determine the processing methods that directly optimize the criteria of interest – methods which may no longer
be linear filters. This theory and its practical applications to pattern recognition
will be the topic of the sequel of this book.

APPENDIX 2.A: Definitions and notation

In this appendix, we introduce some notations and definitions which are used throughout the
book. We will consider discrete images, which are functions of with possibly complex values
on a discrete 2-dimensional grid G. Each node of the grid will be defined by two indices m and

n, and will be called a *pixel*. The sensors that we will consider in this book provide real-valued measures, however, we will introduce complex images as results of Fourier transforms.

We will consider that the grid on which the images are defined has M lines and N columns. These images can then be defined as matrices of dimension $M \times N$ and will be denoted by bold-font symbols. The value of the image s corresponding to the pixel (m, n) will be denoted $s_{m,n}$. However, when possible, it may be preferable to represent an image by a vector rather than a matrix. In particular, this makes it possible to represent linear operators on images by matrices. There exist several ways to order the elements of a matrix in a vector. We will choose the lexicographic order, which consists in reading the matrix from left to right, downwards. Thus the first N elements of the vector will contain the first line, the N next ones the second line, and so on. The obtained vector is of length $L = M \times N$.

In Chapters 2 and 3, we will use the vector notation whenever possible, since this makes the notations simpler. In the other chapters, we will use the 2-D notation. In all cases, we will use the same bold-font symbol to denote the matrix and the vector image, in order to simplify notation. Between these two representations, we will have the following relation:

$$s_{m,n} = s_{mN+n} \tag{2.A.1}$$

We will first describe some notions of linear signal processing on 1-D vectors, then we will generalize them to 2-D images. We will show that linear operations can be formally denoted in the same way for 1-D and 2-D vectors, so that as far as linear filtering is concerned, the two notations are equivalent. We will also define the notion of cyclostationary noise and analyze the particular case of Gaussian cyclostationary noise.

Linear processing on images

Let us first consider a 1-D signal s of length N. The discrete Fourier transform is an example of linear operator on a discrete signal. The Fourier transform $\tilde{s}$ of vector s with N components is defined as:

$$\tilde{s}_k = \frac{1}{\sqrt{N}} \sum_{n=0}^{N-1} s_n \exp\left[-2i\pi \frac{nk}{N}\right] \tag{2.A.2}$$

where $i^2 = -1$. The Fourier transform can be viewed as a linear operation consisting of applying to s the following matrix F:

$$F_{n,n'} = \frac{1}{\sqrt{N}} \exp\left[-2i\pi \frac{nn'}{N}\right] \tag{2.A.3}$$

It can be shown that F is a unitary matrix, that is, $FF^\dagger = I$ where the symbol $\dagger$ represents the transconjugation operator and I is the identity matrix. One can thus write:

$$\tilde{s} = F s \ , \ s = F^\dagger \tilde{s}, \ \text{and} \ \tilde{A} = F A F^\dagger \tag{2.A.4}$$

where s is an N-component image vector, A a square matrix, and $\tilde{s}$ and $\tilde{A}$ their respective Fourier transforms. Thanks to the fact that F is unitary, the discrete Fourier transform verifies the Parseval relation :

$$s^\dagger r = s^\dagger F^\dagger F r = (Fs)^\dagger (Fr) = \tilde{s}^\dagger \tilde{r} \tag{2.A.5}$$

One of the operations that we will apply most frequently to images is the correlation with a filter. This operation is defined as:

$$c_n = [\mathbf{h} \star \mathbf{s}]_n = \sum_{u=0}^{N-1} h_{u-n} s_u \tag{2.A.6}$$

where **h** represents the impulse response of the filter. One can remark that in order for this definition to be valid for every $u \in [0, N-1]$, one has to define what happens at the borders of the image. We will choose circular – or periodic – conditions, that is, in Eq. 2.A.6, $u - n$ should be read $(u - n)$ modulo N. The so-defined operation is called *circular correlation*.

The operation of circular correlation can be written in the form of a matrix transformation $\mathbf{c} = H\mathbf{s}$ with:

$$H = \begin{bmatrix} h_0 & h_1 & \dots & h_{N-1} \\ h_{N-1} & h_0 & \dots & h_{N-2} \\ \vdots & \vdots & & \vdots \\ h_1 & h_2 & \dots & h_0 \end{bmatrix} = \text{circ}(h_0, h_1, \dots, h_{N-1}) \qquad (2.A.7)$$

Such a matrix is termed *circulant*. More precisely, a square matrix M with dimension $N \times N$ is termed *circulant* if $M_{u,v} = M_{0,(v-u)\text{modulo}N}$. In other terms, each line of the matrix is deduced from the previous one by a circular translation. Such a matrix is thus entirely defined by its first line.

If the correlation operation is defined in this way, the expression of the Fourier transform of the correlation product is very simple, since it is the product of the complex conjugate of the Fourier transform of the filter impulse response $\tilde{h}^*$ and the Fourier transform of the image $\tilde{s}$:

$$\tilde{c}_k = \tilde{h}_k^* \tilde{s}_k \qquad (2.A.8)$$

In matrix notation, one has:

$$\tilde{\mathbf{c}} = F H \mathbf{s} = F H F^\dagger F \mathbf{s} = F H F^\dagger \tilde{\mathbf{s}} \qquad (2.A.9)$$

and $F H F^\dagger = \text{diag}(\tilde{h}_0^*, \dots, \tilde{h}_{N-1}^*)$. In other words, if H is a circulant matrix, it is diagonalizable in the space spanned by the column vectors of matrix F and its eigenvalues are the Fourier transform of its first row.

In the Fourier space, circular correlation thus amounts to computing N products. It can be shown that for filters whose support is large enough, it is more efficient to perform the computation in the Fourier space, provided a Fast Fourier Transform (FFT) algorithm is used to compute the Fourier transforms. Choosing circular correlation is thus mainly a way of simplifying the computations, at the price of a physical inconsistency. Indeed, there is no reason why the pixels outside the right side border should look like those of the left part of the signal.

Let us now consider that the signal **s** and the filter **h** are 2-D images of dimension $M \times N$. The definition of correlation is then:

$$c_{m,n} = \sum_{u=0}^{M-1} \sum_{v=0}^{N-1} h_{u-m,v-n}\, s_{u,v} \qquad (2.A.10)$$

If the images **s** and **h** are put in vector form by lexicographic ordering, the correlation operation can be expressed in the form of a matrix-vector product $\mathbf{c} = H\mathbf{s}$ with

$$H = \begin{bmatrix} H^0 & H^1 & \dots & H^{M-1} \\ H^{M-1} & H^0 & \dots & H^{M-2} \\ \vdots & \vdots & & \vdots \\ H^1 & H^2 & \dots & H^0 \end{bmatrix} = \text{circ}(H^0, H^1, \dots, H^{M-1}) \qquad (2.A.11)$$

where the matrices H^i are circulant and defined as follows: $H^i = \text{circ}(h_{i,0}, h_{i,1}, \dots, h_{i,N-1})$. The matrix H is termed *doubly block-circulant*.

The 2-D Fourier transform is defined as:

$$\tilde{s}_{k,l} = \frac{1}{\sqrt{MN}} \sum_{m=0}^{P-1} \sum_{n=0}^{Q-1} s_{m,n} \exp\left[-2i\pi\left(\frac{mk}{M} + \frac{nl}{N}\right)\right] \tag{2.A.12}$$

This operation can be expressed as a product of the lexicographic vector s with a matrix $\mathcal{F}$, so that: $\tilde{s} = \mathcal{F}s$. It can be shown that doubly block-circulant matrices are diagonal in the space spanned by the column vectors of $\mathcal{F}$ [76]. Consequently, the Fourier transform of a correlation product is equal to $\tilde{c}_k = \tilde{h}_k^* \tilde{c}_k$. This is identical to the relation obtained in the case of 1-D vectors, but now, the coefficients $\tilde{h}_k$ and $\tilde{s}_k$ are those of the 2-D Fourier transform of the 2-D image reordered in vector form with lexicographic ordering.

This result shows that as far as correlation and Fourier transform operations are concerned, the same notations can be used for 1-D vectors and 2-D images put in vector form. We use this property in Chapters 2 and 3 of this book.

Cyclostationary noise

Let us now define the notion of cyclostationary noise. We define a random vector b as a 1-D vector whose N elements are random variables. We will use this notion to model noise in discrete images. A random vector is termed *second-order cyclostationary* if it possesses the following properties:

1 The vector is periodic in the sense that $b_{n-n'} = b_{(n-n')\text{modulo}N}$.

2 One has the following property:

$$\langle b_i \rangle = m \tag{2.A.13}$$
$$S_{n,n'} = \langle [b_n - m][b_{n'} - m] \rangle = S_{0,(n-n')\text{modulo}N} \tag{2.A.14}$$

where the symbol $<>$ denotes ensemble averaging. Equation 2.A.13 means that the average is the same at each position of the signal. Equation 2.A.14 means that the covariance matrix S is symmetric and circulant. In this case, it is easily seen that its Fourier transform $\tilde{S}$ is diagonal. The diagonal coefficients $\tilde{S}_k$, $k \in [1, N]$ define the *power spectral density* (PSD) of the noise. If the PSD is constant, that is, it does not depend on k, then the noise is termed *white*.

The notion of cyclostationary noise can be generalized to 2-D random images, which we will call *random fields*, in the same way as the correlation operation.

Gaussian cyclostationary noise

As an example of cyclostationary noise, we now consider a Gaussian random vector. It is defined by its mean m and its covariance matrix S, which is symmetric and circulant in order to ensure cyclostationarity. Its probability density function (pdf) is by definition:

$$P_b(b) = \frac{1}{(2\pi)^{N/2}[\det S]^{1/2}} \exp-\left[\frac{1}{2}(b-m)^T S^{-1}(b-m)\right] \tag{2.A.15}$$

where T denotes transposition, m is a vector whose N elements are m, and det S is the determinant of matrix S. Since the matrix S is circulant, its Fourier transform $\tilde{S}$ is diagonal. Using the Parseval relation and the fact that the determinant of a matrix is the product of its eigenvalues, the pdf of the Gaussian random field can also be written as a function of the Fourier transform $\tilde{b}$ of b:

$$P_b(b) = \frac{1}{(2\pi)^{N/2} \prod_{k=0}^{N-1} \tilde{S}_k^{1/2}} \exp\left[-\frac{1}{2}\sum_{k=0}^{N-1} \frac{|\tilde{b}_k - m\,\delta_k|^2}{S_k}\right] \tag{2.A.16}$$

where δ_k is the Kronecker symbol: $\delta_k = 1$ if $k = 0$ and $\delta_k = 0$ otherwise.

As an example, let us consider a Gaussian random field with the following PSD (in 2-D notation):

$$\tilde{S}^p_{k,l} = \frac{1}{(k^2 + l^2)^{p/2}} \qquad (2.A.17)$$

This type of random field is often called $1/k^p$ *noise*. We have represented in Figure 2.A.1 realizations of such a noise for different values of p. If $p = 0$, the PSD is uniform and the noise is totally uncorrelated, or "white." We can see that as p increases, the texture of the random field becomes more and more correlated.

APPENDIX 2.B: Lagrange multipliers

The method of Lagrange multipliers is a powerful tool for searching an extremum in a function under some constraints. In this appendix, we present this method and limit ourselves to the case where the function to be optimized has a finite number of variables.

Let $E(x_1, x_2, \ldots, x_N)$ be a function of N variables defined on a domain $\mathcal{D}$. One searches for the point $(x_1^m, x_2^m, \ldots, x_N^m)$ which minimizes $E(x_1, x_2, \ldots, x_N)$ on the subset $\mathcal{D}$ defined by:

$$g(x_1, x_2, \ldots, x_N) = 0$$

It is easily verified that this formulation is very general.

One will assume that in the neighborhood of each point $(x_1, x_2, \ldots, x_N)$ of $\mathcal{D}$, $g(x_1, x_2, \ldots, x_N)$ defines an implicit function:

$$x_N = X_n(x_1, x_2, \ldots, x_{N-1})$$

Since x_N does not play any particular role in the following, if the condition to define an implicit function with respect to x_N is not verified, one may use another coordinate.

The minimum value of $E(x_1, x_2, \ldots, x_N)$ under the constraint $g(x_1, x_2, \ldots, x_N) = 0$ can thus be obtained by injecting $x_N = X_n(x_1, x_2, \ldots, x_{N-1})$ and by minimizing:

$$E(x_1, x_2, \ldots, X_n(x_1, x_2, \ldots, x_{N-1}))$$

with respect to $x_1, x_2, \ldots, x_{N-1}$:

$$\frac{\partial E}{\partial x_i} = 0 \ , \ \forall i \in [1, N-1]$$

which can also be written as:

$$\frac{\partial E(x_1, x_2, \ldots, x_N)}{\partial x_i} + \frac{\partial E(x_1, x_2, \ldots, x_N)}{\partial x_N} \frac{\partial X_n(x_1, x_2, \ldots, x_{N-1})}{\partial x_i} = 0$$

Moreover, the constraint gives us the following relation:

$$\frac{\partial g(x_1, x_2, \ldots, x_N)}{\partial x_i} + \frac{\partial g(x_1, x_2, \ldots, x_N)}{\partial x_N} \frac{\partial X_n(x_1, x_2, \ldots, x_{N-1})}{\partial x_i} = 0$$

These two relations can be rewritten in matrix form:

$$\begin{pmatrix} \frac{\partial E}{\partial x_i} & \frac{\partial E}{\partial x_N} \\ \frac{\partial g}{\partial x_i} & \frac{\partial g}{\partial x_N} \end{pmatrix} \begin{pmatrix} 1 \\ \frac{\partial X_n}{\partial x_i} \end{pmatrix} = \begin{pmatrix} 0 \\ 0 \end{pmatrix}$$

This relation means that the two vectors formed by the lines of the matrix are proportional:

$$\begin{pmatrix} \frac{\partial E}{\partial x_i} \\ \frac{\partial E}{\partial x_N} \end{pmatrix} = \lambda_i \begin{pmatrix} \frac{\partial g}{\partial x_i} \\ \frac{\partial g}{\partial x_N} \end{pmatrix} \ , \ \forall i \in [1, N-1]$$

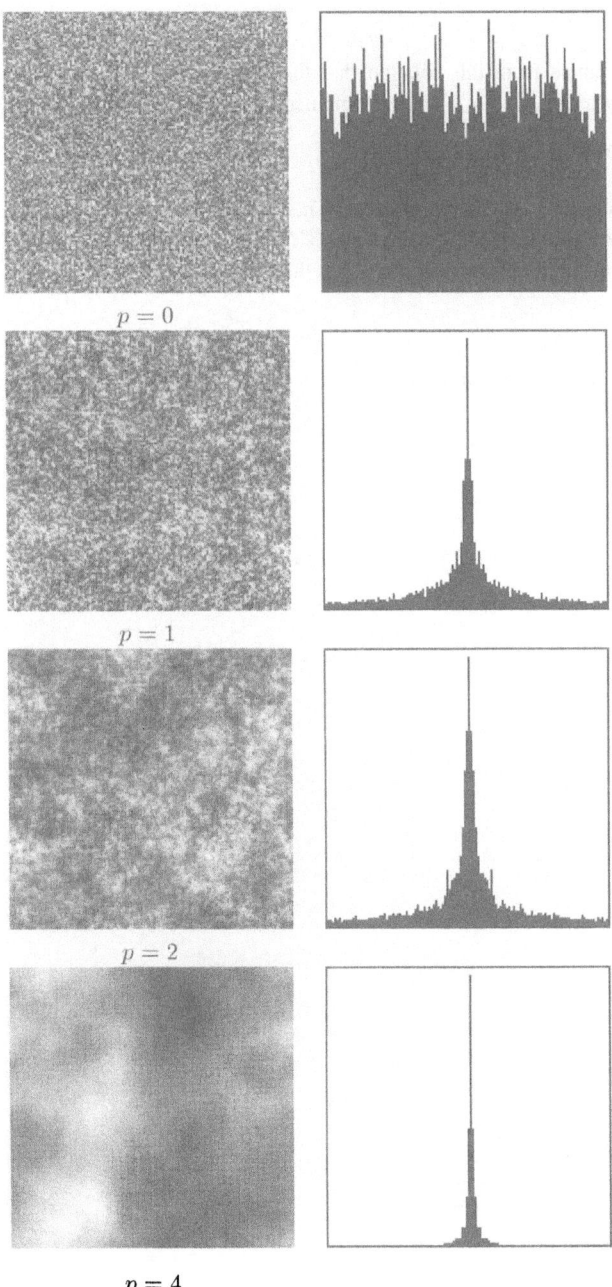

$p = 0$

$p = 1$

$p = 2$

$p = 4$

Figure 2.A.1. First column: realizations of $1/k^p$ noises (see Eq. 2.A.17) for different values of p. Second column: cross-section of the estimate of the PSD from these realizations.

However, since $\frac{\partial E}{\partial x_N} = \lambda_i \frac{\partial g}{\partial x_N}$, $\forall i \in [1, N-1]$, one has necessarily $\lambda_i = \lambda$, $\forall i \in [1, N-1]$ and thus:

$$\frac{\partial E}{\partial x_i} - \lambda \frac{\partial g}{\partial x_i} = 0$$

One can thus deduce from this relation that the minimum of $E(x_1, x_2, \ldots, x_N)$ under the constraint that $g(x_1, x_2, \ldots, x_N) = 0$ is obtained by minimizing the following function:

$$\Psi(x_1, x_2, \ldots, x_N) = E(x_1, x_2, \ldots, x_N) - \lambda \, g(x_1, x_2, \ldots, x_N)$$

where λ is called the *Lagrange multiplier*. The minimization of $\Psi(x_1, x_2, \ldots, x_N)$ leads to a parametric form of the minimum $(x_1^m(\lambda), x_2^m(\lambda), \ldots, x_N^m(\lambda))$. The actual value λ_0 of λ is obtained by ensuring that the solution satisfies the constraint:

$$g\left(x_1^m(\lambda_0), x_2^m(\lambda_0), \ldots, x_N^m(\lambda_0)\right) = 0$$

The solution of the optimization problem is then $(x_1^m(\lambda_0), x_2^m(\lambda_0), \ldots, x_N^m(\lambda_0))$.

Chapter 3

STATISTICAL CORRELATION TECHNIQUES

The image processing techniques introduced in the previous chapter optimize heuristic criteria. This method of algorithm design presents some advantages. It makes it possible to introduce very naturally additional constraints, such as regularization terms, and to handle large training sets in an efficient way. However, there is no proof that these algorithms optimize the parameters which are of interest for the user. These parameters may be the probability of recognition for classification applications, estimator variance for angle estimation, or detection probability for detection tasks. The appropriate framework for designing image processing algorithms which optimize such parameters is statistical decision and estimation theory.

Using this approach requires one to clearly specify the sources of randomness in the image. This is done by choosing a probability density function (pdf) to represent the perturbations that affect the image, although it is not necessary to specify all the parameters of this pdf. For example, if the input noise is additive, Gaussian, stationary, and ergodic with a known power spectral density (PSD), it is well known that the classical matched filter optimizes the probability of correct location of a target with known gray levels [43]. For such a noise model, the techniques presented in the previous chapter are thus optimal in the sense of statistical decision and estimation theory, which means that for this type of noise, any other filter, either linear or nonlinear, will result in a lower probability of detection or of correct location than the matched filter. However, if one or more of the above assumptions do not hold, there may exist a better solution than the matched filter. Indeed, the main advantage of statistical decision and estimation theory is that it takes into account the actual characteristics of the noise present in the image to design optimal algorithms.

In the present chapter, we will first discuss in Section 3.1 the limits of the additive, Gaussian, ergodic assumption for the noise in images. We will then

49

introduce in Section 3.2 the basics of statistical decision and estimation theory, which will be used throughout the book. In the rest of the chapter, we will give several examples of the efficiency of this approach to solve target location problems in a variety of image processing situations. In Section 3.3, we will address the conditions for which matched filtering is optimal. We will then tackle a more original problem in Section 3.4, where we will design filters which are optimal when the PSD of the noise is *a priori* unknown. Finally, in Section 3.5, we will apply statistical decision theory to the design of location algorithms optimal in the presence of nonoverlapping noise, which is of prime importance in image processing. Further examples of the power of the decision-theoretic approach for solving image processing applications involving different physical noise models will be given in Chapter 4.

3.1.   Some sources of noise in images

The detection and location of a target in a scene are classical problems, pervasive to many image processing applications. However, the matched filter, as well as the improved linear filtering techniques introduced in the previous chapter, may be shown to perform poorly on certain types of real-world images [44, 45]. This is because such images do not belong to the class for which linear filtering is optimal.

In general, in real-world images, the main source of noise is not the additive detector noise, but the whole background of the scene (clutter), which does not overlap with the target [44]. Second, the gray levels of the reference object can be unknown *a priori*, or may fluctuate. Third, the PSD of the noise is also often unknown *a priori*. In each of these three cases, one important assumption which is necessary to demonstrate the optimality of the matched filter is not fulfilled. As a practical consequence, it has been frequently observed that the matched filter yields poor performance in some real applications of image processing.

In this section, we will discuss in more detail the three cases mentioned above, in which one of the assumptions necessary for the optimality of the matched filter is not fulfilled [46].

3.1.1   Nonoverlapping noise

If the main source of noise in the input image is not the detector noise, which can be considered as additive, but the whole background of the scene, the noise is nonadditive since it does not affect the pixels of the target. It can thus be called *nonoverlapping* [44]. The difference between additive and nonoverlapping noise is illustrated in Figure 3.1. In this figure, it is also pointed out that many real-world scenes can be more accurately represented by a nonoverlapping noise model than by an additive noise model.

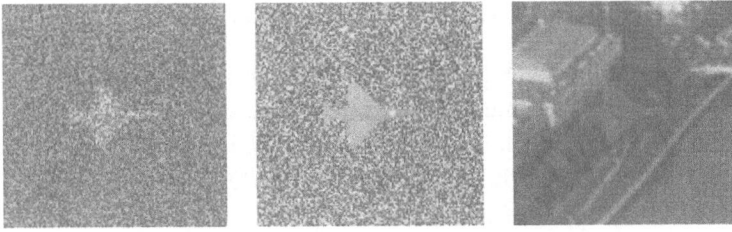

Figure 3.1. Left: additive noise. Center: nonoverlapping noise. Right: realistic image; it can be noticed that the main source of noise is constituted by the background, and is thus nonoverlapping. ©1997 SPIE.

In the presence of nonoverlapping noise, the important task is to discriminate the target with the background clutter and classical linear filters often fail to correctly locate or detect the target in this case [45]. We show in Figure 3.2 (bottom line) an example of such a situation, in which we can clearly observe that the matched filter, the Optimal Trade-off filter and the POF filter, which are all linear filters, are unable to locate a target that can be seen very easily on the image. The same linear filters are efficient (when slightly regularized) in the presence of additive noise (Figure 3.2, top row). We will discuss in Section 3.5 an approach which can solve this problem.

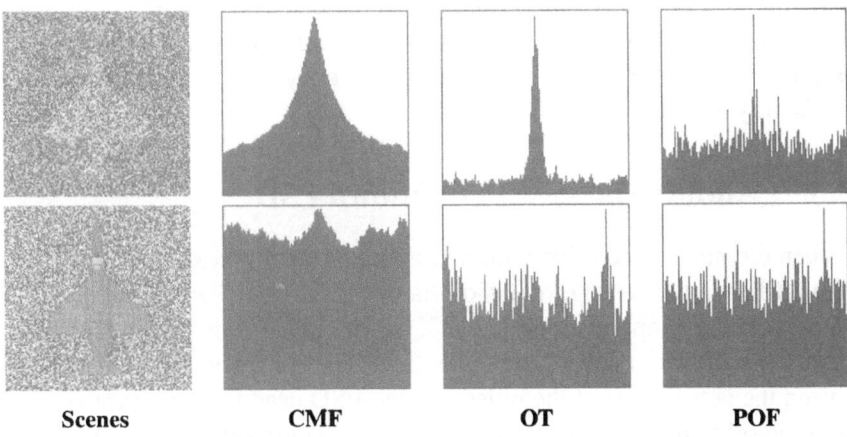

Scenes   **CMF**   **OT**   **POF**

Figure 3.2. Top row: scene with additive noise and result of the processing of this scene with the matched filter (Eq. 2.11), an Optimal Trade-off filter (Eq. 2.33), and a Phase-Only filter (Eq. 2.15) slightly regularized. Bottom row: scene containing the same object (airplane) as above, but corrupted with nonoverlapping noise, and result of the processing of this scene with the same linear filters as above. The curves in the bottom row represent the maximum value of each line of the modulus square of the correlation plane obtained with the corresponding filter. ©1997 SPIE.

3.1.2   Fluctuations of the target's gray levels

Most optimal filtering techniques, such as those introduced in the previous chapter, assume that the internal structure of the target, that is, its gray level distribution, is deterministic and known. This is in particular the case of the matched filter, the MACE, the Optimal Trade-off (OT), and the MINACE filters. However, in many applications, this assumption is not realistic. For example, the target may be subject to sun reflections in optical images or to temperature changes in infrared images: The gray levels of the target then fluctuate in an unpredictable way.

In order to illustrate this problem, let us look at Figure 3.3. The two scenes in the first column represent an object with a given shape in the presence of nonoverlapping noise. Both scenes are processed with a linear filter (second column) and the maximum likelihood algorithm that will be introduced in Section 3.5.2 (third column). The images in the first row are the reference objects from which the corresponding filters have been designed. It can be seen that the linear filter fails to locate the target in the scene of the second row, since it is not adapted to nonoverlapping noise. However, the algorithm adapted to this type of noise correctly locates the target. In the third row, the scene also contains nonoverlapping noise, but its gray level internal structure is different from that of the reference object utilized to build the filters. It can be seen that the optimal filter for nonoverlapping noise also fails on this image. An optimal maximum likelihood algorithm adapted to targets with fluctuating graylevels will be derived in Section 3.5.3. Its behavior is represented in the fourth column of Figure 3.3: It is robust to nonoverlapping noise and to fluctuations of the target's gray levels.

3.1.3   Additive noise with unknown PSD

When the noise is additive with a known PSD, we will show below that the optimal filter in the sense of decision theory is the well-known matched filter. However, in many applications, the PSD of the noise is unknown *a priori* and can vary rapidly from one image to another. The consequences of a difference between the actual PSD of the noise and the PSD used for filter synthesis have been precisely studied in Ref. 18 for Optimal Trade-off filters and nonlinear filters. It has been shown that the matched filter is in general very sensitive to such a difference. It has been demonstrated in Ref. 18 that Optimal Trade-off filters designed with a white noise model are less sensitive to that problem than the matched filter, since they are stable while the matched filter can be unstable. However, Optimal Trade-off filters are not optimal when the PSD of the noise is unknown. It would thus be useful to find out the optimal algorithm in that case. We will address this issue in Section 3.4.

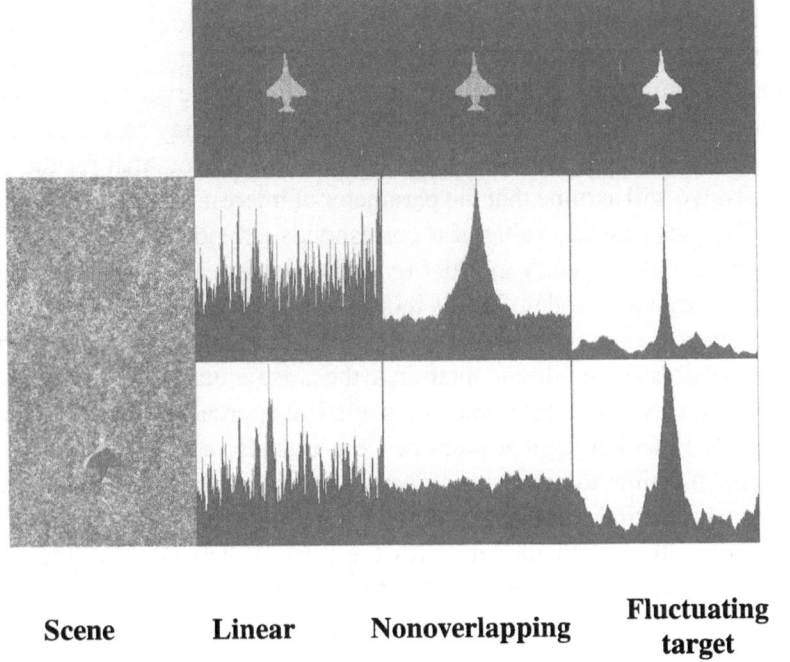

Scene   **Linear**   **Nonoverlapping**   **Fluctuating target**

Figure 3.3. Behavior of several filters on a target with fluctuating graylevels (see text). ©1997 SPIE.

3.2. Background on statistical decision and estimation theory

This brief review of the cases where the matched filter fails shows that in order to design efficient pattern recognition techniques, it is necessary to take into account the nature of the perturbations that affect the image. Statistical decision and estimation theory provides efficient tools for solving this problem. There exist many classical and authoritative books on this topic [29,47-49] and we will in the present section only briefly describe the notions and the tools that will be of direct use in the sequel of this book.

3.2.1 Decision and estimation theory without nuisance parameters

Decision and estimation theory consists of extracting some information from an image, in the form of parameter values, by assuming that this image is the realization of a random process. Let us denote the image s and the parameter to estimate from this image θ. The nature of this parameter will vary according to the type of information that has to be extracted from the image. For example, in a detection application, θ will be a binary variable: $\theta = 1$ if the object of interest is present and $\theta = 0$ if it is absent. In target location application, θ will represent the coordinates of the target's position in the scene. For in-plane orientation estimation, θ represents the angle of the target with respect to a reference axis. In target recognition applications, θ may represent the class of target: for example, the class *tank*, *truck*, or *jeep*. Note that for the sake of simplicity, we will assume that the parameter of interest θ can only take discrete values. To each possible value of θ corresponds a hypothesis H_θ.

The first point to specify in order to apply decision and estimation theory is the image formation model. This is indeed an important property of this theory to allow one to design optimal methods adapted to each image formation model. It is precisely because of this adaptation to the noise actually present in the image that better results can be obtained with statistical approaches than with heuristic ones. From a methodological point of view, it is also an important property of decision and estimation theory to clearly specify – in a statistical sense – the problem which will be solved.

The image formation model states the relation between the image s, the reference object r, and the noise n for a given hypothesis H_θ:

$$H_\theta \quad : \quad \mathbf{s} = F_\theta(\mathbf{r}, \mathbf{n}) \tag{3.1}$$

For example, for an estimation problem such as target location in the presence of additive noise, this model is:

$$H_\theta \quad : \quad s_i = r_{i-\theta} + n_i \; ; \quad \forall i \in [0; N-1] \tag{3.2}$$

where θ represents the unknown location of the target (it is a scalar value if the signal is one dimensional) and $r_{i-\theta}$ is a version of the reference r translated of the value θ. On the other hand, for target location in the presence of multiplicative noise, one has:

$$H_\theta \quad : \quad s_i = n_i \, r_{i-\theta} \; \forall i \in [0; N-1] \tag{3.3}$$

In the general case, the noise is not independent from the target and the image formation model is deduced from the conditional probability $P[\mathbf{s}|\mathbf{r}, \theta]$, which represents the probability density of observing s assuming that the value of the unknown parameter is θ and the gray levels of the reference are r. $P[\mathbf{s}|\mathbf{r}, \theta]$ plays a central role in decision and estimation theory and is called the *likelihood* of hypothesis H_θ. Indeed, this probability represents the likelihood of the hypothesis that the value of the parameter of interest is equal to θ, assuming that the target is known and equal to r. For this reason, $P[\mathbf{s}|\mathbf{r}, \theta]$ will be denoted $L[\mathbf{s}|\mathbf{r}, \theta]$ in the following.

The likelihood can be determined very easily from the image model itself. For that purpose, let $P_\mathbf{n}(\mathbf{n})$ denote the pdf of the noise and assume that Eq. 3.1

can be rewritten in the form:

$$\mathbf{n} = F_{\boldsymbol{\theta}}^{-1}(\mathbf{r}, \mathbf{s}) \tag{3.4}$$

The expression of the likelihood is then:

$$L[\mathbf{s}|\mathbf{r}, \boldsymbol{\theta}] = P_{\mathrm{n}}\left(F_{\boldsymbol{\theta}}^{-1}(\mathbf{r}, \mathbf{s})\right) \tag{3.5}$$

As explained in Section 3.1, the target graylevel reference $\mathbf{r}$ may not be known, but only the target shape $\mathbf{w}$ ($\mathbf{w}$ is a binary mask whose value is 1 within the target and 0 outside). In this case, the likelihood may be written: $L[\mathbf{s}|\mathbf{w}, \boldsymbol{\theta}]$. We can now describe the Maximum Likelihood (ML) principle which is of prime importance in statistical inference problems.

3.2.1.1   Maximum likelihood principle

The ML estimate of the unknown parameter $\boldsymbol{\theta}$ is the value which maximizes $L[\mathbf{s}|\mathbf{p}, \boldsymbol{\theta}]$ where $\mathbf{p}$ is any known parameter. The parameter $\mathbf{p}$ can be for example $\mathbf{r}$ or $\mathbf{w}$. The ML estimate of $\boldsymbol{\theta}$ can thus be written:

$$\widehat{\boldsymbol{\theta}}^{ML} = \arg\max_{\boldsymbol{\theta}} L[\mathbf{s}|\mathbf{p}, \boldsymbol{\theta}] \tag{3.6}$$

Since the likelihood $L[\mathbf{s}|\mathbf{p}, \boldsymbol{\theta}]$ is the probability of observing $\mathbf{s}$ under the assumption that the value of the unknown parameter is $\boldsymbol{\theta}$, the basic physical idea behind the ML principle is to choose the value of $\boldsymbol{\theta}$ which makes very probable (or likely) the data that has been observed. This is a very reasonable approach but it can also appear arbitrary at this level.

3.2.1.2   Maximum *a posteriori* principle

The Maximum likelihood approach provides an efficient decision rule for parameter estimation, but its optimality properties are not always clearly defined in the context of hypothesis testing. In order to go further, one can specify a numerical criterion which indicates the goodness of any decision rule, and determine the decision rule which optimizes it.

Let us first define some notations. In all the applications considered in the following, we will assume that the parameter $\boldsymbol{\theta}$ can only take discrete values. The problem we address is thus classification between a finite number K of hypotheses. The different hypotheses $H_{\boldsymbol{\theta}_1}, \ldots, H_{\boldsymbol{\theta}_K}$ correspond to the different possible values $\boldsymbol{\theta}_1, \ldots, \boldsymbol{\theta}_K$ of the parameter. The classification is done by assigning to each realization $\mathbf{s}$ of the data one of the possible values of the parameter. If we let S denote the set of possible values of the image $\mathbf{s}$, performing classification amounts to partitioning this space into K regions $\Omega_1, \ldots, \Omega_K$, so that if a given realization $\mathbf{s}$ of the image belongs to region Ω_k, then hypothesis

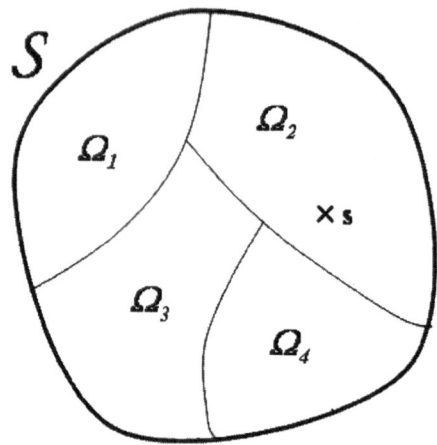

Figure 3.4. Regions $\Omega_1, \ldots, \Omega_4$ defined by a decision rule over the set of possible data realizations S. In this example, the value of the parameter chosen for the data realization s will be θ_2.

H_{θ_k} is chosen (see Figure 3.4). The purpose of the design of a decision rule is to determine the best delimitation of regions Ω_k.

In order to solve this problem, one first has to define a criterion which indicates the goodness of a decision rule. We will consider in the following the probability of error. In order to define this criterion, one considers the *posterior probability density* $P[\theta_k|\mathbf{s}, \mathbf{p}]$. This is the probability that a given realization of the data $\mathbf{s}$ corresponds to the hypothesis H_{θ_k}. It is directly related to the likelihood through the Bayes rule:

$$P[\theta|\mathbf{s}, \mathbf{p}] = \frac{L[\mathbf{s}|\mathbf{p}, \theta]P[\theta]}{P[\mathbf{s}]} \tag{3.7}$$

$P[\theta]$ is called the *prior probability density* for the unknown parameter θ and is frequently simply referred to as *prior*. For example, when tracking a target in an image sequence, this prior can be obtained from the previous estimated location using a location prediction algorithm [50].

A classification error is obtained when the decision rule assigns a realization s belonging in the class θ_j to the class θ_i, with $i \neq j$. The probability P_{ij} of obtaining such an error is easily defined from the posterior density:

$$P_{ij} = \int_{\Omega_i} P[\theta_j|\mathbf{s}, \mathbf{p}]P[\mathbf{s}]ds \tag{3.8}$$

where $\int F[\mathbf{s}]ds$ is a formal notation for: $\int_{s_0} \int_{s_1} \ldots \int_{s_{N-1}} F[\mathbf{s}]ds_0 ds_1 \ldots ds_{N-1}$ and $ds = ds_0 \ldots ds_{N-1}$. The global probability of error associated with a decision rule, or, which is equivalent, with the corresponding partition $\Omega_k, k \in$

$[1 \dots K]$ of the parameter space $\mathcal{S}$, is the sum of all the P_{ij}:

$$P_e = \sum_{i=1}^{K} \sum_{j \neq i} P_{ij} = \sum_{i=1}^{K} \sum_{j \neq i} \int_{\Omega_i} P[\boldsymbol{\theta}_j|\mathbf{s},\mathbf{p}]P[\mathbf{s}]d\mathbf{s} \qquad (3.9)$$

P_e can be rewritten in a simpler form if one recalls that $P[\boldsymbol{\theta}_j|\mathbf{s},\mathbf{p}]$ is a pdf and thus $\sum_{j=1}^{K} P[\boldsymbol{\theta}_j|\mathbf{s},\mathbf{p}] = 1$. Consequently:

$$\sum_{j \neq i} P[\boldsymbol{\theta}_j|\mathbf{s},\mathbf{p}] = 1 - P[\boldsymbol{\theta}_i|\mathbf{s},\mathbf{p}] \qquad (3.10)$$

Injecting this relation in Eq. 3.9, and using the fact that $\sum_{i=1}^{K} \int_{\Omega_i} P[\mathbf{s}]d\mathbf{s} = \int_{\mathcal{S}} P[\mathbf{s}]d\mathbf{s} = 1$, one obtains:

$$P_e = 1 - \sum_{i=1}^{K} \int_{\Omega_i} P[\boldsymbol{\theta}_i|\mathbf{s},\mathbf{p}]P[\mathbf{s}]d\mathbf{s} \qquad (3.11)$$

We now have to find the partition Ω_k which minimizes the probability of error in Eq. 3.11. It is easily seen that in order to minimize this expression, one has to associate to any realization $\mathbf{s}$ of the data the class Ω_i such that $P[\boldsymbol{\theta}_i|\mathbf{s},\mathbf{p}]P[\mathbf{s}]$ is maximal. Thus the decision rule which minimizes the probability of error P_e consists in choosing $\boldsymbol{\theta}$ as [29, 47]:

$$\widehat{\boldsymbol{\theta}}^{MAP} = \arg \max_{\boldsymbol{\theta}} P[\boldsymbol{\theta}|\mathbf{s},\mathbf{p}] \qquad (3.12)$$

This decision rule is called the *Maximum a Posteriori* (MAP) estimator since it consists in choosing the value of the parameter $\boldsymbol{\theta}$ which maximizes the *posterior* pdf. Using Eq. 3.7, this estimator can also be expressed as a function of the likelihood and of the prior:

$$\widehat{\boldsymbol{\theta}}^{MAP} = \arg \max_{\boldsymbol{\theta}} \{L[\mathbf{s}|\mathbf{p},\boldsymbol{\theta}]P[\boldsymbol{\theta}]\} \qquad (3.13)$$

If we introduce the *loglikelihood*, defined as follows:

$$l[\mathbf{s}|\mathbf{p},\boldsymbol{\theta}] = \log[L[\mathbf{s}|\mathbf{p},\boldsymbol{\theta}]] \qquad (3.14)$$

the MAP estimate can also be written as:

$$\widehat{\boldsymbol{\theta}}^{MAP} = \arg \max_{\boldsymbol{\theta}} \{l[\mathbf{s}|\mathbf{p},\boldsymbol{\theta}] + \log(P[\boldsymbol{\theta}])\} \qquad (3.15)$$

which clearly shows that when the loglikelihood is known, it is in general simple to obtain the MAP estimate. In particular, it can be noted that if the prior is uniform, that is, it takes the same value for all possible values of $\boldsymbol{\theta}$, then $P[\boldsymbol{\theta}]$ is constant and the Maximum Likelihood approach is the optimal decision rule. In the following, we will thus consider in general ML estimation of the parameter $\boldsymbol{\theta}$, the generalization to MAP estimation being straightforward as soon as the prior is known.

3.2.2 Decision and estimation theory in presence of nuisance parameters

In the general case, the likelihood $L[\mathbf{s}|\mathbf{p}, \boldsymbol{\theta}]$ may depend on unknown parameters, which will be denoted $\boldsymbol{\mu}$ in the following. The parameters $\boldsymbol{\mu}$ are called *nuisance parameters* since we are generally not interested in their values. A classical example of such a situation is the location of a target with unknown illumination β in the presence of additive noise:

$$s_i = \beta \, r_{i-\theta} + n_i \ , \ \ \forall i \in [0; N-1] \tag{3.16}$$

In this example, the nuisance parameter is the scalar value β. We will consider more complex examples in the following.

In the presence of nuisance parameters, the likelihood is denoted $L[\mathbf{s}|\mathbf{p}, \boldsymbol{\mu}, \boldsymbol{\theta}]$, since it is conditional on the value of $\boldsymbol{\mu}$. There exist several methods to deal with nuisance parameters. We propose to briefly describe three of them, which are frequently used in practice and which will be useful in the following developments.

3.2.2.1 ML estimation of the nuisance parameter

In this first approach, the nuisance parameter is estimated in the ML sense, and then inserted in the expression of the likelihood. The ML estimation of the nuisance parameter $\boldsymbol{\mu}$ consists in determining:

$$\widehat{\boldsymbol{\mu}}^{ML}(\boldsymbol{\theta}) = \arg \max_{\boldsymbol{\mu}} L[\mathbf{s}|\mathbf{p}, \boldsymbol{\mu}, \boldsymbol{\theta}] \tag{3.17}$$

Note that the estimate $\widehat{\boldsymbol{\mu}}^{ML}(\boldsymbol{\theta})$ is also a function of $\mathbf{s}$, but this will not be explicitly denoted in the following in order to simplify the notation. The ML estimate of $\boldsymbol{\theta}$ is:

$$\widehat{\boldsymbol{\theta}}_{ML}^{ML} = \arg \max_{\boldsymbol{\theta}} \left\{ L[\mathbf{s}|\mathbf{p}, \widehat{\boldsymbol{\mu}}^{ML}(\boldsymbol{\theta}), \boldsymbol{\theta}] \right\} = \arg \max_{\boldsymbol{\theta}} \left\{ \mathcal{L}(\mathbf{s}, \boldsymbol{\theta}) \right\} \tag{3.18}$$

where

$$\mathcal{L}(\mathbf{s}, \boldsymbol{\theta}) = L[\mathbf{s}|\mathbf{p}, \widehat{\boldsymbol{\mu}}^{ML}(\boldsymbol{\theta}), \boldsymbol{\theta}] \tag{3.19}$$

It must be noted that the function $\mathcal{L}(\mathbf{s}, \boldsymbol{\theta})$ obtained after injection of the ML estimate $\widehat{\boldsymbol{\mu}}^{ML}(\boldsymbol{\theta})$ (which is also a function of $\mathbf{s}$) into the likelihood is no longer a pdf since $\int \mathcal{L}(\mathbf{s}, \boldsymbol{\theta}) d\mathbf{s} \neq 1$. We will thus call it a *pseudo-likelihood*.

In practice, the maximum likelihood estimate may appear to be unstable. The likelihood is said to be unstable if a small variation of the initial conditions (here $s \to s + ds$) can lead to a large variation of $\mathcal{L}(s, \theta)$. [1] The MAP approach can be used to overcome this problem.

3.2.2.2 MAP estimation of the nuisance parameter

The MAP approach is interesting when some values of the nuisance parameter are very unlikely or lead to unstable estimates of the likelihood and thus of the parameter θ. The MAP estimate of the nuisance parameter μ is obtained by maximizing the posterior pdf:

$$\widehat{\mu}^{MAP}(\theta) = \arg\max_{\mu} P[\mu|s, p, \theta] = \arg\max_{\mu} \{L[s|p, \mu, \theta]P[\mu]\} \quad (3.20)$$

The ML estimate of θ can thus be written as:

$$\widehat{\theta}_{MAP}^{ML} = \arg\max_{\theta} \left\{ L\left[s|p, \widehat{\mu}^{MAP}(\theta), \theta\right] P\left[\widehat{\mu}^{MAP}(\theta)\right] \right\} \quad (3.21)$$

The prior $P[\mu]$ is thus useful to penalize undesirable values of μ. It is also very easy to apply this approach when computing the MAP estimate of θ, which is:

$$\widehat{\theta}_{MAP}^{MAP} = \arg\max_{\theta} \left\{ L\left[s|p, \widehat{\mu}^{MAP}(\theta), \theta\right] P\left[\widehat{\mu}^{MAP}(\theta)\right] P[\theta] \right\} \quad (3.22)$$

It is clear that if the prior $P[\mu]$ is uniform, the MAP approach is equivalent to the ML method.

3.2.2.3 Marginal Bayesian approach

In the ML and the MAP approaches, the nuisance parameters are considered as parameters of interest which are estimated and then injected back into the expression of the likelihood. A more rigorous approach consists in considering the following relation, which is obtained by applying the Bayes rule:

$$P[\theta|s, p] = \int P[\theta, \mu|p, s]d\mu = \int P[\theta|s, p, \mu]P[\mu]d\mu \quad (3.23)$$

where μ is a real vector of dimension n. This relation shows that the posterior pdf which appears in the expression of the MAP decision rule (see Eq. 3.12) can be considered as the marginal pdf of the joint distribution of the parameter of interest θ and the nuisance parameter: $P[\theta, \mu|p, s]$. According to Eq. 3.12,

---

[1] One can also consider stability against reference variation $r \to r + dr$ as in Chapter 2. Both approaches should be analyzed, but for statistical models considered here, the stability requirement against input image variation implies stability against reference variation.

the MAP decision rule then consists in determining:

$$\widehat{\theta}_{Bay}^{MAP} = \arg \max_{\theta} \int P[\theta|s, p, \mu] P[\mu] d\mu \qquad (3.24)$$

This approach is called *marginal Bayesian* and has been the subject of many investigations and theoretical studies [47, 51]. The motivation for this approach in the applications considered here is the same as for the MAP. The main difference is that in the marginal Bayesian method, the nuisance parameters are really considered of no interest while in the MAP approach, they are considered as parameters of interest.

With both the MAP and the marginal Bayesian approaches, the value of the estimate depends on the expression of the prior $P[\mu]$. This point is sometimes considered as a limitation of these methods [51]. However, if the goal is mainly to suppress the cases where the likelihood is unstable, from a practical point of view, the choice of $P[\mu]$ generally results from the following considerations:

1 To penalize the undesirable values of μ,

2 To lead to MAP or marginal Bayesian estimates that are close to the ML estimate for the values of the nuisance parameter for which the likelihood is stable,

3 To lead to tractable mathematical equations.

In Section 3.4, we will illustrate these different approaches in a particular case for which the PSD of the noise is unknown *a priori* and can thus be considered as a nuisance parameter.

3.2.3   Two-hypothesis testing

The case where one has to decide between two hypotheses is particularly interesting, since it corresponds to detection problems. Let H_{θ_0} and H_{θ_1} be the two considered hypotheses, and let us first assume that there is no nuisance parameter. Then according to Section 3.2.1.2, the minimum probability of error is obtained with the MAP decision rule (see Eq. 3.12). When there are only two hypotheses, this decision rule amounts to considering the *likelihood ratio*:

$$R(s) = \frac{L[s|p, \theta_1]}{L[s|p, \theta_0]} \qquad (3.25)$$

and to choosing between the two hypotheses according to the following rule:

$$\begin{cases} \text{Choose } H_{\theta_1} & \text{if } R(s) > \frac{P[\theta_0]}{P[\theta_1]} \\ \text{Choose } H_{\theta_0} & \text{otherwise} \end{cases} \qquad (3.26)$$

where $P[\boldsymbol{\theta}_1]$ ($P[\boldsymbol{\theta}_0]$) is the prior probability of hypothesis $H_{\boldsymbol{\theta}_1}$ ($H_{\boldsymbol{\theta}_0}$). $R(\mathbf{s})$ is called a *decision statistics*, since it is a random variable, function the data $\mathbf{s}$, which is compared to a threshold in order to make a decision.

As explained above, $R(\mathbf{s})$ is the decision statistics which minimizes the global probability of error P_e. However, in detection applications, it is often more interesting to consider the probability of detection P_d and the probability of false alarm P_{fa} separately. These probabilities are defined as follows:

- P_d: probability of choosing $H_{\boldsymbol{\theta}_1}$ when it is true.

- P_{fa}: probability of choosing $H_{\boldsymbol{\theta}_1}$ when $H_{\boldsymbol{\theta}_0}$ is true.

One can also define the probability of miss P_m, which is the probability of choosing $H_{\boldsymbol{\theta}_0}$ when $H_{\boldsymbol{\theta}_1}$ is true. It is clear that the probability of error is equal to:

$$P_e = P_m + P_{fa} = 1 - P_d + P_{fa} \qquad (3.27)$$

This means that the two possible types of errors are weighted equally in the expression of P_e. In many instances, such a weighting is inadequate. Indeed, when performing target detection with a radar or an imaging system, it is most important to maintain a sufficiently low false alarm probability in order not to overload the subsequent processing steps with too many spurious detections. In such cases, going from a 10^{-3} to a 10^{-6} value of P_{fa} may have very benefic influence in practice, whereas it has a marginal influence on the global probability of error P_e. What is needed here is to be able to optimize the probability of detection for a given value of the probability of false alarm. The Neyman-Pearson theory that we describe in the following makes this possible [29].

Let us denote Ω_0 and Ω_1 the regions of acceptance of hypotheses $H_{\boldsymbol{\theta}_0}$ and $H_{\boldsymbol{\theta}_1}$. The probabilities of detection and of false alarm can then be written as:

$$
\begin{aligned}
P_d &= \int_{\Omega_1} P[\boldsymbol{\theta}_1|\mathbf{s},\mathbf{p}]P[\mathbf{s}]d\mathbf{s} \\
P_{fa} &= \int_{\Omega_1} P[\boldsymbol{\theta}_0|\mathbf{s},\mathbf{p}]P[\mathbf{s}]d\mathbf{s}
\end{aligned}
\qquad (3.28)
$$

Our objective is to maximize P_d for a given value of P_{fa}. The problem can be solved by using the Lagrange parameters method. This consists in maximizing the following function:

$$F(\Omega_1) = P_d - \lambda P_{fa} = \int_{\Omega_1} \{P[\boldsymbol{\theta}_1|\mathbf{s},\mathbf{p}] - \lambda P[\boldsymbol{\theta}_0|\mathbf{s},\mathbf{p}]\} P[\mathbf{s}]d\mathbf{s} \qquad (3.29)$$

It is clear that in order to maximize $F(\Omega_1)$, one has to associate a realization $\mathbf{s}$ of the data to Ω_1 if the argument of the integral in Eq. 3.29, that is, $P[\boldsymbol{\theta}_1|\mathbf{s},\mathbf{p}] -$

$\lambda P[\boldsymbol{\theta}_1|\mathbf{s}, \mathbf{p}]$, is positive, and to Ω_0 if it is negative. Consequently, the optimal decision rule is the following:

$$\begin{cases} \text{Choose } H_{\boldsymbol{\theta}_1} & \text{if } R(\mathbf{s}) > \lambda' \\ \text{Choose } H_{\boldsymbol{\theta}_0} & \text{otherwise} \end{cases} \tag{3.30}$$

where $\lambda' = \frac{P[\boldsymbol{\theta}_0]}{P[\boldsymbol{\theta}_1]}\lambda$. This decision rule is called the Likelihood Ratio Test (LRT). The Lagrange parameter λ is identified by imposing that P_{fa} has a given value. The interpretation of λ' is very clear: It plays the role of a detection threshold and large values of λ' lead to low values of P_d, but also of P_{fa}.

Since $R(\mathbf{s})$ is a random variable, one can define its conditional pdf in the two hypotheses $H_{\boldsymbol{\theta}_0}$ and $H_{\boldsymbol{\theta}_1}$, that will be denoted $P_{R|\boldsymbol{\theta}_0}(R|\boldsymbol{\theta}_0)$ and $P_{R|\boldsymbol{\theta}_1}(R|\boldsymbol{\theta}_1)$. The probabilities P_d and P_{fa} can be expressed as a function of these pdf and of the detection threshold λ':

$$P_d(\lambda') = \int_{\lambda'}^{+\infty} P_{R|\boldsymbol{\theta}_1}(R|\boldsymbol{\theta}_1)dR \tag{3.31}$$

$$P_{fa}(\lambda') = \int_{\lambda'}^{+\infty} P_{R|\boldsymbol{\theta}_0}(R|\boldsymbol{\theta}_0)dR \tag{3.32}$$

Of particular interest is the parametric curve which represents $P_d(\lambda')$ as a function of $P_{fa}(\lambda')$ when the detection threshold λ' varies. This curve is called the *Receiver Operating Characteristic* (ROC) and an example of such a curve is represented in Figure 3.5. The fact that the LRT is optimal means that the ROC of this test lies above the ROC of any other detection test based on a different decision statistics $D(\mathbf{s})$.

In detection problems, it is also frequent to have to deal with nuisance parameters. The rigorous way to handle the presence of such parameters is the marginal Bayesian approach described in Section 3.2.2.3. According to the results obtained in this section, in particular Eq. 3.12, the likelihood ratio is defined as follows:

$$R(\mathbf{s}) = \frac{\int L[\mathbf{s}|\mathbf{p}, \boldsymbol{\mu}, \boldsymbol{\theta}_1]P[\boldsymbol{\mu}]d\boldsymbol{\mu}}{\int L[\mathbf{s}|\mathbf{p}, \boldsymbol{\mu}, \boldsymbol{\theta}_0]P[\boldsymbol{\mu}]d\boldsymbol{\mu}} \tag{3.33}$$

However, the integration over the nuisance parameter required in the marginal Bayesian approach is often intractable. In practice, one then preferably uses the maximum likelihood estimation of these parameters as described in Section 3.2.2.1, and form the ratio of the two obtained pseudo-loglikelihoods (see Eq. 3.18). This approach leads to the *Generalized Likelihood Ratio Test* (GLRT):

$$\mathcal{R}(\mathbf{s}) = \frac{\max\limits_{\boldsymbol{\mu}} L[\mathbf{s}|\mathbf{p}, \boldsymbol{\mu}, \boldsymbol{\theta}_1]}{\max\limits_{\boldsymbol{\mu}} L[\mathbf{s}|\mathbf{p}, \boldsymbol{\mu}, \boldsymbol{\theta}_0]} = \frac{\mathcal{L}(\boldsymbol{\theta}_1)}{\mathcal{L}(\boldsymbol{\theta}_0)} \underset{H_{\boldsymbol{\theta}_0}}{\overset{H_{\boldsymbol{\theta}_1}}{\gtrless}} \lambda' \tag{3.34}$$

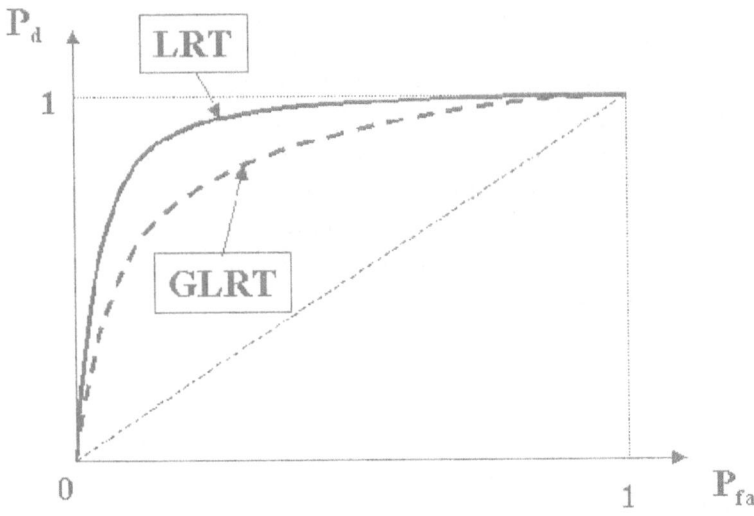

Figure 3.5. Examples of ROCs of the LRT and of the GLRT.

It must be noted that the GLRT does not possess the optimality properties of the LRT, and as a consequence its ROC lies below that of the LRT (see Figure 3.5). Extensive study of the properties of the GLRT can be found in Ref. 49. In practice, the GLRT gives good results and we will make wide use of this test to solve detection problems in the sequel of this book.

In practice, we will often use the log-GLRT defined as follows:

$$r(\mathbf{s}) = \log \mathcal{R}(\mathbf{s}) = \ell(\mathbf{s}, \boldsymbol{\theta}_1) - \ell(\mathbf{s}, \boldsymbol{\theta}_0) \tag{3.35}$$

where $\ell(\mathbf{s}, \boldsymbol{\theta}) = \log[\mathcal{L}(\mathbf{s}, \boldsymbol{\theta})]$ is the log-pseudolikelihood of hypothesis $H_{\boldsymbol{\theta}}$.

3.3. Matched filtering and statistical decision theory

As a first example of the application of statistical decision theory to a practical pattern recognition problem, let us consider image processing tasks in the presence of Gaussian additive noise with known PSD. Let s denote the input image and $\mathbf{r}^{\theta}$ the reference objects which depend on a parameter θ. As stated above, the nature of this parameter depends on the tasks to perform. Let us consider some examples. If we want to locate a reference $\mathbf{r}$ in the input image, θ denotes the different possible locations of the target. In that case $r_i^{\theta} = r_{i-\theta}$. If we try to estimate the angle of a rotation around a given axis, $\mathbf{r}^{\theta} = R_{\theta}\mathbf{r}$ where R_{θ} is the rotation operator with an angle θ. If we are in the presence of a detection problem, there are two possibilities: $\theta = 0$ or $\theta = 1$. $\theta = 1$ corresponds to the hypothesis that the target is present and $\theta = 0$ that it is absent. More generally, if P types of target are possible (for example, P different objects or P different attitudes of the object) this is a classification problem and θ can be a label associated with each type of targets ($\theta = 1, 2, ..., P$).

Let us consider that the input noise on the image is additive, Gaussian, stationary and has a known spectral density. The input image s is thus the addition of a noise and of a target with an unknown value of θ and, in general, with an unknown illumination β.[2] Let H_θ denote the hypothesis that the searched object is in the input image s with parameter θ. The global illumination factor β is a nuisance parameter. Under hypothesis H_θ and for a given value of the nuisance parameter β, we can write:

$$\mathbf{s} = \beta \, \mathbf{r}^\theta + \mathbf{n} \tag{3.36}$$

where $\mathbf{n}$ is a Gaussian additive noise with a known covariance matrix S and power spectral density (PSD) $\tilde{S}$.

The probability density function $P_n(\mathbf{n})$ of the noise can be written:

$$P_n(\mathbf{n}) = \frac{1}{\sqrt{(2\pi)^N \det S}} \, \exp\left[-\frac{1}{2}\mathbf{n}^T S^{-1}\mathbf{n}\right] \tag{3.37}$$

where the noise covariance matrix S is assumed to be circulant (see Appendix 2.A), and $\det S$ is its determinant. In this case, the Fourier transform $\tilde{S}$ of S is diagonal with values equal to the PSD. The likelihood associated with this problem thus has the following expression:

$$L(\mathbf{s}|\beta, \theta) = \frac{1}{\sqrt{(2\pi)^N \det S}} \, \exp[-\frac{1}{2}(\mathbf{s} - \beta \, \mathbf{r}^\theta)^T S^{-1}(\mathbf{s} - \beta \, \mathbf{r}^\theta)] \tag{3.38}$$

The ML estimation of the nuisance parameter β is obtained by maximizing this likelihood. Indeed, annulling the derivative of the likelihood with respect to β, one obtains [29]:

$$\widehat{\beta}^{ML} = \frac{[\mathbf{r}^\theta]^T S^{-1}\mathbf{s}}{[\mathbf{r}^\theta]^T S^{-1}\mathbf{r}^\theta} \tag{3.39}$$

The ML estimate of θ (Eq. 3.18) is then obtained by injecting $\widehat{\beta}^{ML}$ in Eq. 3.38 and taking the value of θ which maximizes $L(\mathbf{s}|\widehat{\beta}^{ML}, \theta)$. One obtains [29]:

$$\widehat{\theta}_{ML}^{ML} = \arg\max_\theta \frac{\left|[\mathbf{r}^\theta]^T S^{-1}\mathbf{s}\right|^2}{[\mathbf{r}^\theta]^T S^{-1}\mathbf{r}^\theta} \tag{3.40}$$

---

[2]The parameter β is introduced with its specific notation since its ML estimation is very easy.

In order to be somewhat more concrete, let us assume that the target can belong to one of P possible classes. Furthermore, let us consider that this recognition task has to be performed with translation invariance.

The parameter $\boldsymbol{\theta} = (p, \tau)$ is now composed of a value p indicating the class and a value τ for the location. The estimation of p and τ is thus given by:

$$(\hat{p}^{ML}, \hat{\tau}^{ML}) = \underset{p, \tau}{\text{argmax}} \frac{|\sum_{i,j=0}^{N-1} r_{i-\tau}^{p}[S^{-1}]_{i,j} s_j|^2}{\sum_{i,j=0}^{N-1} r_{i-\tau}^{p}[S^{-1}]_{i,j} r_{j-\tau}^{p}} \tag{3.41}$$

Since S is assumed circulant, it holds that $[S^{-1}]_{i,j} = [S^{-1}]_{i-\tau, j-\tau}$, and thus:

$$\sum_{i,j=0}^{N-1} r_{i-\tau}^{p}[S^{-1}]_{i,j} r_{j-\tau}^{p} = \sum_{i,j=0}^{N-1} r_i^{p}[S^{-1}]_{i,j} r_j^{p}$$

Furthermore, the expression of Eq. 3.41 can be simplified if the matched filter is introduced:

$$h_j^{p} = \sum_{i=0}^{N-1} r_i^{p}[S^{-1}]_{i,j}$$

If τ is considered as a nuisance parameter, Eq. 3.41 becomes:

$$\hat{p}^{ML} = \underset{p}{\arg\max} \left[\frac{C_{max}[p]}{A[p]} \right] \tag{3.42}$$

with

$$C_{max}[p] = \max_{\tau} \left\{ \left| \sum_{j=0}^{N-1} h_{j-\tau}^{p} s_j \right|^2 \right\}$$

and

$$A[p] = \sum_{i,j=0}^{N-1} r_i^{p}[S^{-1}]_{i,j} r_j^{p}$$

$C_{max}[p]$ is the maximum value of the modulus square of the correlation function between $\mathbf{s}$ and $\mathbf{h}^p$. The ML classification is thus performed in two steps. In the first step, the maximum value of the modulus squares of the correlation functions between $\mathbf{s}$ and $\mathbf{h}^p$, $C_{max}[p]$, is determined for each p. In the second step, these values are normalized (i.e., $C_{max}[p]/A[p]$) and the class p for which this normalized value is maximal is chosen.

It is interesting to note that this result theoretically justifies the use of linear filtering techniques (or correlation techniques) for pattern recognition. It also clearly shows their optimality when the noise is additive, Gaussian and has a known PSD.

However, the assumption of additive noise is not completely realistic in image processing, as discussed above. We will describe solutions to this problem in Section 3.5 and in Chapter 4. The second important point is that the matched filter assumes that the PSD of the noise $\tilde{S}$ is known. This is not the case in general in image processing, where in contrast to radar processing, it is very difficult to estimate. Then, an important question in the context of pattern recognition is the determination of an appropriate model for $\tilde{S}$. Furthermore, as mentioned above, there is no reason to consider that the realizations of noise can be precisely described with a temporally stationary statistical distribution.

It appears also clearly that if the number of classes P is large, determining the correlation function with all the $\mathbf{h}^p$ can be prohibitive in terms of the computer memory necessary to store all the $\mathbf{h}^p$ and of the computational power required to determine all the correlation products. This problem is an important motivation for using the SDF filters described in Chapter 2, even though they are suboptimal in the sense of statistical decision theory.

3.4.   Optimal filter for unknown PSD

In this section, we address a first deviation from the standard noise model used above. We still consider the noise Gaussian and additive, but we no longer assume that its PSD is known. To simplify matters, we will consider target location applications. We will show that under the considered hypotheses, the optimal location algorithm is very close to the adaptive nonlinear filter introduced in Section 2.4.2.4. We will also have the opportunity to discuss some classical forms of *a priori* pdf which are useful for dealing with nuisance parameters. More details on this problem can be found in Refs. 52 and 53.

In the problem we consider, the parameter θ to estimate is the position of the searched object in the image. This position will be denoted j. The image model can thus be described as:

$$s_i = r_i^j + n_i \tag{3.43}$$

where $r_i^j = r_{i-j}$ is a version of $\mathbf{r}$ centered on the position j. For this image model, let us determine the optimal location method using the statistical approach described above.

First, we can notice that since we consider sampled random signals with a finite number of pixels, it is possible to define their Fourier transforms:

$$\tilde{s}_k = \tilde{r}_k^j + \tilde{n}_k \tag{3.44}$$

Let us now compute the likelihood of $\tilde{\mathbf{s}}$, that is, the probability of observing the input image $\tilde{\mathbf{s}}$ with the hypothesis that the spectral density is Γ and the object location is j. Since $\mathbf{n}$ is a cyclostationary Gaussian noise, its pdf can be expressed in a simple way as a function of its Fourier transform $\tilde{\mathbf{n}}$ (see

Eq. 2.A.16):

$$P(\mathbf{n}|\mathbf{\Gamma}) = \prod_{k=0}^{N-1} \frac{1}{\sqrt{2\pi\Gamma_k}} \exp\left[-\frac{|\tilde{n}_k|^2}{2\Gamma_k}\right] \tag{3.45}$$

where Γ_k denotes the component of the PSD $\mathbf{\Gamma}$ at frequency k (note that in this section, we will denote the PSD $\mathbf{\Gamma}$ instead of $\tilde{S}$, to emphasize the fact that it is now an unknown parameter). The likelihood, which is the probability of observing the input image s knowing j and $\mathbf{\Gamma}$, thus has the following expression:

$$L(\mathbf{s}|\mathbf{\Gamma}, j) = \prod_{k=0}^{N-1} \frac{1}{\sqrt{2\pi\Gamma_k}} \exp\left[-\frac{|\tilde{s}_k - \tilde{r}_k^j|^2}{2\Gamma_k}\right] \tag{3.46}$$

In this problem, the parameter of interest is j and $\mathbf{\Gamma}$ is a nuisance parameter. We will now determine the optimal location algorithms obtained by estimating $\mathbf{\Gamma}$ in the ML and in the MAP sense.

3.4.1 ML estimation of the spectral density

Let us estimate the spectral density $\mathbf{\Gamma}$ in the ML sense. The solution is obtained by solving the following equation:

$$\frac{\partial L(\mathbf{s}|\mathbf{\Gamma}, j)}{\partial \Gamma_k} = 0 \tag{3.47}$$

which leads to the following estimate:

$$\widehat{\Gamma}_k^{ML}(j) = |\tilde{s}_k - \tilde{r}_k^j|^2 \tag{3.48}$$

After injecting this estimate into Eq. 3.46 and taking its logarithm, we obtain the following expression of the pseudo-loglikelihood:

$$\ell_{ML}(j) = -\sum_{k=0}^{N-1} \frac{1}{2} \log\left[|\tilde{s}_k - \tilde{r}_k^j|^2\right] + K \tag{3.49}$$

where K is a constant independent of j. One can remark that $\ell_{ML}(j)$ diverges if $|\tilde{s}_k - \tilde{r}_k^j| = 0$ for some frequency k: This estimate can thus be unstable. In order to have a stable estimate, it will be necessary to consider a prior pdf $P(\mathbf{\Gamma})$ which penalizes null values for Γ_k. This will be the subject of the next sections.

For now, let us consider Eq. 3.49, which represents the location method which is optimal in the maximum likelihood sense for the considered image model. It is very computation intensive, since it cannot be expressed as a correlation operation, and thus cannot benefit from an implementation with the Fast Fourier Transform (FFT). However, we are going to show that a first-order development of this expression leads to an algorithm which has the same computational complexity as a linear filter.

Let us first assume that for every frequency k, we always have $|\tilde{s}_k - \tilde{r}_k^j|^2 > \sigma^2$ with $\sigma^2 > 0$, which can be fulfilled if noise is present at each frequency k. In order to simplify the following analysis, let us introduce the notations:

$$\Delta_k^j = |\tilde{s}_k - \tilde{r}_k^j|^2 \tag{3.50}$$

$$U_k^j = [\tilde{s}_k]^* \, \tilde{r}_k^j + [\tilde{r}_k^j]^* \, \tilde{s}_k \tag{3.51}$$

$$D_k = |\tilde{s}_k|^2 + |\tilde{r}_k|^2 \tag{3.52}$$

We have used the fact that $\tilde{r}_k^j = \tilde{r}_k \exp(-i2\pi jk)$, where $\tilde{r}_k$ is the Fourier transform of the target when it is centered at the origin. With these definitions, one has: $\Delta_k^j = D_k - U_k^j$ (U_k^j is real) and the loglikelihood can be written:

$$\ell_{ML}(j) = K - \frac{1}{2} \sum_{k=0}^{N-1} \left(\log[D_k] + \log \left[1 - \frac{U_k^j}{D_k} \right] \right) \tag{3.53}$$

Since D_k is independent of j and since the assumption $\Delta_k^j > \sigma^2$ leads to $\frac{U_k^j}{D_k} < 1 - \frac{\sigma^2}{D_k}$, one can consider the first-order development of the loglikelihood:

$$\ell_{ML}(j) \simeq K' + \frac{1}{2} \sum_{k=0}^{N-1} \frac{U_k^j}{D_k} \tag{3.54}$$

where K' is a constant independent of j. Let us define:

$$\tilde{F}_k = \frac{\tilde{s}_k (\tilde{r}_k)^*}{|\tilde{s}_k|^2 + |\tilde{r}_k|^2} \tag{3.55}$$

By using the fact that $\tilde{r}_k^j = \tilde{r}_k e^{-i2\pi jk}$, one can write:

$$\sum_{k=0}^{N-1} \frac{U_k^j}{D_k} = \sum_{k=0}^{N-1} \tilde{F}_k \, \exp(+i2\pi jk) + \sum_{k=0}^{N-1} (\tilde{F}_k)^* \, \exp(-i2\pi jk)$$

$$= F_j + (F_j)^* = 2 \, F_j \tag{3.56}$$

where F_j is the inverse Fourier transform of $\tilde{F}_k$, which is real-valued since r and s are real-valued. Thus maximizing $\ell_{ML}(j)$ is equivalent to maximizing the inverse Fourier transform of $\tilde{F}_k$.

The Fourier transform of the optimal nonlinear filter introduced in Section 2.4.2.4 is:

$$\tilde{C}_k = \frac{\tilde{s}_k (\tilde{r}_k)^*}{a + \mu \, |\hat{r}_k|^2 + (1 - \mu) \, |\tilde{s}_k|^2} \tag{3.57}$$

One can remark that $\tilde{F}_k$ is a particular case of $\hat{C}_k$ with the regularization term a equal to 0 and $\mu = 1 - \mu = 1/2$. We have thus determined noise conditions

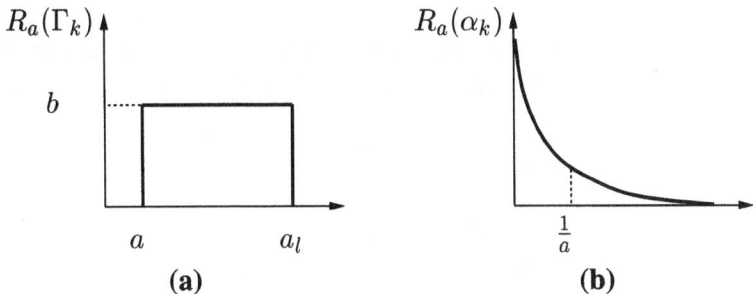

Figure 3.6. Examples of possible priors. **(a)** Uniform prior (on Γ_k). **(b)** Exponential prior (on $\alpha_k = 1/\Gamma_k$).

in which this nonlinear filter, which had been derived on a heuristic basis in Section 2.4.2.4, is optimal: It is when the PSD of the Gaussian additive noise is unknown and must be considered as a nuisance parameter. However, this result comes at the price of a linear approximation of the likelihood function.

If the assumption $\Delta_k^j > \sigma^2$ is not fulfilled for σ^2 sufficiently large, the nonlinear filter defined by $\tilde{F}_k$ can be unstable. We thus analyze in the following respectively the MAP and the marginal Bayesian approaches which allow one to regularize the ML solution.

3.4.2 MAP estimation of the spectral density

Let us now consider a prior $P[\mathbf{\Gamma}]$ so as to estimate $\mathbf{\Gamma}$ in the MAP sense. We have seen in Section 3.2.2.1 that this amounts to maximizing the posterior density $P(\mathbf{\Gamma}, j|\mathbf{s})$, or equivalently, to maximizing the product $L(\mathbf{s}|\mathbf{\Gamma}, j)P[\mathbf{\Gamma}]$. The likelihood $L(\mathbf{s}|\mathbf{\Gamma}, j)$ becomes unstable for null values of $\Gamma_k^{ML}(j)$, so from a practical point of view, the main goal of $P[\mathbf{\Gamma}]$ is to penalize small values of Γ_k. There are many types of such functions. In the following, we will analyze two of them: the uniform prior and the exponential prior (see Figure 3.6).

3.4.2.1 Uniform prior

The expression of the uniform prior on $\mathbf{\Gamma}$ is $P[\mathbf{\Gamma}] = \prod_{k=0}^{N-1} R_a[\Gamma_k]$ where:

$$\begin{cases} R_a[\Gamma_k] = b & \text{if } a \leq \Gamma_k \leq a_l \\ R_a[\Gamma_k] = 0 & \text{otherwise} \end{cases} \tag{3.58}$$

a_l is determined so that on the current scene $\Gamma_k < a_l$ for all k. Moreover $b = 1/(a_l - a)$. This prior fulfills the second requirement of Section 3.2.2: It is obvious that it will lead to PSD estimates close – in fact equal – to the ML estimate values when the likelihood is stable.

The optimal filter in the MAP sense corresponding to this prior is derived in Appendix 3.A. The computation is complex, and so is its final expression. However, after one approximation, it can be shown that its first-order term is:

$$\tilde{F}_k = \frac{\tilde{s}_k (\tilde{r}_k)^*}{\max \left[|\tilde{s}_k|^2 + |\tilde{r}_k|^2 , \, a \right]} \tag{3.59}$$

This expression is quite satisfying intuitively, since it penalizes the very small values of $|\tilde{s}_k|^2 + |\tilde{r}_k|^2$, which could make the expression of the filter unstable. This prior leads to a solution which presents strong analogies with the truncature method discussed in Section 2.4.1.

3.4.2.2   Exponential prior

In this section, it will be interesting to write the pdf of Eq. 3.45 in the canonical form of the exponential family [51] by introducing $\alpha_k = 1/(\Gamma_k)$. We thus define a vector $\boldsymbol{\alpha}$ with components α_k and we obtain the following expression for the probability of observing the input image s (i.e., the likelihood):

$$L(\mathbf{s}|\boldsymbol{\alpha}, j) = \prod_{k=0}^{N-1} \frac{\sqrt{\alpha_k}}{\sqrt{2\pi}} \exp \left[-\frac{\alpha_k}{2} \Delta_k^j \right] \tag{3.60}$$

As discussed in Section 3.2.2, an interesting property for the choice of the prior is to lead to tractable mathematical equations and to penalize small values of Γ_k or, in the present case, large values of α_k. The exponential prior fulfills this requirement:

$$P[\boldsymbol{\alpha}] = \prod_{k=0}^{N-1} a \, exp(-a \, \alpha_k) \tag{3.61}$$

If this prior is used, we have:

$$L(\mathbf{s}|\boldsymbol{\alpha}, j)P(\boldsymbol{\alpha}) = \prod_{k=0}^{N-1} \frac{a \sqrt{\alpha_k}}{\sqrt{2\pi}} \exp \left[-\frac{\alpha_k}{2} (\Delta_k^j + 2a) \right] \tag{3.62}$$

and it is easily shown that the MAP estimate of the spectral density has the following expression [52]:

$$\hat{\Gamma}_k^{MAP}(j) = 2\, a + \Delta_k^j \tag{3.63}$$

After injecting this estimate into Eq. 3.62 and taking its logarithm, we obtain the following expression of the log-posterior pdf:

$$\ell_{MAP}(j) = - \sum_{k=0}^{N-1} \frac{1}{2} \log[2\, a + |\tilde{s}_k - \tilde{r}_k^j|^2] + K' \tag{3.64}$$

where K' is a constant independent of j. Reasoning in the same manner as in Eqs. 3.54 and 3.55, the first order of the Taylor expansion of the logarithm shows that the MAP estimate of j is obtained by maximizing the inverse Fourier transform F_j of:

$$\tilde{F}_k = \frac{[\tilde{s}_k]^* \, \tilde{r}_k}{2 \, a + |\tilde{s}_k|^2 + |\tilde{r}_k|^2} \tag{3.65}$$

Within this approximation, the optimal estimate of the location is the value of j which maximizes F_j, the inverse Fourier transform of $\tilde{F}_k$. This expression is identical to that of the nonlinear filter introduced in Section 2.4.2.4 (see also Eq. 3.57) with $\mu = 1 - \mu = 1/2$. The MAP approach has thus allowed us to determine the optimal value of a parameter which remained free in the developments of Section 2.4.2.4. This is an example of the practical advantage of decision theory versus heuristic approaches.

3.4.3 Marginal Bayesian approach

Let us again consider the exponential prior of Eq. 3.61. In the marginal Bayesian approach, one has to determine:

$$P(\mathbf{s}|j) = \int L(\mathbf{s}|\alpha, j) P[\alpha] d\alpha \tag{3.66}$$

where the integral has to be interpreted with the same notations as in Section 3.2.2.3. It is easily shown that:

$$P(\mathbf{s}|j) = \prod_{k=0}^{N-1} \frac{2a}{\sqrt{\pi}} \frac{1}{[2 \, a + \Delta_k^j]^{\frac{3}{2}}} \tag{3.67}$$

Its logarithm is:

$$\ell_{Bay}(j) = - \sum_{k=0}^{N-1} \frac{3}{2} \log[2 \, a + |\tilde{s}_k - \tilde{r}_k^j|^2] + K'' \tag{3.68}$$

where K'' is a constant independent of j. One can remark that the j-dependent term of $\ell_{Bay}(j)$ is proportional to the j-dependent term of $\ell_{MAP}(j)$ (see Eq. 3.64). The maximization of $\ell_{Bay}(j)$ with respect to j is thus equivalent to the maximization of $\ell_{MAP}(j)$. In other terms, for the exponential prior, the optimal marginal Bayesian solution is equivalent to the MAP approach. Of course, this is also true for the first-order developments, and thus the filter defined in Eq. 3.65 is also a good approximation of the marginal Bayesian solution. This filter is thus close to the optimal decision-theoretical solution for an exponential prior, whatever the technique which is used to deal with the nuisance parameters.

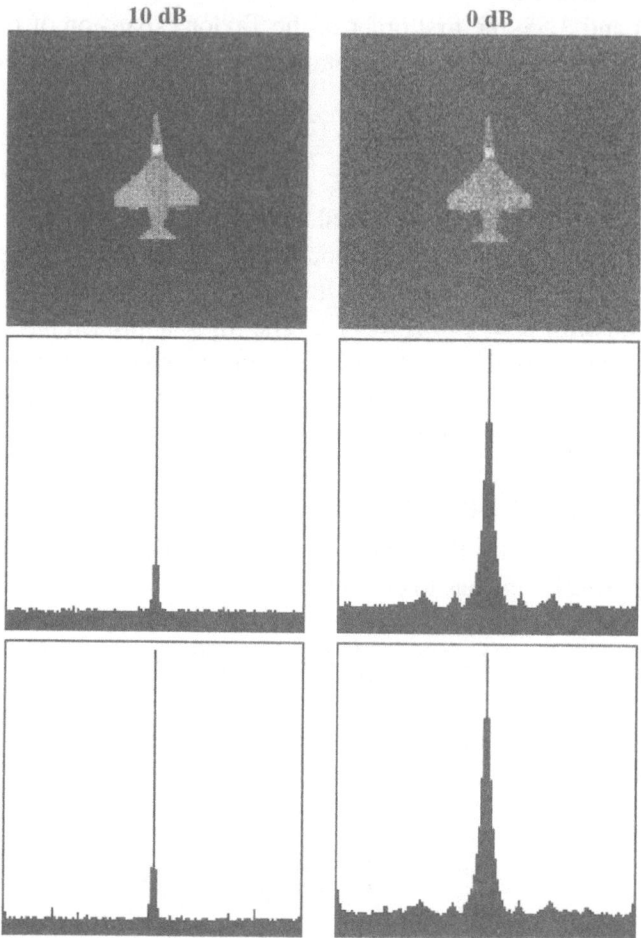

Figure 3.7.   First row: scenes with additive white uniform noise (low levels of noise). Second row: result of the application of the first-order approximation in Eq. 3.65 to the corresponding scene. Third row: result of the application of the MAP optimal algorithm of Eq. 3.64. NB: the graphs of the second and third lines are plots of the maximum of each line of the output plane obtained with the corresponding algorithm. ©1997 SPIE.

3.4.4   Examples of application

In order to illustrate the previous results, we have constructed two scenes containing an object embedded in additive white noise. These scenes are represented in Figure 3.7. The SNR_{in} is respectively equal to 10 dB (very low noise) and 0 dB. In the second line appears the correlation plane obtained by applying Eq. 3.65 to the corresponding scene. In the third line appears the result of the MAP filter without approximation as defined in Eq. 3.64. We can

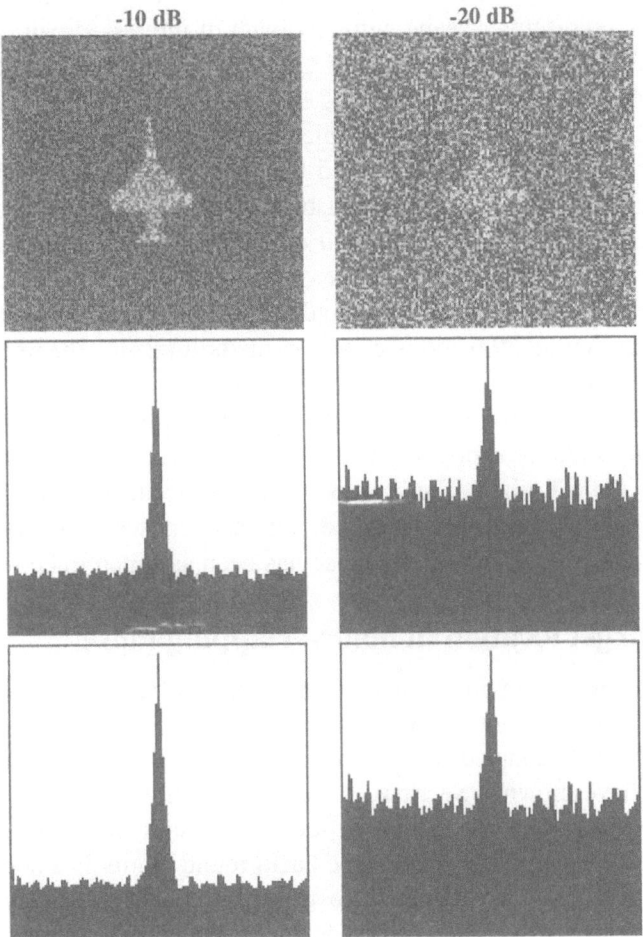

Figure 3.8. First row: scenes with additive white uniform noise (higher levels of noise). Second row: result of the application of the first-order approximation in Eq. 3.65 to the corresponding scene. Third row: result of the application of the MAP optimal algorithm of Eq. 3.64. ©1997 SPIE.

see that the correlation planes corresponding to both filters are almost identical. This shows that the higher-order terms of Eq. 3.64 are negligible with respect to the first-order approximation. Figure 3.8, which represents the same data computed from noisier images (SNR_{in} of -10 and -20 dB), leads to the same conclusions. Whatever the level of noise present in the image, the approximated nonlinear filter is thus a good alternative to the rigorous MAP – or Marginal Bayesian – algorithm.

Let us now illustrate the efficiency of the described approach on real images. Figure 3.9.a represents a scene with an object of interest (a pair of pliers) and Figure 3.9.b represents a reference of the searched object. The reference is a binary shape since one will detect and localize the object only from its shape, not from its graylevels which are subject to changes due to the illumination. Two algorithms will be tested to find the object in the scene: the OT filter optimizing the trade-off between robustness to white noise and peak sharpness (see Eq. 2.33), and the approximation of the MAP filter (see Eq. 3.64). The procedure consists in searching for the object in all its possible orientations, by applying the algorithms based on rotated versions of the reference object (every 2 degrees) in Figure 3.9.b and in choosing the best result. The results obtained with both algorithms with the best estimated angle are shown in Figure 3.9. One can see that the performance of the approximation of the MAP algorithm is better, the peak corresponding to the object appearing more clearly than in the case of the OT filter. This can be attributed to the fact that in the MAP algorithm, the PSD of the scene is taken into account whereas it is not with the considered OT filter, in which the noise has been assumed to be white.

3.5.   Target location in nonoverlapping noise

In this section, we will consider another case in which statistical decision theory makes it possible to go beyond matched filtering, because the processing algorithms can be adapted to the kind of noise actually present in the scene. Indeed, we have shown in Section 3.1.1 that linear filters can fail to detect or locate a target if the noise perturbing the scene is not additive but *nonoverlapping*. For example, in real-world scenes, the background clutter is a nonoverlapping noise since, by definition, it does not overlap with the target. We will see in the present section how to design target location techniques adapted to this type of noise by using the tools of statistical decision theory. In particular, we will consider successively the cases where the graylevel structure of the searched target is totally known, totally unknown, and partially known.

3.5.1   The SIR image model and optimal location algorithms

In order to describe nonoverlapping noise, let us define the following image model. The observed scene (or subscene, if the image has been subdivided into several regions) is composed of two different zones: the target zone and the background zone. These two zones are assumed to have known shapes, which are respectively the shape of the target, and its complementary. The graylevels in these two zones are assumed to be random.

In the following developments, we will consider images as N-dimensional vectors, N being their total number of pixels. Let s denote the analyzed scene

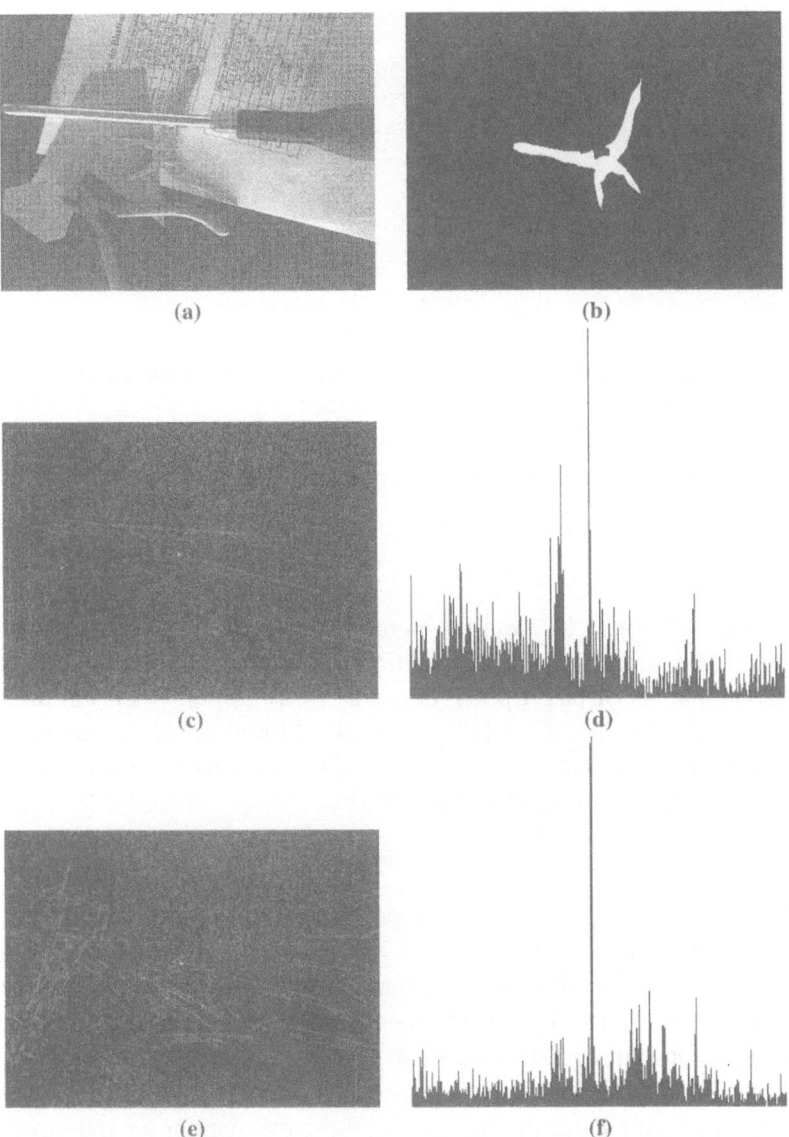

Figure 3.9. **(a)** Scene. **(b)** Object to be searched in the scene (arbitrary orientation). **(c)** Correlation plane obtained with an OT filter (see Eq. 2.33), $\mu = 0.99$, built from reference (b) rotated 102 degrees. **(d)** Maximum of each line of plane (d). **(e)** Correlation plane obtained with the approximation of the MAP algorithm (see Eq. 3.64), $\alpha = 0.5$, built from reference (b) rotated 102 degrees. **(f)** Maximum of each line of plane (e).

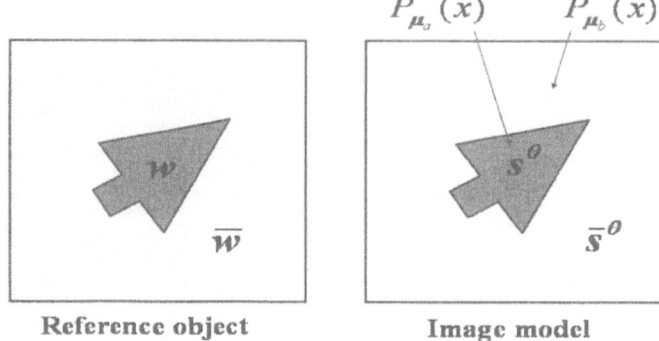

$$P_{\mu_a}(x) \qquad P_{\mu_b}(x)$$

Reference object   Image model

Figure 3.10.   Illustration of the SIR image model (see Eq. 3.70). **w** defines the object region and $\bar{\mathbf{w}}$ the background region.

and **w** the window function of the target, defined as:

$$\left\{ \begin{array}{ll} w_i = 1 & \text{if pixel } i \text{ belongs to the target} \\ w_i = 0 & \text{otherwise} \end{array} \right. \qquad (3.69)$$

We consider an object location problem and thus the parameter θ to be estimated is a scalar representing the position of the object in the image. We will also denote the set of target pixel values $\mathbf{s}^\theta = \{s_i | w_{i-\theta} = 1\}$ and the set of background pixel values $\bar{\mathbf{s}}^\theta = \{s_i | w_{i-\theta} = 0\}$. Moreover, N_w is the number of pixels of the mask **w** and $N_{\bar{w}}$ the number of pixels of its complementary.

The input scene is then defined as follows:

$$s_i = a_i \, w_{i-\theta} + b_i \, (1 - w_{i-\theta}) \ , \quad i \in [0, N-1] \qquad (3.70)$$

a and **b** represent the graylevels of, respectively, the target and the background zone. θ is the position of the object in the scene. In order to proceed, it is necessary to specify the statistical distributions $P_{\mu_a}(\mathbf{x})$ of the set of target graylevels **a** and $P_{\mu_b}(\mathbf{x})$ of the background graylevels **b**. μ_a (μ_b) represents the parameters of the pdf of **a** (**b**). Note that in most real images, the luminances of the target and of the background correspond to different physical processes. It is thus natural to assume that the statistics of these two regions are independent of each other. This is the fundamental hypothesis we will employ in the following. The image model represented by Eq. 3.70, where it is assumed that the statistics of the target and background regions are independent, is called the *Statistically Independent Region* (SIR) image model. It is illustrated in Figure 3.10.

Let us determine the ML algorithm for the estimation of θ. Given the hypothesis of independence between the two regions, the likelihood is written:

$$L(\mathbf{s} | \mu_a, \mu_b, \theta) = P_{\mu_a}(\mathbf{s}^\theta) \, P_{\mu_b}(\bar{\mathbf{s}}^\theta) \qquad (3.71)$$

The log-likelihood corresponding to the image model can thus be written as follows:

$$l(\boldsymbol{\mu}_a, \boldsymbol{\mu}_b, \theta) = \log\left[L(\mathbf{s}|\boldsymbol{\mu}_a, \boldsymbol{\mu}_b, \theta)\right] = \log P_{\boldsymbol{\mu}_a}(\mathbf{s}^\theta) + \log P_{\boldsymbol{\mu}_b}(\bar{\mathbf{s}}^\theta) \quad (3.72)$$

In this target location application, we are only interested in θ. The parameters $\boldsymbol{\mu}_a$ and $\boldsymbol{\mu}_b$ of the pdf's $P_{\boldsymbol{\mu}_a}(\mathbf{x})$ and $P_{\boldsymbol{\mu}_b}(\mathbf{x})$ thus constitute *nuisance parameters*. In the following, we will choose to estimate them in the ML sense. That is, we will determine the values $\widehat{\boldsymbol{\mu}}_a(\theta)$ and $\widehat{\boldsymbol{\mu}}_b(\theta)$ that maximize $l(\boldsymbol{\mu}_a, \boldsymbol{\mu}_b, \theta)$ for a given value of θ. These estimates are then injected back into Eq. 3.72 so as to yield the pseudo-loglikelihood $\ell(\theta) = l\left(\widehat{\boldsymbol{\mu}}_a(\theta), \widehat{\boldsymbol{\mu}}_b(\theta), \theta\right)$. This function only depends on θ, which is the parameter of interest. The ML estimate of the position of the object will thus be given by:

$$\widehat{\theta} = \arg\max_\theta \left\{ l\left(\widehat{\boldsymbol{\mu}}_a(\theta), \widehat{\boldsymbol{\mu}}_b(\theta), \theta\right) \right\} = \arg\max_\theta \left\{ \ell(\theta) \right\} \quad (3.73)$$

In order to proceed, one has to specify parametric expressions for the pdf's $P_{\boldsymbol{\mu}_a}(x)$ and $P_{\boldsymbol{\mu}_b}(x)$. Of course, the precise expression of the ML location algorithm will depend on this choice. In the following, we will specialize the SIR image model to two types of target's and background's probability density functions. In each case, we will determine the expression of the ML optimal location algorithm.

3.5.2 Targets with known graylevels: The D2SIR algorithm

We assume that the target's internal structure $\mathbf{r}$ is known to within a scalar factor β, which can account for global brightness changes in the image [54, 55]. Note that $\mathbf{r}$ is zero outside $\mathbf{w}$. The whole scene is assumed to be covered with a white Gaussian additive noise $\mathbf{n}$ with zero mean and unknown standard deviation σ_n. The graylevels in the background region are assumed to be random variables with Gaussian pdf of mean m_b and variance σ_b^2. This distribution will account for the background clutter plus the detector noise. The image model is thus:

$$s_i = (\beta\, r_{i-\theta} + n_i)\, w_{i-\theta} + b_i\, (1 - w_{i-\theta}) \quad (3.74)$$

Recall that $\mathbf{r}$ is defined to be *zero outside* $\mathbf{w}$. In this model, there are four nuisance parameters: $\boldsymbol{\mu} = [\beta, \sigma_n, m_b, \sigma_b]^T$. The loglikelihood corresponding to this model is, after removing the constant terms:

$$\begin{aligned}
\mathcal{L}[\boldsymbol{\mu}, \theta] \;=\; & -\frac{1}{2\sigma_n^2} \sum_{i=0}^{N-1} \left[(s_i - \beta\, r_{i-\theta})^2\, w_{i-\theta} \right] - N_w \log \sigma_n \\
& -\frac{1}{2\sigma_b^2} \sum_{i=0}^{N-1} [s_i - m_b]^2\, (1 - w_{i-\theta}) - N_{\bar{w}} \log \sigma_b \quad (3.75)
\end{aligned}$$

Let us estimate the nuisance parameters in the ML sense:

$$\frac{\partial \mathcal{L}}{\partial \beta} = 0 \quad \Longleftrightarrow \quad \hat{\beta} = \frac{\sum_{i=0}^{N-1} s_i \, r_{i-\theta}}{E_r} \tag{3.76}$$

$$\frac{\partial \mathcal{L}}{\partial \sigma_n^2} = 0 \quad \Longleftrightarrow \quad \hat{\sigma}_n^2(\theta) = \frac{1}{N_w} \sum_{i=0}^{N-1} [s_i - \hat{\beta} r_{i-\theta}]^2 \, w_{i-\theta} \tag{3.77}$$

$$\frac{\partial \mathcal{L}}{\partial m_b} = 0 \quad \Longleftrightarrow \quad \hat{m}_b(\theta) = \frac{1}{N_{\tilde{w}}} \sum_{i=0}^{N-1} s_i \, (1 - w_{i-\theta}) \tag{3.78}$$

$$\frac{\partial \mathcal{L}}{\partial \sigma_b^2} = 0 \quad \Longleftrightarrow \quad \hat{\sigma}_b^2(\theta) = \frac{1}{N_{\tilde{w}}} \sum_{i=0}^{N-1} [s_i - \hat{m}_b(\theta)]^2 \, (1 - w_{i-\theta}) \tag{3.79}$$

where $E_r = \sum_{i=0}^{N-1} r_i^2$. After injecting these parameters in the expression of the likelihood, one obtains [54, 55]:

$$F_\theta^{(d2)} = -\frac{N_w}{2} \log\left[\hat{\sigma}_n^2(\theta)\right] - \frac{N_{\tilde{w}}}{2} \log\left[\hat{\sigma}_b^2(\theta)\right] \tag{3.80}$$

This algorithm will be called D2SIR, since the term $\hat{\sigma}_n^2(\theta)$ amounts to computing the quadratic difference between the scene **s** and the reference **r** multiplied by the estimate of the luminance $\hat{\beta}$. Consequently, the image model defined by Eqs. 3.74 will also be called the *D2SIR image model*. Note that in the following, we will call an image model and the location algorithm optimal in the ML sense for this model with the same name.

If $\hat{\beta}$ is replaced by its expression in Eq. 3.76, the expression of $\hat{\sigma}_n^2$ can be put in the following form:

$$\hat{\sigma}_n^2(\theta) = \frac{1}{N_w} \left[\sum_{i=0}^{N-1} s_i^2 \, w_{i-\theta} - \frac{1}{E_r} \left(\sum_{i=0}^{N-1} s_i \, r_{i-\theta} \right)^2 \right] \tag{3.81}$$

Thus the expressions of $\hat{\sigma}_n^2(\theta)$ and $\hat{\sigma}_b^2(\theta)$ can be written as:

$$\hat{\sigma}_n^2(\theta) = \frac{1}{N_w}[\mathbf{s}^2 \star \mathbf{w}]_\theta - \frac{1}{N_w} \frac{[\mathbf{s} \star \mathbf{r}]_\theta^2}{E_r}$$

$$\hat{\sigma}_b^2(\theta) = \frac{1}{N_{\tilde{w}}} \left(\sum_{i=0}^{N-1} s_i^2 - [\mathbf{s}^2 \star \mathbf{w}]_\theta \right) - \frac{1}{N_{\tilde{w}}^2} \left(\sum_{i=0}^{N-1} s_i - [\mathbf{s} \star \mathbf{w}]_\theta \right)^2$$

where $\mathbf{s}^2 = \{s_i^2 \, , \, i \in [0, N-1]\}$ and $\star$ denotes the correlation operation. From Eq. 3.80, we see that the D2SIR algorithm can be decomposed into three steps. First, preprocess the input image **s**. Here, the preprocessing is quite simple since it consists in taking the square of all pixel values. Second, perform three correlation products $[s \star w]_\theta$, $[s \star r]_\theta$, and $[s^2 \star w]_\theta$. Finally, postprocess the obtained correlation planes according to Eq. 3.80. The pre- and postprocessings are pointwise operations. Their computational complexity is thus N (N being the total number of pixels in the image) whereas the complexity of correlations is $N \log_2 N$ if they are implemented with the FFT. The whole algorithm thus has the same complexity as a linear correlation algorithm. Despite this fact, its properties are different from those of linear filters, because the correlations are performed on nonlinearly preprocessed versions of the scene, and they are nonlinearly postprocessed. It must be noted that these nonlinear pre- and postprocessings are not empirical, but follow from optimality considerations.

The efficiency of this filter can be verified in Figure 3.3, where it is shown that in the presence of nonoverlapping noise, a linear filter fails (see second column) whereas the D2SIR filter correctly locates the object (see third column). An example of application of the D2SIR filter on a realistic image is shown in Figure 3.11.

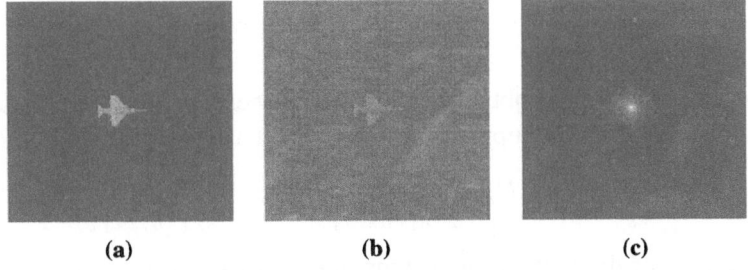

    (a)   (b)   (c)

Figure 3.11. **(a)** Reference object. **(b)** Scene. **(c)** Result of processing image (b) with the D2SIR algorithm (see Eq. 3.80)

3.5.3   Targets with fluctuating graylevels: The WGSIR algorithm

As pointed out in Section 3.1.2, the assumption that the target's internal structure r is known to within a scalar factor can be too strong in some cases. For example, while tracking an object in an image sequence, the graylevels of the object can fluctuate. In order to model this kind of situation, one can modify the SIR image model by assuming that the graylevels of both the target zone and the background zone are random. More precisely, one assumes that the target's graylevels are distributed with a white Gaussian pdf with mean m_a and standard deviation σ_a, and, as in the previous case, the background's graylevels are distributed with a white Gaussian distribution with mean m_b and standard deviation σ_b [56]. This image model will be the *White Gaussian SIR* (WGSIR).

In this model, there are four nuisance parameters: $\boldsymbol{\mu} = [m_a, m_b, \sigma_a, \sigma_b]^T$. From Eq. 3.72, and taking into account that the pdf in the target and the background regions are white and Gaussian, the loglikelihood has the following expression:

$$\mathcal{L}(\boldsymbol{\mu}, \theta) = -N_w \log \sigma_a - N_{\bar{w}} \log \sigma_b - \frac{1}{2\sigma_a^2} \sum_{i=0}^{N-1} [s_i - m_a]^2 w_{i-\theta}$$

$$-\frac{1}{2\sigma_b^2} \sum_{i=0}^{N-1} [s_i - m_b]^2 (1 - w_{i-\theta}) \tag{3.82}$$

The nuisance parameters m_a, m_b, σ_a, and σ_b are estimated in the ML sense. The estimates of m_b and σ_b are given by Eqs. 3.78 and 3.79. Those of m_a and σ_a are similar:

$$\widehat{m}_a(\theta) = \frac{1}{N_w} \sum_{i=0}^{N-1} s_i \, w_{i-\theta} \tag{3.83}$$

$$\widehat{\sigma}_a^2(\theta) = \frac{1}{N_w} \sum_{i=0}^{N-1} [s_i - \widehat{m}_a(\theta)]^2 \, w_{i-\theta} \tag{3.84}$$

The optimal algorithm is obtained by injecting these values in the expression of the loglikelihood, and suppressing the terms that do not depend on θ. One obtains:

$$F_\theta^{(wg)} = -\frac{N_w}{2} \log \left[\widehat{\sigma}_a^2(\theta)\right] - \frac{N_{\bar{w}}}{2} \log \left[\widehat{\sigma}_b^2(\theta)\right] \tag{3.85}$$

As in the case of the D2SIR algorithms, the estimates of the variances $\widehat{\sigma}_a^2(\theta)$ and $\widehat{\sigma}_b^2(\theta)$ can be expressed as linear combinations of two correlation products $[\mathbf{s} \star \mathbf{w}]_\theta$ and $[\mathbf{s}^2 \star \mathbf{w}]_\theta$:

$$\widehat{\sigma}_a^2(\theta) = \frac{1}{N_w} [\mathbf{s}^2 \star \mathbf{w}]_\theta - \frac{1}{N_w^2} [\mathbf{s} \star \mathbf{w}]_\theta^2 \tag{3.86}$$

$$\widehat{\sigma}_b^2(\theta) = \frac{1}{N_{\bar{w}}} \left(\sum_{i=0}^{N-1} s_i^2 - [\mathbf{s}^2 \star \mathbf{w}]_\theta \right)$$

$$-\frac{1}{N_{\bar{w}}^2} \left(\sum_{i=0}^{N-1} s_i - [\mathbf{s} \star \mathbf{w}]_\theta \right)^2 \tag{3.87}$$

The computational complexity of the WGSIR filter is thus equivalent to that of linear filtering.

The efficiency of this filter can be verified in Figure 3.3, where it is shown that in the presence of fluctuating target, the D2SIR filter fails whereas the WGSIR filter correctly locates the object (see last column). An example of application of the WGSIR on a realistic image is shown in Figure 3.12.

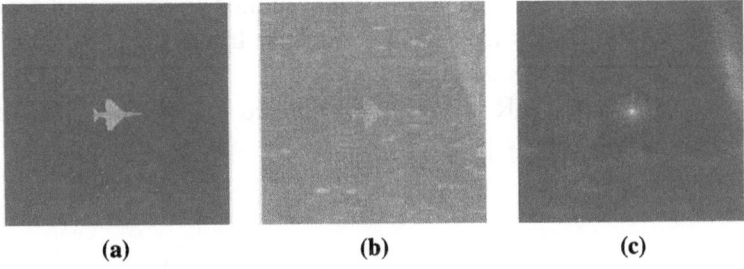

(a) **(b)** **(c)**

Figure 3.12. **(a)** Reference object. **(b)** Scene. **(c)** Result of processing image (b) with the WGSIR algorithm (see Eq. 3.85).

3.5.4   Partially fluctuating targets: The USIR algorithm

We have seen that the D2SIR algorithm is adapted to the case where the target's graylevels are known to within a scalar factor, and the WGSIR is adapted to the case where these graylevels are totally random. In real applications, the situation is often a balance between these two extreme cases since one often has a partial knowledge of the target's graylevels. For example, in tracking applications, one may know the graylevels of the object at the beginning of the process (after the initialization), but as time goes by, these graylevels may change.

To overcome this problem, one can model the texture of the target as a weighted sum of the deterministic texture of the target reference **r** and of a random white Gaussian texture **a** [55, 57]. The weights are assumed unknown and they will be estimated from the data itself in the ML sense. The model we consider is thus the following:

$$s_i = [\beta \, r_{i-\theta} + a_i] \, w_{i-\theta} + b_i \, (1 - w_{i-\theta}) \tag{3.88}$$

with **a** and **b** random white Gaussian vectors with respective means m_a and m_b, and standard deviations σ_a and σ_b. The parameter β determines the weight of the deterministic model in the target's texture, and the mean m_a the weight of the random part. One can introduce a parameter d which synthesizes the balance between these two contributions:

$$d = \frac{\beta m_r}{m_a + \beta m_r} \tag{3.89}$$

with $m_r = 1/N_w \sum_{i=0}^{N-1} r_i$. If both m_a and m_r are zero, we set $d = 0$.

In the case of a target with known texture and only perturbed by an additive noise with zero mean, we have $m_a = 0$ (and $\beta \neq 0$) and $d = 1$. In these conditions, the model of Eq. 3.88 is similar to the D2SIR model (see Eq. 3.74). On the other hand, in the case of a target with an unknown texture, we have

$\beta = 0$ (and $m_a \neq 0$) and $d = 0$. This model can thus be considered a weighted mixing between the D2SIR and WGSIR models. It will thus be called USIR (for "unified" SIR) in the following.

Let us consider the USIR image model defined in Eq. 3.88. The vector of nuisance parameters is $\boldsymbol{\mu} = [\beta, m_a, m_b, \sigma_a, \sigma_b]^T$ and the loglikelihood can be written as follows:

$$
\begin{aligned}
\mathcal{L}(\boldsymbol{\mu}, \theta) \;=\; & -\frac{N}{2} \log(2\pi) - N_w \log \sigma_a - N_{\bar{w}} \log \sigma_b \\
& -\frac{1}{2\sigma_a^2} \sum_{i=0}^{N-1} [s_i - \beta r_{i-\theta} - m_a]^2 \, w_{i-\theta} \\
& -\frac{1}{2\sigma_b^2} \sum_{i=0}^{N-1} [s_i - m_b]^2 \, (1 - w_{i-\theta})
\end{aligned}
\tag{3.90}
$$

The nuisance parameters β, m_a, m_b, σ_a, and σ_b can be estimated in the ML sense. We obtain:

$$
\hat{\beta} \;=\; \frac{1}{N_w} \frac{\displaystyle\sum_{i=0}^{N-1} [s_i(r_{i-\theta} - m_r)] \, w_{i-\theta}}{\displaystyle\sum_{i=0}^{N-1} (r_i - m_r)^2 \, w_{i-\theta}}
\tag{3.91}
$$

$$
\hat{m}_a(\theta) \;=\; \frac{1}{N_w} \sum_{i=0}^{N-1} \left(s_i - \hat{\beta}\, r_{i-\theta}\right) w_{i-\theta}
\tag{3.92}
$$

$$
\hat{m}_b(\theta) \;=\; \frac{1}{N_{\bar{w}}} \sum_{i=0}^{N-1} s_i (1 - w_{i-\theta})
\tag{3.93}
$$

$$
\hat{\sigma}_a^2(\theta) \;=\; \frac{1}{N_w} \sum_{i=0}^{N-1} \left[s_i - \hat{\beta}\, r_{i-\theta} - \hat{m}_a(\theta)\right]^2 w_{i-\theta}
\tag{3.94}
$$

$$
\hat{\sigma}_b^2(\theta) \;=\; \frac{1}{N_{\bar{w}}} \sum_{i=0}^{N-1} [s_i - \hat{m}_b(\theta)]^2 (1 - w_{i-\theta})
\tag{3.95}
$$

If we inject these estimates into the expression of the likelihood, the function to be maximized becomes:

$$
F_\theta^{(mlu)} \;=\; -\frac{N_w}{2} \log \hat{\sigma}_a^2(\theta) - \frac{N_{\bar{w}}}{2} \log \hat{\sigma}_b^2(\theta)
\tag{3.96}
$$

This algorithm is based on computing the three correlation products $[\mathbf{s} \star \mathbf{w}]_\theta$, $[\mathbf{s}^2 \star \mathbf{w}]_\theta$, and $[\mathbf{s} \star \mathbf{r}]_\theta$ since we can write:

$$\hat{\sigma}_a^2(\theta) = \frac{1}{N_w}[\mathbf{s}^2 \star \mathbf{w}]_\theta - \frac{1}{N_w^2}[\mathbf{s} \star \mathbf{w}]_\theta^2 - \frac{1}{N_w^2}\frac{(m_r[\mathbf{s} \star \mathbf{w}]_\theta - [\mathbf{s} \star \mathbf{r}]_\theta)^2}{\displaystyle\sum_{i=0}^{N-1}(r_i - m_r)^2 w_{i-\theta}}$$

$$\hat{\sigma}_b^2(\theta) = \frac{1}{N_{\tilde{w}}}\left(\sum_{i=0}^{N-1} s_i^2 - [\mathbf{s}^2 \star \mathbf{w}]_\theta\right) - \frac{1}{N_{\tilde{w}}^2}\left(\sum_{i=0}^{N-1} s_i - [\mathbf{s} \star \mathbf{w}]_\theta\right)^2$$

We will denote this algorithm USIR in the following. Let us now illustrate its performance. The scene in Figure 3.13.a contains a graylevel target (the airplane) to which the D2SIR and the USIR algorithms are matched. One can see in the first row that the D2SIR correctly locates the target in the scene, but not the WGSIR. Figure 3.13.e contains a target which has the same shape as in Figure 3.13.a, but with different graylevels. The D2SIR and USIR are still matched to the graylevels of the target appearing in Figure 3.13.a. One can see in the second line of Figure 3.13 that the WGSIR correctly locates the target in this scene, but not the D2SIR. On the other hand, one can see in Figures 3.13.b and 3.13.f that the USIR enables one to locate the target in both scenes. Thus the USIR algorithm performs an efficient unification of the D2SIR and WGSIR, and allows one to process scenes for which two different algorithms would have been necessary.

We have introduced in this section the SIR image model, which allows for dealing with images perturbed with nonoverlapping noise. We have seen how the tools of decision and estimation theory enable one to define optimal target location algorithms according to the knowledge which is available on the graylevel structure of the target. In Chapter 4, we will generalize the SIR image model to a variety of pattern recognition applications, such as detection, classification, or segmentation. We will then illustrate its usefulness in cases where the noise cannot be considered white Gaussian.

One can remark that all the optimal algorithms presented in this section, as well as those presented in Section 3.4, are based on correlations. Some of them consist of nonlinear preprocessings of the input scene followed by correlations (see Figure 3.14 for an illustration). Others are based on nonlinearities in the Fourier plane. Their computational complexities are thus equivalent to that of linear filters, although their performance is higher since they are adapted to the nature of the noise present in the image. It is important to notice that this property is not postulated *a priori*, as in the heuristic approaches, but obtained *a posteriori* through the optimization of statistical criteria.

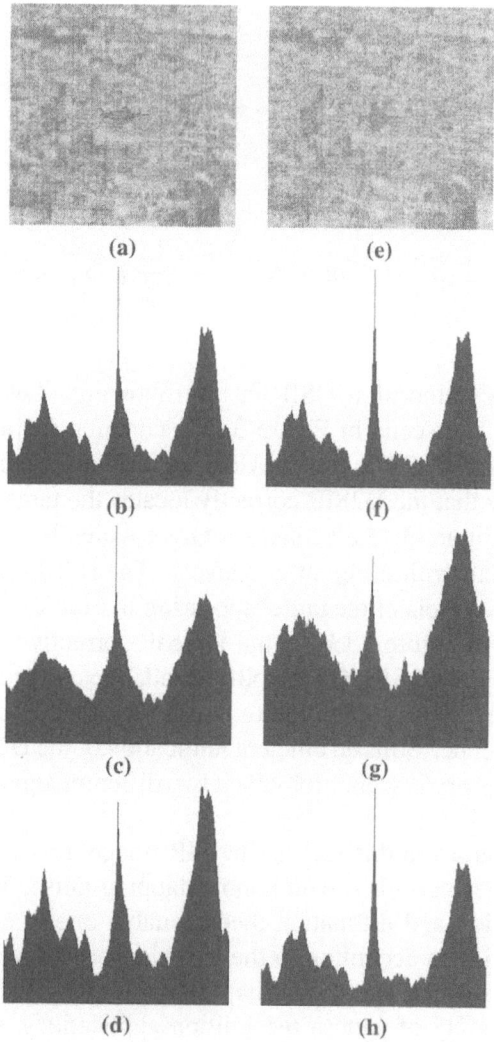

Figure 3.13. **(a)** Scene (256×256 pixels). **(b)** Result of processing (a) with the USIR algorithm. **(c)** Result of processing (a) with the D2SIR algorithm. **(d)** Result of processing (a) with the WGSIR algorithm. **(e)** Scene (256×256 pixels). **(f)** Result of processing (e) with the USIR algorithm. **(g)** Result of processing (e) with the D2SIR algorithm. **(h)** Result of processing (e) with the WGSIR algorithm. In all cases, the D2SIR and USIR algorithms are matched to the graylevel target appearing in scene (a). All the curves represent the maximum of each column of the output plane of the corresponding algorithm.

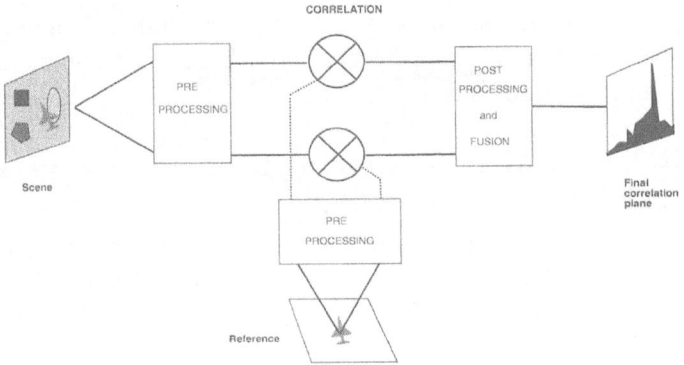

Figure 3.14. Principle of target location algorithm consisting of preprocessing, correlation, and postprocessing modules.

3.6. Conclusion

In this chapter, we have analyzed how statistical decision and estimation theory can be applied to solve image processing problems. We have first presented the general theory, with the ML and the MAP approaches. We have also discussed the different ways of handling nuisance parameters, i.e., the ML, the MAP, and the marginal Bayesian methods. We have then illustrated these different approaches on object location applications. These results show that the statistical decision and estimation theory approach makes it possible to solve problems that could not be solved by heuristic approaches.

The main advantage of statistical decision and estimation theory compared to heuristic techniques is that it provides algorithms which are optimally adapted to the noise perturbing the images. The reason is that with statistical approaches, the starting point is the image model itself, which represents – as accurately as possible – the characteristics of real scenes. We will see further examples of this property in Chapter 4. The other advantage of this approach is that the formulation of a problem in terms of likelihood makes it possible to solve a variety of applications. The algorithms for detection, location, and classification all rely on the same likelihood function, which uniquely depends on the image model. In Chapter 5, we will see that the same approach leads very naturally to active contour algorithms for object segmentation.

The main drawback of the statistical approach is that it can lead to computationally intensive algorithms. This is the case for example for invariant pattern recognition, for which the optimal procedure (in the sense of statistical decision and estimation theory) consists in computing the likelihood of any possible target configuration and in choosing the configuration which leads to a maximum. For problems with many degrees of freedom, this approach rapidly becomes cumbersome and the SDFs introduced in Chapter 2, although suboptimal, may be more appropriate. However, even when it is operationally impractical, the statistical decision and estimation theory approach yields a benchmark to which suboptimal strategies can be compared.

APPENDIX 3.A: MAP location algorithm in the presence of uniform prior

In this appendix, we determine the ML location algorithm with MAP estimation of the nuisance parameter Γ, assuming a uniform prior on Γ (cf. Section 3.4.2 and Figure 3.6).

The criterion to optimize is the following:

$$\mathcal{L}(s|j,\Gamma)P[\Gamma] = -\frac{1}{2}\sum_k \log \Gamma_k - \frac{1}{2}\sum_k \frac{\Delta_k^j}{\Gamma_k} + \log P[\Gamma] \qquad (3.A.1)$$

In order to maximize this expression with respect to Γ_k, one first computes the ML estimate (without taking the term $\log P[\Gamma]$ into account). According to Eq. 3.48, its expression is:

$$\hat{\Gamma}_k^{ML} = \Delta_k^j \qquad (3.A.2)$$

The MAP estimate of Γ is then defined by:

$$\Delta_k^j \geq a \implies \hat{\Gamma}_k^{MAP} = \Delta_k^j \qquad (3.A.3)$$
$$\Delta_k^j < a \implies \hat{\Gamma}_k^{MAP} = a \qquad (3.A.4)$$

Let us define the following two sets of integer values:

$$\mathcal{D} = \left\{ k \in [0, N-1] \mid \Delta_k^j \geq a \right\} \qquad (3.A.5)$$
$$\bar{\mathcal{D}} = \left\{ k \in [0, N-1] \mid \Delta_k^j < a \right\} \qquad (3.A.6)$$

After injecting the estimates of Γ_k^{MAP} into Eq. 3.A.1, one obtains the following pseudo-likelihood:

$$\ell(j) = -\frac{1}{2}\sum_{k \in \mathcal{D}} \log \Delta_k^j - \frac{1}{2}\sum_{k \in \bar{\mathcal{D}}} (\log a - 1) - \frac{1}{2}\sum_{k \in \bar{\mathcal{D}}} \frac{\Delta_k^j}{a} + K \qquad (3.A.7)$$

where K is independent of j. Let us now perform on $\ell(j)$ the same first-order development as in Section 3.4.1. One obtains:

$$\begin{aligned} d\ell(j) &= \frac{1}{2}\sum_{k \in \mathcal{D}} \frac{U_k^j}{D_k} - \frac{1}{2}\sum_{k \in \mathcal{D}} \log D_k - \frac{1}{2}\sum_{k \in \bar{\mathcal{D}}} (\log a - 1) \\ &\quad -\frac{1}{2}\sum_{k \in \bar{\mathcal{D}}} \frac{D_k}{a} + \frac{1}{2}\sum_{k \in \bar{\mathcal{D}}} \frac{U_k^j}{a} + K \end{aligned} \qquad (3.A.8)$$

where U_k^j and D_k are defined in Eqs. 3.52. This expression is quite involved since $\mathcal{D}$ and $\bar{\mathcal{D}}$ depend on j itself. In order to obtain a more tractable equation, one has to do a further approximation, which is consistent with the hypothesis of validity of the first-order development: Let us suppose that $D_k >> U_k^j$ so that $\Delta_k^j \simeq D_k$. We can then replace the set $\mathcal{D}$ and $\bar{\mathcal{D}}$ with:

$$\mathcal{D}' = \{ k \in [0, N-1] \mid D_k \geq a \} \qquad (3.A.9)$$
$$\bar{\mathcal{D}}' = \{ k \in [0, N-1] \mid D_k < a \} \qquad (3.A.10)$$

The advantage of this operation is that $\mathcal{D}'$ and $\bar{\mathcal{D}}'$ do not depend on j, so that the expression of $d\ell(j)$ becomes:

$$d\ell'(j) = \frac{1}{2} \sum_{k \in \mathcal{D}'} \frac{U_k^j}{D_k} + \frac{1}{2} \sum_{k \in \bar{\mathcal{D}}'} \frac{U_k^j}{a} + K' \qquad (3.A.11)$$

where K' is a constant which does not depend on j. Consequently, following the same reasoning as in Section 3.4.1, maximizing $d\ell'(j)$ with respect to j is equivalent to maximizing the inverse Fourier transform of:

$$\tilde{F}_k = \frac{\tilde{s}_k(\tilde{r}_k)^*}{\max[D_k, a]} \qquad (3.A.12)$$

Chapter 4

APPLICATIONS OF STATISTICAL CORRELATION TECHNIQUES TO DIFFERENT PHYSICAL NOISES

We have introduced in Chapter 3 the basics of statistical decision and estimation theory as a tool for designing image processing algorithms adapted to the noise that actually affects the images. The efficiency of this approach stems from the fact that it is based on a physical model of the image and of the noise: The derived algorithms are thus, by construction, adapted to these perturbations. This approach has many applications nowadays, as more and more new imagery systems based on various physical phenomena are developed. Indeed, each of these systems may have different noise characteristics and statistical decision and estimation theory provides an efficient way of designing adapted processing algorithms.

Consequently, the crucial point to apply the statistical theory to practical image processing problems is to determine a good model of the noise present in the image. In making such a choice, the algorithm designer will be guided by two goals, which are often antagonist. The first one is of course that the noise model describe as accurately as possible the image characteristics. Unfortunately, images are very complex signals, and the sources of perturbations that affect them are of various types, as briefly illustrated in Section 3.1. Consequently, realistic image models tend to become complex and the corresponding statistical decision-based algorithms are also complex and time consuming. This is in opposition with the second goal of the algorithm designer – which is also very important to the algorithm user – that the algorithms be as simple and as computationally fast as possible. Indeed, the image processing applications that we are considering (target detection, location, segmentation) are often the first levels of the image processing chain. They must be accurate, but also fast since they usually deal with all the pixels in the image. This issue becomes even more crucial with the development of multichannel imagery, such as multispectral, hyperspectral, or polarimetric systems, in which the quantity of data increases proportionally to the number of channels. Simplicity and rapidity are thus key parameters for the applicability of image processing techniques.

The challenge for the algorithm designer is thus to choose a noise model which will represent accurately enough the image perturbation to yield good performance, while being simple enough to lead to sufficiently fast processing algorithms. We have provided examples of this approach in Section 3.5, where we have shown that considering the nonoverlapping nature of the noise in images leads to a considerable increase of performance compared to the additive noise model, for a similar computational complexity since only correlation and pointwise operations are required.

We present in this chapter further applications of this approach to nonconventional types of images. We begin in Section 4.1 by generalizing in two different ways the SIR image model that was introduced in the previous chapter. First, we show that this model can be used to solve not only target location applications, but also detection, segmentation, and recognition tasks. Second, we introduce a family of noise statistics, the exponential family [58], which has the double advantage of accurately describing the noise present in a variety of different image modalities and to lead to simple and tractable expressions of the likelihood, so that the resulting optimal algorithms are computationally attractive. The rest of the chapter is devoted to applications of this approach to different types of imagery. We analyze in Section 4.2 the problem of target location in the presence of spatially correlated noise and nonhomogeneous backgrounds. We then consider target location in binary images in Section 4.3 and edge detection/location in SAR images in Section 4.4.

4.1. A general framework for designing image processing algorithms

The SIR image model has been introduced in Section 3.5 in order to model nonoverlapping noise, that is, the fact that in most images, the main perturbation is caused by the background clutter rather than additive noise. This model has been used to solve target location problems. However, its field of application is much more general. We show in this section that it can be used to model a wide variety of image processing problems and of noise statistics.

4.1.1   Generalization of the SIR image model

Let us write the model of the observed scene s as:

$$s(x,y) = a(x,y)\, w^{\boldsymbol{\theta}}(x,y) + b(x,y)\left[1 - w^{\boldsymbol{\theta}}(x,y)\right] \qquad (4.1)$$

Note that from now on, we will use two-dimensional notation for the images. Equation 4.1 means that the scene consists of two regions: region a, which corresponds to the object of interest and will be called the *target region*, and region b, which is the *background region*. The mask $\mathbf{w}^{\theta} = \left\{ w^{\theta}(x,y)|(x,y) \in [1,P] \times [1,Q] \right\}$ is a binary window function that defines the shape of the target, so that:

$$w^{\theta}(x,y) = 1 \qquad \text{within the target (region } a)$$
$$w^{\theta}(x,y) = 0 \qquad \text{outside the target (region } b) \qquad (4.2)$$

The parameter θ corresponds to a certain hypothesis H_{θ} on some attributes of the target: It is the parameter of interest which has to be estimated from the model. In the previous chapter, since we considered target location applications, θ represented the position of the target in the image. However, this parameter can have other meanings depending on the applications which have to be solved:

- In detection applications, θ is a binary value, which is 1 if a target of shape $\mathbf{w}$ is present in the image, and 0 otherwise. In other words, $w^{1}(x,y) = w(x,y)$ and $w^{0}(x,y) = 0$.

- In recognition (or more precisely discrimination) applications, θ is a value belonging to a discrete set, each value corresponding to one of the objects to recognize. For example, in a vehicle recognition application, $\theta = 1$ if a truck is present in the image (the mask $\mathbf{w}^{1}$ will represent the shape of the truck), $\theta = 2$ if it is a jeep, etc.

- For target location application, θ is the position of some characteristic points of the object (for example the center of gravity) in the image.

- For orientation estimation, θ is a set of possible orientations of the object. θ is a scalar value if rotation is estimated in a plane, and a two-component vector if one considers 3-D rotations.

- For segmentation purpose, that is, when one has to determine the whole shape of the object, θ represents this shape, which can be parametrized in different ways, using B-splines [59], Fourier descriptor [60], or polygons [61, 62]. In the latter case, if the shape is approximated by a polygonal contour, θ is the set of coordinates of the nodes of the polygon.

The definition of θ corresponding to each of these applications is summarized in Table 4.1.

The target's gray levels $\mathbf{a}$ and the background noise $\mathbf{b}$ defined in Eq. 4.2 are modeled as homogeneous random fields, which means that each value $a(x,y)$ or $b(x,y)$ is a random variable. For simplicity's sake, we will assume that these random variables are spatially uncorrelated (i.e., the graylevels of each pixel are statistically independent). In other words, we consider that $\mathbf{a}$ and $\mathbf{b}$ are white random fields. Of course, this assumption may be limiting in some

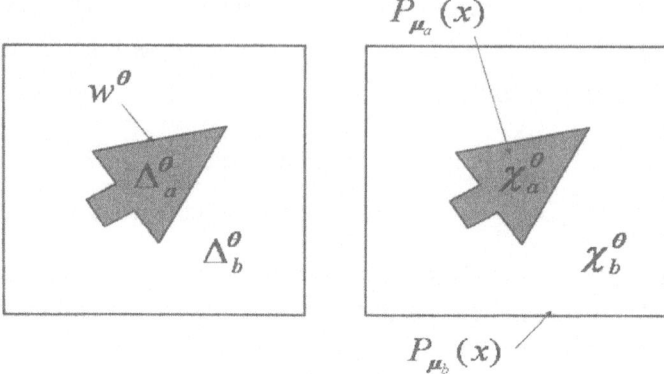

Figure 4.1. Left: Geometrical description of the SIR image model. Right: Statistical description of the SIR image model.

cases, and we will describe in Section 4.2.1 a simple method to generalize the present approach to spatially correlated graylevels. Under the assumption of spatial independence, the random fields **a** and **b** are thus characterized by their respective pointwise pdf $P_{\mu_a}(x)$ and $P_{\mu_b}(x)$. μ_a and μ_b are the parameters of the pdf, which will be considered as *a priori* unknown and thus constitute *nuisance parameters*. These parameters can be scalars or vectors if more than one scalar parameter is needed to determine the pdf.

Finally, let us define the following notation corresponding to regions a and b, which will be used throughout the chapter:

- $\Delta_a^\theta = \{(x,y) \mid w^\theta(x,y) = 1\}$ denotes the set of pixels corresponding to region a.

- $\Delta_b^\theta = \{(x,y) \mid w^\theta(x,y) = 0\}$ denotes the set of pixels corresponding to region b.

- $\chi_u^\theta = \{s(x,y) \mid (x,y) \in \Delta_u^\theta\}$, $u = a$ or b, denotes the set of the graylevel values of the pixels belonging in Δ_u^θ. Since these values are modeled as random variables, χ_u^θ is a statistical sample.

- $N_u(\theta) = card(\Delta_u^\theta)$ denotes the number of pixels in region u. This number may depend on θ, for example in segmentation applications, where θ represents the shape of the object.

Figure 4.1 schematically represents the layout of the SIR image model.

Let us now express the likelihood corresponding to the SIR image model. We will consider that the parameter θ can take discrete values, so that each

Application	Nature of θ
Detection	0 or 1
Discrimination	discrete set
Location	(x, y)
Attitude estimation	angles
Polygonal segmentation	node coordinates

Table 4.1.   Examples of image processing applications involving the SIR image model, and the corresponding nature of the parameter θ.

value of θ corresponds to a hypothesis H_θ. The likelihood of this hypothesis depends on θ, but also on the nuisance parameters μ_a and μ_b. It can be written:

$$L[s|\theta, \mu_a, \mu_b] = \prod_{(x,y) \in \Delta_a^\theta} P_{\mu_a}[s(x,y)] \prod_{(x,y) \in \Delta_b^\theta} P_{\mu_b}[s(x,y)] \qquad (4.3)$$

where we have made use of the fact that the graylevels are statistically independent. The loglikelihood has the following expression:

$$\mathcal{L}[s|\theta, \mu_a, \mu_b] = \sum_{(x,y) \in \Delta_a^\theta} \log P_{\mu_a}[s(x,y)] + \sum_{(x,y) \in \Delta_b^\theta} \log P_{\mu_b}[s(x,y)] \quad (4.4)$$

All the methods presented in Section 3.2 can be used to deal with nuisance parameters μ_a and μ_b. In the following, for simplicity's sake, we will concentrate on the ML approach, which consists in choosing the values of the nuisance parameters which maximize the loglikelihood so as to obtain the following pseudo-loglikelihood:

$$\ell(s, \theta) = \max_{\mu_a, \mu_b} \mathcal{L}[s|\theta, \mu_a, \mu_b]$$

$$= \mathcal{L}[s|\theta, \hat{\mu}_a(\theta), \hat{\mu}_b(\theta)] \qquad (4.5)$$

where $\hat{\mu}_a(\theta)$ and $\hat{\mu}_b(\theta)$ are the ML estimates of μ_a and μ_b. The ML estimate of θ is obtained by:

$$\hat{\theta}^{ML} = \arg\max_\theta \ell(s, \theta) \qquad (4.6)$$

We have established the general properties of the image model which will be considered in the following. Its main characteristics are the randomness and the statistical independence of the target and background pixel graylevels. In order to proceed, we have to choose the expressions of the pdf $P_{\mu_a}(x)$ and $P_{\mu_b}(x)$.

4.1.2   The exponential family

In order to represent the statistical properties of the pixel graylevels in the SIR image model, we will use pdf that belong to the exponential family [58]. Members of this family include the Bernoulli, Gamma, Gaussian, Poisson, Rayleigh, and many other familiar statistical distributions. As we shall see, these distributions fulfill the requirements for efficient algorithm design that we have indicated in the introduction of this chapter: They accurately represent the fluctuations induced by several widely used imaging modalities, and their mathematical form leads to computationally efficient algorithms.

For example, the Bernoulli pdf is adapted to handle the case of binary images [63]. It is also well known that at low photon levels, the noise present in images is described by Poisson pdf [64] since the arrival of photons on the sensor can be considered as discrete random events which follow a Poisson law. This situation occurs, for example, in medical or astronomical imagery when the exposure time of the sensor is short. Synthetic Aperture Radar (SAR) intensity images are corrupted by a multiplicative noise, also known as speckle [65], which can be described in first approximation with Gamma pdf. This issue has been widely studied and it is now well known that in order to obtain efficient algorithms, the statistical properties of speckle noise have to be taken into account in the design of image processing algorithms (see Ref. 66 and references therein). Ultrasonic medical images correspond to amplitude detection of the incident acoustic field and in these images, the speckle noise can be described with a Rayleigh pdf [67]. We will also discuss the case of optronic images and the relevance of Gaussian pdf when a whitening preprocessing is used [68].

4.1.2.1   Exponential family and ML parameter estimation

Probability density functions (pdf) which belong to the exponential family are defined by:

$$P_{\boldsymbol{\mu}}(x) = \exp\left[\boldsymbol{\alpha}(\boldsymbol{\mu}).\boldsymbol{\beta}(x) + \phi(\boldsymbol{\mu}) + \kappa(x)\right] \tag{4.7}$$

where $\boldsymbol{\mu} = [\mu_1, \mu_2, ..., \mu_n]^T$ is the vector of parameters of the pdf, $\kappa(x)$ is a scalar function of x while $\boldsymbol{\alpha}(\boldsymbol{\mu})$ and $\boldsymbol{\beta}(x)$ are p-component vector functions of respectively $\boldsymbol{\mu}$ and x, and ".." stands for scalar product. The domain of definition of x is assumed independent of $\boldsymbol{\mu}$. We summarize in Table 4.2 the pdf of some of the members of the exponential family which will be discussed in the following.

Law	pdf P(x)	Parameters μ
Bernoulli	$p\delta(x) + (1-p)\delta(1-x)$	p
Gamma	$\left(\frac{L}{\mu}\right)^L \frac{x^{L-1}}{\Gamma(L)} exp[-\frac{L}{\mu}x]$, $x \geq 0$	$\mu, (L)$
Gaussian	$\frac{1}{\sqrt{2\pi}\sigma} exp[-\frac{(x-m)^2}{2\sigma^2}]$	m, σ
Poisson	$\sum_{n \in \mathbf{N}} \delta(x-n)e^{-p}\frac{p^n}{n!}$	p
Rayleigh	$\frac{x}{\mu} exp[-\frac{x^2}{2\mu}]$, $x \geq 0$	μ

Table 4.2. pdf of some members of the exponential family and their corresponding parameters (the parenthesized parameters will not be estimated). $\delta(x)$ is the Dirac distribution, $\mathbf{N}$ is the set of integers, and $n! = n(n-1)\ldots 2.1$.

The exponential family possesses some important properties in terms of parameter estimation which make it interesting for the development of fast and efficient image processing algorithms. Let us consider a sample χ_u of N_u random variables distributed with a pdf $P_{\mu_u}(x)$: $\chi_u = x_1, x_2, ..., x_{N_u}$. If the pdf belongs to the exponential family, the likelihood is:

$$L(\chi_u \,|\boldsymbol{\mu}_u) = \prod_{i=1}^{N_u} exp\left\{\boldsymbol{\alpha}(\boldsymbol{\mu}_u).\boldsymbol{\beta}(x_i) + \phi(\boldsymbol{\mu}_u) + \kappa(x_i)\right\} \qquad (4.8)$$

The ML estimate of $\boldsymbol{\mu}_u$ is thus:

$$\widehat{\boldsymbol{\mu}}_u^{ML} = \arg\max_{\boldsymbol{\mu}} \left[\prod_{i=1}^{N_u} \exp\left\{\boldsymbol{\alpha}(\boldsymbol{\mu}).\boldsymbol{\beta}(x_i) + \phi(\boldsymbol{\mu}) + \kappa(x_i)\right\}\right] \qquad (4.9)$$

which can also be written:

$$\widehat{\boldsymbol{\mu}}_u^{ML} = \arg\max_{\boldsymbol{\mu}} \left[\exp\left\{\boldsymbol{\alpha}(\boldsymbol{\mu}).\mathbf{T}[\chi_u] + N_u\,\phi(\boldsymbol{\mu})\right\}\right] \qquad (4.10)$$

with

$$\mathbf{T}[\chi_u] = \sum_{i=1}^{N_u} \beta(x_i) \tag{4.11}$$

It is seen from Eq. 4.10 that the ML estimate of parameter μ depends on the data only through $\mathbf{T}[\chi_u]$. This function is called a *sufficient statistic* (see Appendix 4.A). As will be seen in the following, this property is of great interest for the design of image processing algorithms.

The exponential family possesses another important property regarding parameter estimation. Indeed, let us consider a pdf belonging to the exponential family and for which the domain of definition of x is independent of μ. Then, if the ML estimate of μ is unbiased, it is optimal in the sense that it has the smallest variance that any estimator can have for the estimation problem at hand (see Appendix 4.A for a definition of these notions). This property gives a theoretical justification to the use of the ML principle to estimate the nuisance parameters of the SIR image model (see Eq. 4.5).

4.1.2.2   Some examples of pdf belonging in the exponential family

As an illustration, let us first take the example of the Gamma pdf defined in Table 4.2. We will assume that the order L is known and that the only parameter of interest is the scalar parameter μ. The pdf can be written in the following form:

$$P_\mu(x) = \exp\left[-L\frac{x}{\mu} - L\log\mu + (L-1)\log x + L\log L - \log\Gamma(L)\right] \tag{4.12}$$

By comparing this expression to Eq. 4.7, it clearly seen that $\beta(x)$ is a scalar function equal to $\beta(x) = x$ and the sufficient statistics $T(\chi_u) = \sum_{i=1}^{N_u} x_i$ is simply the sum of the elements of the sample.

Let us now consider the case of the Gaussian pdf. It involves two parameters so that the parameter vector is $\mu = (m, \sigma^2)^T$. The pdf can be written in the following form:

$$P_\mu(x) = \exp\left[-\frac{x^2}{2\sigma^2} + \frac{m}{\sigma^2}x - \frac{m^2}{2\sigma^2} - \log\left(\sqrt{2\pi}\sigma\right)\right] \tag{4.13}$$

It is clearly seen that in this case, β is a 2-D vector equal to $\beta = (x, x^2)^T$ and the sufficient statistics is also a vector:

$$\mathbf{T} = \left(\begin{array}{c} \sum_{i=1}^{N_u} x_i \\ \sum_{i=1}^{N_u} x_i^2 \end{array}\right) \tag{4.14}$$

We provide in Table 4.3 the sufficient statistics for the other members of the exponential family which will be discussed in the following. It should be noted that in all cases but the Gaussian pdf, the sufficient statistic will be a scalar.

Law	Parameter estimates	Sufficient statistics $\mathbf{T}[\chi_u]$
Bernoulli	$p = T_1/N_u$	$T_1 = \sum_{i=1}^{N_u} s_i$
Gamma	$\mu = T_1/N_u$	$T_1 = \sum_{i=1}^{N_u} s_i$
Gaussian	$m = T_1/N_u$ $m^2 + \sigma^2 = T_2/N_u$	$T_1 = \sum_{i=1}^{N_u} s_i$ $T_2 = \sum_{i=1}^{N_u} (s_i)^2$
Poisson	$p = T_1/N_u$	$T_1 = \sum_{i=1}^{N_u} s_i$
Rayleigh	$\mu = T_2/N_u$	$T_2 = \sum_{i=1}^{N_u} (s_i)^2$

Table 4.3. Mathematical expressions of the sufficient statistics for the parameters defined in Table 4.2. One has considered that the parameters are estimated in the ML sense.

4.1.2.3   Exponential family and SIR model

In the SIR image model, the two samples of interest are the $N_a(\boldsymbol{\theta})$ target pixel values which form the sample $\chi_a^{\boldsymbol{\theta}}$ and the $N_b(\boldsymbol{\theta})$ background pixel values which form the sample $\chi_b^{\boldsymbol{\theta}}$. ML estimation can be used for dealing with the nuisance parameters and possesses optimal properties if it is unbiased and if the pdf belongs to the exponential family, as discussed in Section 4.1.2.1. If one has prior knowledge about their values, they can be processed with the MAP or the marginal Bayesian approach. In all cases, the obtained loglikelihood (in the case of the marginal Bayesian approach) or pseudo-loglikelihoods (in the case of the ML and the MAP approach) have the following expression [69]:

Property:

Whatever the adopted approach to deal with the nuisance parameters – ML,

MAP, or marginal Bayesian – the loglikelihood of hypothesis H_θ is:

$$\ell(\mathbf{s}, \boldsymbol{\theta}) = -N_a(\boldsymbol{\theta})\, f_a\left(\frac{\mathbf{T}[\chi_a^{\boldsymbol{\theta}}]}{N_a(\boldsymbol{\theta})}\right) - N_b(\boldsymbol{\theta})\, f_b\left(\frac{\mathbf{T}[\chi_b^{\boldsymbol{\theta}}]}{N_b(\boldsymbol{\theta})}\right) + G(\mathbf{s}) \quad (4.15)$$

with

$$\mathbf{T}[\chi_u^{\boldsymbol{\theta}}] = \sum_{(x,y)\in\Delta_u^{\boldsymbol{\theta}}} \beta[s(x,y)] \quad \text{with } u = a, b \quad (4.16)$$

$$G(\mathbf{s}) = \sum_{i\in\Delta} \kappa(s_i) \quad (4.17)$$

where $\Delta = \Delta_a^{\boldsymbol{\theta}} \bigcup \Delta_b^{\boldsymbol{\theta}}$. This equation means that after elimination of the nuisance parameters, the likelihood of H_θ is a function of the image graylevels s through the sufficient statistics.

In the case of the marginal Bayesian and MAP approaches, the functions $f_a(x)$ and $f_b(x)$ depend on the considered pdf and on the prior densities of the nuisance parameters. In the case of ML estimation of the nuisance parameters, and if the type of pdf is the same in regions a and b, $f_a(x)$ and $f_b(x)$ are equal. However, this is not always the case. For example, in sonar images of submarine mines, the pdf in the mine region can be described with a Rayleigh distribution because of the amplitude speckle, but noise in the shadow zone created by the mine may be more accurately modeled as a Gaussian noise since there is no reflected acoustic wave but only detector noise. This type of situation can be naturally handled with the SIR model, by assuming a different type of noise pdf in regions a and b. In this case, the functions $f_a(x)$ and $f_b(x)$ are different even if the nuisance parameters are estimated in the ML sense.

In Eq. 4.17, the last term $G(\mathbf{s})$ is independent of $\boldsymbol{\theta}$. In Table 4.4 are listed the expressions of the $\boldsymbol{\theta}$-dependent part of the pseudo-loglikelihood defined in Eq. 4.15 when the nuisance parameters are estimated with the ML method. We propose to illustrate in the sequel of this chapter the introduced concepts by using the example of target location in different types of images. The case of object segmentation, which is of prime interest in image processing, needs further developments to be efficiently implemented and will constitute the subject of Chapter 5.

4.2.   Performing object location with algorithms based on the SIR image model

The problem of target location using a SIR model and Gaussian pdf has already been addressed in Section 3.5. In this case, $\boldsymbol{\theta}$ represents the position of

Law	$\mathbf{f_u(z)}$	$\mathbf{z}$
Bernoulli	$-z \, \log[z] - (1 - z) \, \log[1 - z]$	$z = T_1/N_u$
Gamma	$L \, \log[z]$	$z = T_1/N_u$
Gaussian	$\frac{1}{2} \, \log[z]$	$z = T_2/N_u - [T_1/N_u]^2$
Poisson	$-z \, \log[z]$	$z = T_1/N_u$
Rayleigh	$\log[z]$	$z = T_2/N_u$

Table 4.4. Mathematical expressions which define the varying part of the loglikelihod (see Eq. 4.15) in terms of the sufficient statistics defined in Table 4.3. $\log(z)$ is the neperian logarithm. L is the order of the Gamma pdf (see Table 4.2).

the target in the image, which will be denoted as $\boldsymbol{\tau} = [\tau_x, \tau_y]$ so that:

$$w^{\boldsymbol{\theta}}(x, y) = w^{\boldsymbol{\tau}}(x, y) = w(x - \tau_x, y - \tau_y) \tag{4.18}$$

With the concepts introduced in the previous section, this target location procedure can be generalized to any image statistics belonging to the exponential family. The general ML solution for the estimation of the location τ can be written (see Eq. 4.15):

$$\hat{\boldsymbol{\tau}} = \arg \max_{\boldsymbol{\tau}} \ \ell(\boldsymbol{\tau})$$

$$= \arg \max_{\boldsymbol{\tau}} \ \left\{ -N_a \, f_a \left(\frac{1}{N_a} \sum_{(x,y) \in \Delta_a^{\boldsymbol{\tau}}} \beta[s(x, y)] \right) \right.$$

$$\left. -N_b \, f_b \left(\frac{1}{N_b} \sum_{(x,y) \in \Delta_b^{\boldsymbol{\tau}}} \beta[s(x, y)] \right) \right\} \tag{4.19}$$

Note that in order to simplify the notation, we have denoted the pseudo-likelihood $\ell(\boldsymbol{\tau})$ instead of $\ell(\mathbf{s}, \boldsymbol{\tau})$. The number of pixels in regions a and b have been denoted N_a and N_b instead of $N_a(\boldsymbol{\theta})$ and $N_b(\boldsymbol{\theta})$ since they do not depend on $\boldsymbol{\theta}$ in this application. This location algorithm will be called *SIR filter* in the following.

We now specialize Eq. 4.19 to particular pdf's belonging to the exponential family when the ML estimation of the nuisance parameters is considered. For simplicity, let us introduce the following notations:

$$t_k^{(u)}(\boldsymbol{\tau}) \;=\; \frac{1}{N_u} \sum_{(x,y)\in\Delta_u^{\boldsymbol{\tau}}} [s(x,y)]^k \qquad (4.20)$$

$$\sigma_u^2(\boldsymbol{\tau}) \;=\; t_2^{(u)}(\boldsymbol{\tau}) - [t_1^{(u)}(\boldsymbol{\tau})]^2 \qquad (4.21)$$

where $k = 1$ or 2 and $u = a, b$. Using Table 4.4 and Eq. 4.19, we obtain the following expressions of the location algorithm for different pdf belonging to the exponential family:

- **Bernoulli:**

$$\ell_{ber}(\boldsymbol{\tau}) = -N_a \, f[t_1^{(a)}(\boldsymbol{\tau})] - N_b \, f[t_1^{(b)}(\boldsymbol{\tau})] \qquad (4.22)$$

with

$$f(x) = -x\log(x) - (1-x)\log(1-x) \qquad (4.23)$$

- **Gaussian:**

$$\ell_{gau}(\boldsymbol{\tau}) = -\frac{N_a}{2} \, \log[\sigma_a^2(\boldsymbol{\tau})] - \frac{N_b}{2} \, \log[\sigma_b^2(\boldsymbol{\tau})] \qquad (4.24)$$

- **Gamma:**

$$\ell_{gam}(\boldsymbol{\tau}) = -N_a \, L \log[t_1^{(a)}(\boldsymbol{\tau})] - N_b \, L \log[t_1^{(b)}(\boldsymbol{\tau})] \qquad (4.25)$$

- **Rayleigh:**

$$\ell_{ray}(\boldsymbol{\tau}) = -N_a \log\left[t_2^{(a)}(\boldsymbol{\tau})\right] - N_b \log\left[t_2^{(b)}(\boldsymbol{\tau})\right] \qquad (4.26)$$

- **Poisson:**

$$\ell_{poi}(\boldsymbol{\tau}) = -N_a \, f[t_1^{(a)}(\boldsymbol{\tau})] - N_b \, f[t_1^{(b)}(\boldsymbol{\tau})] \qquad (4.27)$$

with

$$f(x) = -x\log(x) \qquad (4.28)$$

It must be noted that the SIR image model implies two assumptions which can be considered as limits. The first one is that we have assumed the pixel graylevels to be statistically independent, in other words, we have modeled the target and background graylevels as white random fields. In some images, the background clutter may be better modeled by correlated random fields. The second limitation is that the background region is assumed statistically homogeneous. This is clearly not a realistic assumption in most image processing applications, where the backgrounds are complex. These two limits will be addressed in the following sections. We will present methods which can improve the performance of the location algorithms in the presence of spatially correlated and nonhomogeneous backgrounds.

4.2.1 The whitening process

In some real-world images, the fluctuations of both the target and the background cannot be represented with good accuracy with uncorrelated random fields. In these situations, the SIR filter is then suboptimal and can fail. As an example, we show in Figure 4.2 two scenes and the maximum of each line of the likelihood planes $\ell_{gau}(\tau)$ (see Eq. 4.24) obtained with the SIR filter adapted to white Gaussian pdf. The maximum of the likelihood plane represents the estimated location of the target. In scene 4.2.a, the pdf's of both the target and the background graylevels are white and Gaussian whereas those of scene 4.2.b are also Gaussian but spatially correlated. One can note that the SIR algorithm adapted to white Gaussian statistics fails on scene 4.2.b whereas it is able to locate the target on scene 4.2.a.

To design an optimal algorithm for the location of a random correlated target appearing on a random correlated background, the main problem consists in finding texture models that characterize real situations and for which the optimal solution is mathematically simple. Correlated texture models are usually complex, in particular in terms of parameter estimation. The simplest ones are probably Markov Random Fields (MRF). Methods for adapting such texture models to the SIR image model have been proposed [70, 71]. Although they have shown interesting results, we will focus in this section on another method, which has the advantage of being adaptive to the correlation properties of the image and computationally efficient. This method consists in applying a whitening preprocessing to the input scene in order to obtain an image with white Gaussian textures and then to apply the SIR method which is optimal in that case [68]. We will show in the next chapter that this preprocessing is also useful for object segmentation algorithms.

The Fourier transform of the scene **s** is denoted $\tilde{s}$ (or $\tilde{s}(k_x, k_y)$ at spatial frequencies k_x and k_y), z^* is the complex conjugate of z and $|z|$ its modulus.

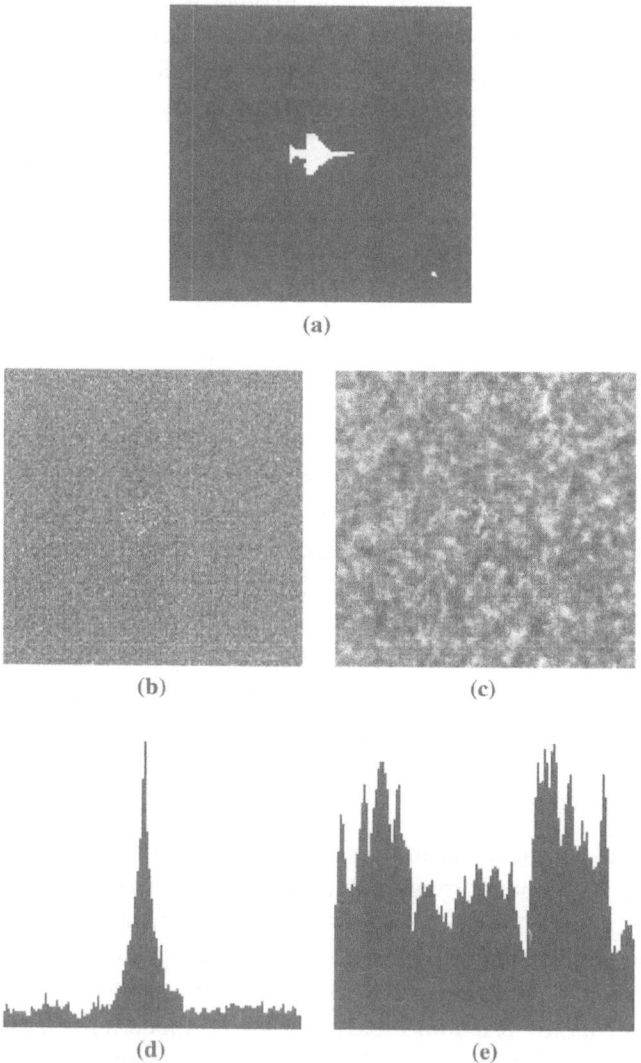

Figure 4.2.   **(a)** Reference shape. **(b)** Example of a scene with white Gaussian textures for the target and the background. The shape of the target is the reference shape in (a). **(c)** Example of a scene with correlated Gaussian textures for the target and the background. The shape of the target is the reference shape in (a). **(d)** Result of processing (b) with the SIR algorithm adapted to white Gaussian statistics $\ell_{gau}(\tau)$. **(e)** Result of processing (c) with the same algorithm. Note that (d) and (e) are plots of the maximum of each line of the output plane of the algorithm. ©1999 SPIE.

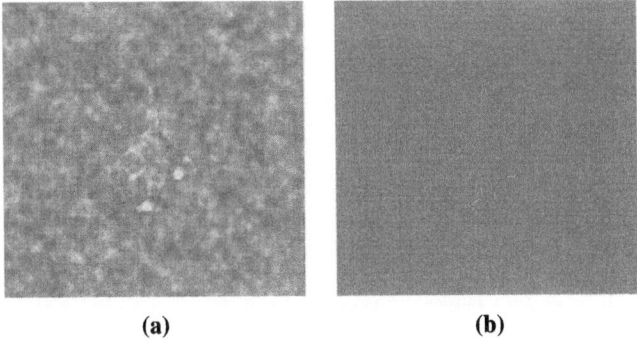

We define the whitening filter in the Fourier domain by [68]:

$$\tilde{h}(k_x, k_y) = \frac{1}{\epsilon + |\tilde{s}(k_x, k_y)|} \qquad (4.29)$$

where ϵ is a small positive constant introduced as a regularization parameter which avoids divergence when $|\tilde{s}(k_x, k_y)|$ is close or equal to zero. The Fourier transform $\tilde{z}$ of the preprocessed image z is thus:

$$\tilde{z}(k_x, k_y) = \tilde{h}(k_x, k_y)\tilde{s}(k_x, k_y) \qquad (4.30)$$

It is easy to see that the square modulus of $\tilde{z}$ is approximately constant. We can conjecture that the pixels of the preprocessed image z are approximately Gaussian uncorrelated variables. In Figure 4.3, we show a target with a correlated texture which appears on a random correlated background and the obtained preprocessed image. One can show that describing the pixel values of the preprocessed image as Gaussian random variables is a good approximation [68]. If we model the preprocessed image with two independent regions and if the nuisance parameters are estimated in the ML sense, the expression of the SIR filter is $\ell_{gau}(\tau)$ in Eq. 4.24.

Figure 4.4 shows an example of application of this method. The synthetic but realistic scene represents an airplane flying over a city and the image is slightly blurred. Image 4.4.b represents the scene after application of the whitening preprocessing. Figure 4.4.d represents the results of applying the SIR filter optimal for Gaussian pdf $\ell_{gau}(\tau)$ constructed with the mask $\mathbf{w}$ appearing in Figure 4.4.c.

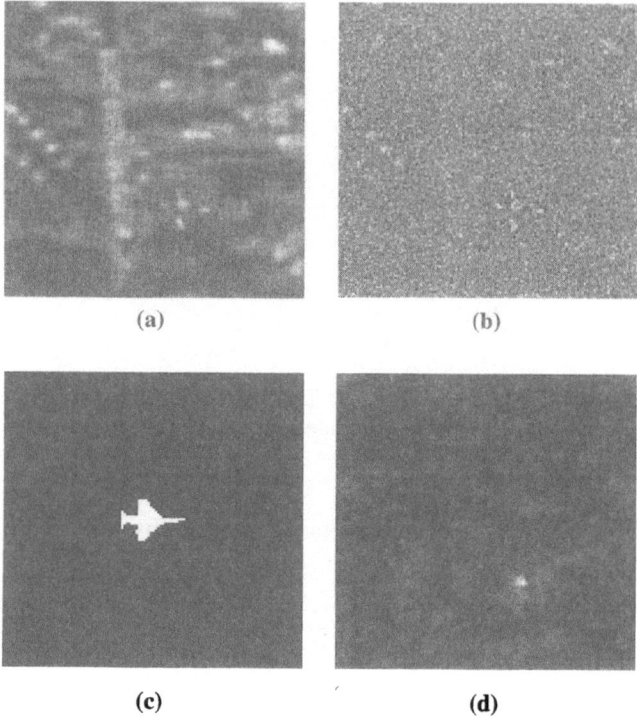

Figure 4.4. (a) Synthetic scene with an airplane appearing on an urban background. The airplane, which is barely visible, is located in the bottom right quarter of the image and the whole scene is blurred and corrupted with white Gaussian additive noise. (b) Whitened version of scene (a). (c) Reference object **w**. (d) Result of processing (b) with the ML algorithm adapted to Gaussian statistics, which shows that the airplane is correctly located. The reference object was (c). ©1999 SPIE.

4.2.2   The generalized likelihood ratio test (GLRT) approach

The other strong hypothesis which is made in the SIR image model is that the target region Δ_a^θ and the background region Δ_b^θ are considered to have homogeneous statistics. This is often a nonrealistic assumption since real-world backgrounds are in general better modeled with several zones having different textures. In order to overcome this problem, instead of estimating the background statistical parameters on the whole image, one will estimate them in a small subwindow $\mathbf{F}^\tau$ centered on the assumed target location τ (see Figure 4.5). If one considers a sufficiently small subwindow, the hypothesis that the background is homogeneous becomes a better approximation.

One way to formalize this problem is to model it as a detection/estimation task. In the framework of the SIR image model, this amounts to considering

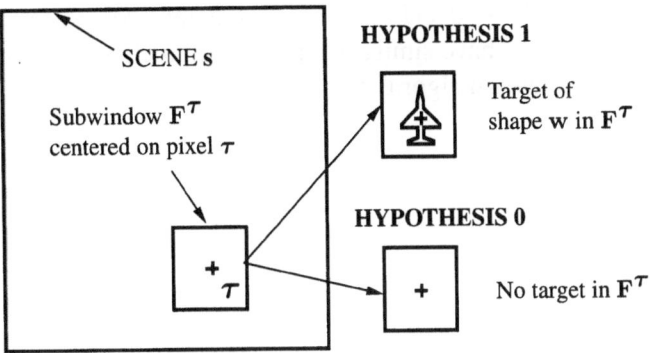

Figure 4.5.   Sketch of the two hypotheses considered in the GLRT scheme.

a compound parameter $\boldsymbol{\theta} = \{h, \boldsymbol{\tau}\}$ where $\boldsymbol{\tau} = (\tau_x, \tau_y)$ represents coordinates in the image and $h \in \{0, 1\}$ is a binary number indicating the presence or not of the target at coordinates $\boldsymbol{\tau}$. Let us consider a binary mask $\mathbf{F}$, which can be for example a square shaped region which is larger than the mask $\mathbf{w}$ defining the target to detect/localize. Let $\mathbf{F}^{\boldsymbol{\tau}}$ denote the version of the mask $\mathbf{F}$ centered on pixel $\boldsymbol{\tau}$ and $\Delta_F^{\boldsymbol{\tau}} = \{(x, y) | F(x - \tau_x, y - \tau_y) = 1\}$, the set of pixels covered by the mask $\mathbf{F}^{\boldsymbol{\tau}}$. Note that in this section, $\Delta_b^{\boldsymbol{\tau}}$ will denote the part of the complementary of $\Delta_a^{\boldsymbol{\tau}}$ belonging to $\mathbf{F}^{\boldsymbol{\tau}}$: In other words, $\Delta_F^{\boldsymbol{\tau}} = \Delta_a^{\boldsymbol{\tau}} \bigcup \Delta_b^{\boldsymbol{\tau}}$. The detection problem at each location $\boldsymbol{\tau}$ then consists of the following hypothesis-testing scheme:

- Hypothesis H_0: the window $\mathbf{F}^{\boldsymbol{\tau}}$ contains only background noise $\mathbf{b}$, so that:

$$\forall (x, y) \in \Delta_F^{\boldsymbol{\tau}} \ , \ s(x, y) = b(x, y) \tag{4.31}$$

- Hypothesis H_1: the target is present in the center of the window $\mathbf{F}^{\boldsymbol{\tau}}$, so that $\forall (x, y) \in \Delta_F^{\boldsymbol{\tau}}$:

$$s(x, y) = a(x, y) w(x - \tau_x, y - \tau_y) + b(x, y)[1 - w(x - \tau_x, y - \tau_y)] \tag{4.32}$$

This type of test can deal with target detection [66, 72], but also with edge detection or region fusion [73]. Indeed, let us consider Figure 4.6, which represents the different configurations of the two region masks $\mathbf{w}$ and $\bar{\mathbf{w}}$ corresponding to different applications. The leftmost image clearly corresponds to the detection of a square-shaped target. In the center of the figure, the configuration corresponds to the detection of a vertical edge. On the right side, the region masks $\mathbf{w}$ and $\bar{\mathbf{w}}$ are adjacent with irregular shapes. This test can be used for deciding to perform a fusion, that is, to decide if these two regions belong to the same part of the image since they have similar statistical parameters. Region fusion is useful in some segmentation algorithms [74].

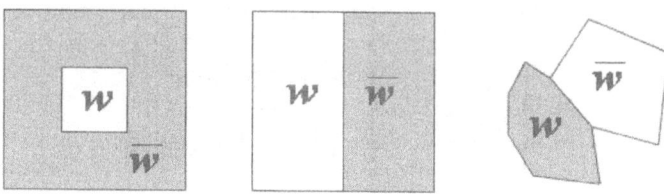

Figure 4.6.   Different configurations of masks **w** and **w̄** (**F** = **w** ⋃ **w̄**). Left: Target detection. Center: Edge detection. Right: Region fusion.

A very classical method for determining the best choice between these two hypotheses is the GLRT which has been described in Section 3.2.3. It consists in computing the likelihoods of both hypotheses, eliminating the nuisance parameters by estimating them in the ML sense, and in taking the ratio of the two obtained pseudo-likelihoods $\mathcal{L}(0, \tau)$ and $\mathcal{L}(1, \tau)$ (see Eq. 3.34):

$$R(\tau) = \frac{\mathcal{L}(1, \tau)}{\mathcal{L}(0, \tau)} \tag{4.33}$$

or, which is equivalent, the difference of the pseudo-loglikelihoods. One then obtains the log-GLRT which will be used in the following:

$$r(\tau) = \ell(1, \tau) - \ell(0, \tau) \tag{4.34}$$

There are two ways of exploiting the loglikelihood ratio plane $r(\tau)$. If one knows that there is only one target in the whole scene, one estimates its position $\widehat{\tau}$ by finding the value of τ which maximizes $r(\tau)$:

$$\widehat{\tau} = \arg\max_{\tau} r(\tau) \tag{4.35}$$

If there are several, or no targets in the images, $r(\tau)$ is compared to a threshold r_0 and the following decision is taken:

- If $r(\tau) > r_0$, one decides that there is a target in the center of $\mathbf{F}^{\tau}$.

- Otherwise one decides that there is no target.

The value of the threshold r_0 sets a compromise between the probability of detection and the probability of false alarm. Using this method, we can estimate if a target is present or not at each location τ in the image.

The general expression of the likelihood ratio $r(\tau)$ for a SIR image when the graylevel statistics belongs to the exponential family is:

$$r(\tau) = -N_a\, f_a\left(\frac{1}{N_a} \sum_{(x,y)\in\Delta_a^{\mathcal{T}}} \beta[s(x,y)]\right) \tag{4.36}$$

$$-N_b\, f_b\left(\frac{1}{N_b} \sum_{(x,y)\in\Delta_b^{\mathcal{T}}} \beta[s(x,y)]\right) \tag{4.37}$$

$$+N_F\, f_b\left(\frac{1}{N_F} \sum_{(x,y)\in\Delta_F^{\mathcal{T}}} \beta[s(x,y)]\right) \tag{4.38}$$

where N_F is the number of pixels of the scanning subwindow $\mathbf{F}^{\mathcal{T}}$ and thus $N_F = N_a + N_b$. Here again, the functions f_a and f_b are dependent on the considered pdf and on the prior on the nuisance parameters for the marginal Bayesian and MAP approaches but are identical in the case of ML estimation of these parameters.

We now specialize Eq. 4.38 to particular pdf's belonging to the exponential family (which are defined in Table 4.2) when the ML estimation of the nuisance parameters are considered. We obtain the following expressions for the log-GLRT:

- **Bernoulli:**

$$r_{ber}(\tau) = -N_a\, f[t_1^{(a)}(\tau)] - N_b\, f[t_1^{(b)}(\tau)] + N_F\, f[t_1^{(F)}(\tau)] \tag{4.39}$$

with

$$f(x) = -x\log(x) - (1-x)\log(1-x) \tag{4.40}$$

- **Gaussian:**

$$r_{gau}(\tau) = -\frac{N_a}{2}\, \log[\sigma_a^2(\tau)] - \frac{N_b}{2}\, \log[\sigma_b^2(\tau)] + \frac{N_F}{2}\, \log[\sigma_F^2(\tau)] \tag{4.41}$$

- **Gamma:**

$$r_{gam}(\tau) = -N_a\, L\log[t_1^{(a)}(\tau)] - N_b\, L\log[t_1^{(b)}(\tau)] + N_F\, L\log[t_1^{(F)}(\tau)] \tag{4.42}$$

- **Rayleigh:**

$$r_{ray}(\tau) = -N_a\, \log\left[t_2^{(a)}(\tau)\right] - N_b\, \log\left[t_2^{(b)}(\tau)\right] + N_F\, \log\left[t_2^{(F)}(\tau)\right] \tag{4.43}$$

- **Poisson:**

$$r_{poi}(\boldsymbol{\tau}) = -N_a \; f[t_1^{(a)}(\boldsymbol{\tau})] - N_b \; f[t_1^{(b)}(\boldsymbol{\tau})] + N_F \; f[t_1^{(F)}(\boldsymbol{\tau})] \qquad (4.44)$$

with

$$f(x) = -x \log(x) \qquad (4.45)$$

See Eqs. 4.20 and 4.21 for the definitions of $t_k^{(u)}(\boldsymbol{\tau})$ and $\sigma_u^2(\boldsymbol{\tau})$.

4.2.3   The implementation issue

An interesting point is that the pseudo-loglikelihood $\ell(\boldsymbol{\tau})$ in the standard location approach and $r(\boldsymbol{\tau})$ in the GLRT approach can easily be rewritten using correlation operations. Let us consider the GLRT approach, and let $[\mathbf{f} \star \mathbf{g}](\boldsymbol{\tau})$ denote the correlation between $\mathbf{f}$ and $\mathbf{g}$:

$$[\mathbf{f} \star \mathbf{g}](\tau_x, \tau_y) = \sum_x \sum_y f(x,y) g(x - \tau_x, y - \tau_y) \qquad (4.46)$$

and let $\mathbf{w} = \mathbf{w}^0$. Equation 4.20 becomes:

$$t_k^{(a)}(\boldsymbol{\tau}) \;\; = \;\; \frac{1}{N_a} [\mathbf{w} \star (\mathbf{s})^k](\boldsymbol{\tau}) \qquad (4.47)$$

$$t_k^{(F)}(\boldsymbol{\tau}) \;\; = \;\; \frac{1}{N_F} [\mathbf{F} \star (\mathbf{s})^k](\boldsymbol{\tau}) \qquad (4.48)$$

$$t_k^{(b)}(\boldsymbol{\tau}) \;\; = \;\; \frac{1}{N_b} [\bar{\mathbf{w}} \star (\mathbf{s})^k](\boldsymbol{\tau})$$

$$\;\; = \;\; \frac{1}{N_b} \left[[\mathbf{F} \star (\mathbf{s})^k](\boldsymbol{\tau}) - [\mathbf{w} \star (\mathbf{s})^k](\boldsymbol{\tau}) \right] \qquad (4.49)$$

where $(s)^k(x,y) = [s(x,y)]^k$ and $\bar{w}(x,y) = 1 - w(x,y)$. Since the most intensive computations involved in computing $\ell(\boldsymbol{\tau})$ or $r(\boldsymbol{\tau})$ are the determination of $t_k^{(u)}(\boldsymbol{\tau})$ with $u = a, b$, or F, this new formulation is very attractive because it is closely connected to the detection architecture described in Figure 3.14. The correlation operations can thus be computed efficiently with FFT-based techniques. This generalizes the result introduced in Section 3.5 for Gaussian pdf to other members of the exponential family.

In the sequel of this chapter, we will address two applications which illustrate how algorithms based on the SIR model make it possible to process images from various types of imaging systems.

4.3.   Application to binary images: Comparison of optimal and linear techniques

In many applications, one has to deal with binary images which may result from preprocessing of grayscale images or from sensor limitations. It is thus of interest to design processing techniques optimal for this type of images. One of the recognized drawbacks of binary images is of course that some information is lost when the graylevels are binarized. Images may be edge enhanced prior to being binarized, since edge enhancement may improve the contrast of the image, making it easier to find a good binarization threshold. However, if the object of interest has a lower contrast than other objects in the background, the binary image will be perturbed by spurious edges or areas. It is thus important to determine robust location algorithms that can locate the object even if it appears on a heavy binary background noise. It is the purpose of this section to address this issue, which is treated in more detail in Refs. 63 and 75.

4.3.1 The GLRT algorithm for binary images

Whatever the used binarization method, a perfectly binarized image will display the object of interest appearing completely white (binary level 1) on a completely black (binary level 0) background, or the inverse if the contrast of the target is negative. However, in the observed grayscale scene, the background on which the object appears has regions with higher graylevel than the object itself. This will cause some parts of the background to appear white after binarization. Moreover, the target itself may contain darker parts, or edges having low contrast with respect to the background. This will cause some parts of the target to appear black after binarization. This fact is illustrated in Figure 4.7, in which a grayscale image containing the object of interest is binarized. We also show the case where the grayscale image is edge enhanced prior to being binarized.

In order to take the presence of noise into account in binary images, one can use the SIR image model where the graylevels of the target and of the background are modeled with Bernoulli random variables. In other words, the image model is the one defined in Eq. 4.1 with:

$$
a(x,y) = \begin{cases} 1 & \text{with probability } \mu_a \\ 0 & \text{with probability } 1 - \mu_a \end{cases}
$$

$$
b(x,y) = \begin{cases} 1 & \text{with probability } \mu_b \\ 0 & \text{with probability } 1 - \mu_b \end{cases} \tag{4.50}
$$

This means that in the noisy scene, target pixels may be white with probability μ_a, and background pixels may by white with probability μ_b. Examples of realizations of this image model are represented in Figure 4.8 for different values of μ_a and μ_b.

However, real-world binary images may be strongly nonhomogeneous (see Figures 4.7.b and 4.7.e). It can thus be more appropriate to use the GLRT approach to locate targets in binary image. The expression of the GLRT adapted to Bernoulli noise is given in Eq. 4.39. Let us write it below with a slightly

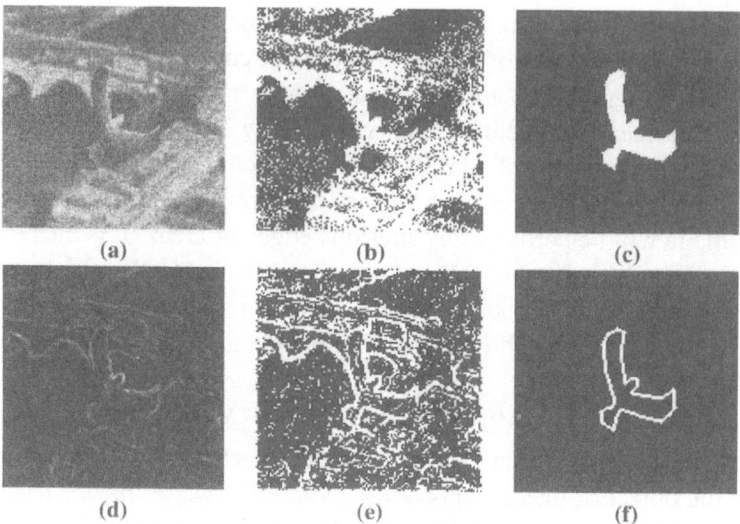

Figure 4.7.   **(a)** Example of a grayscale image containing an object of interest. **(b)** Binarized version of (a). **(c)** Reference object **w** to be searched for in (b). **(d)** Edge enhanced version of (a). **(e)** Binarized version of (d). **(f)** Reference object **w** to be searched for in (e). ©OSA 1998.

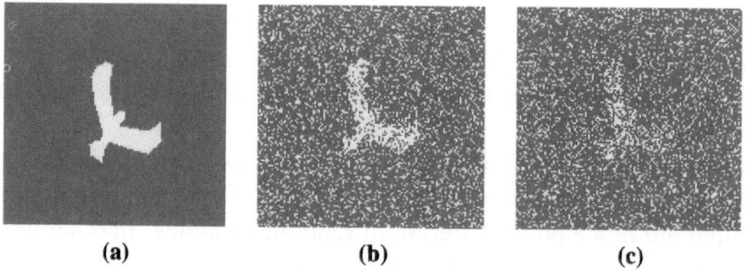

Figure 4.8.   **(a)** Scene without noise. **(b)** Noisy version of (a) with $\mu_a = 0.7$ and $\mu_b = 0.2$. **(c)** Noisy version of (a) with $\mu_a = 0.5$ and $\mu_b = 0.2$. ©OSA 1998.

different notation:

$$r(\boldsymbol{\tau}) = -N_a\, f[\widehat{\mu}_a(\boldsymbol{\tau})] - N_b\, f[\widehat{\mu}_b(\boldsymbol{\tau})] + N_F\, f[\widehat{\mu}_F(\boldsymbol{\tau})] \qquad (4.51)$$

where $f[x]$ is the following one-variable function:

$$f[x] = -x\, \log x - (1 - x)\, \log(1 - x) \qquad (4.52)$$

and

$$\widehat{\mu}_u(\boldsymbol{\tau}) \;=\; \frac{1}{N_u} \sum_{(x,y)\in\Delta_u^{\boldsymbol{\tau}}} s(x, y) \qquad (4.53)$$

with $u = a, b$, or F. It is worth noting that $\widehat{\mu}_a(\boldsymbol{\tau})$ represents the proportion of white pixels within the target window, $\widehat{\mu}_b(\boldsymbol{\tau})$ in the background, and $\widehat{\mu}_F(\boldsymbol{\tau})$ in the search window. In particular, one has:

$$N_F\,\widehat{\mu}_F(\boldsymbol{\tau}) = N_a\,\widehat{\mu}_a(\boldsymbol{\tau}) + N_b\,\widehat{\mu}_b(\boldsymbol{\tau}) \tag{4.54}$$

Considering this expression of the GLRT, it is interesting to note that in terms of computational complexity, the heaviest task involved in computing $r(\boldsymbol{\tau})$ is to determine sums of binary numbers, which can be done very rapidly.

Let us illustrate the efficiency of this algorithm on an example. We have taken a real background, on which we have superimposed a grayscale object to generate a realistic scene which is shown in Figure 4.9.a. It has been edge enhanced with a Sobel operator [76], then binarized using a very low threshold. The binarized image appears in Figure 4.9.b and it can be noted that the object is barely visible. We then applied the GLRT $r(\boldsymbol{\tau})$ built from the reference object represented in Figure 4.9.c, and a search window $\mathbf{F}$ was obtained by morphologically dilating Figure 4.9.c twice with a 3×3 pixel structuring element (see Ref. 74, p. 518). The GLRT plane $r(\boldsymbol{\tau})$ appears in Figure 4.9.d. It can be seen that the target is correctly located. By looking more accurately at the binarized image in Figure 4.9.b, we can see that it is very noisy. This is because a low threshold has been chosen. It is better to have a low threshold since almost all the information-carrying edges are included in the binarized images, but the drawback is that a lot of spurious edges remain after the binarization step. These edges are in general nonhomogeneously distributed over the image, and this is why it is important to use location algorithms robust to nonhomogeneous background noise, such as the GLRT.

4.3.2 A linear approximation to the GLRT algorithm

We have seen that the GLRT algorithm expressed in Eq. 4.51 performs well for target location in nonhomogeneous backgrounds. Indeed, it is optimal for object detection in the type of images we consider. However, we can notice in Eq. 4.53 that its expression is based on computing the number of white pixels in the three regions covered by the masks $\mathbf{w}^T$, $\mathbf{F}^T - \mathbf{w}^T$, and $\mathbf{F}^T$. In other words, the GLRT takes a decision in function of the difference of average values inside and outside the target mask $\mathbf{w}^T$. It has been shown that a wide class of regularized linear filters are based on the same type of target/background contrast information [45]. Consequently, it should be possible to approximate the GLRT with a linear filter.

The derivation of a linear approximation of the GLRT in Eq. 4.51 can be performed as follows [75]. A serial expansion of the function $f[x]$ (see Eq. 4.52) around the point $x = 1/2$ can be written as:

$$f[x] \simeq -4 * (x - \frac{1}{2})^2 + \log 2 \tag{4.55}$$

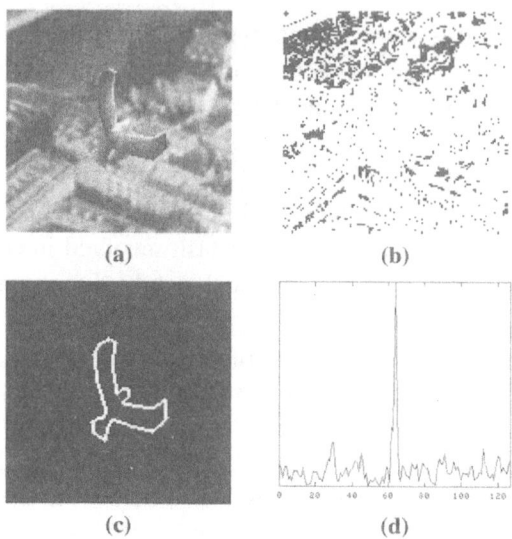

Figure 4.9.   **(a)** Example of realistic scene. **(b)** Image (a) edge-enhanced using a Sobel operator and binarized using a threshold of 0.001 (after having rescaled the edge-enhanced image so that its levels go from 0 to 1). **(c)** Reference object **w** used in the simulation. **(d)** Result of processing image (c) with the GLRT $r(\tau)$ (see Eq. 4.51). The maximum of each column of the GLRT plane is shown. The position of the maximum of this plane corresponds to the true position of the object in scene (a): the object is correctly located. ©OSA 1998.

Replacing $f[x]$ in Eq. 4.51 with this approximation and dividing by 4, one obtains:

$$\frac{1}{4}\,r(\tau) \simeq N_a \left(\widehat{\mu}_a(\tau) - \frac{1}{2} \right)^2 + N_b \left(\widehat{\mu}_b(\tau) - \frac{1}{2} \right)^2 - N_F \left(\widehat{\mu}_F(\tau) - \frac{1}{2} \right)^2 \tag{4.56}$$

Using Eq. 4.54, and after some straightforward calculations, one can show that maximizing the right-hand term of the previous equation amounts to maximizing the following expression:

$$r_{lin}(\tau) = [\widehat{\mu}_a(\tau) - \widehat{\mu}_b(\tau)]^2 \tag{4.57}$$

where $r_{lin}(\tau)$ stands for linear approximation of $r(\tau)$. We note that Eq. 4.53 can be written as:

$$\widehat{\mu}_a(\tau) = \frac{1}{N_a}[\mathbf{s} \star \mathbf{w}](\tau) \tag{4.58}$$

$$\widehat{\mu}_b(\tau) = \frac{1}{N_b}[\mathbf{s} \star \mathbf{F} - \mathbf{s} \star \mathbf{w}](\tau) \tag{4.59}$$

where the symbol $\star$ denotes correlation. Consequently, Eq. 4.57 can be written:

$$r_{lin}(\tau) = [\mathbf{h} \star \mathbf{s}]^2(\tau) \tag{4.60}$$

where $\mathbf{h}$ is the impulse response of a linear filter defined as

$$\mathbf{h} = \frac{\mathbf{w}}{N_a} - \frac{\mathbf{F} - \mathbf{w}}{N_b} \tag{4.61}$$

$r_{lin}(\tau)$ is thus the squared value of a correlation product. It has been shown in Ref. 75 that this approximation can lead to very good results when the target is large enough.

4.4. Application to edge extraction in SAR images

We consider in this section images acquired from a Synthetic Aperture Radar (SAR) [77]. This type of imaging system illuminates the scene with microwaves and constructs an image from the backscattered signal by using the synthetic aperture principle [77], which produces images with a high resolution. The main advantage of SAR compared to optical imaging systems is that it has all-weather capability, since microwaves can pass through clouds. It is also possible to perform 3-D estimation via the interferometric principle. However, since they are formed from coherent radiations, SAR images are perturbed with speckle noise, which makes them very "grainy" [78] (see Figure 4.10). A simple model of speckle noise is the following: At a point in the scene where the mean reflectance is μ, the measured intensity I is a random variable with exponential pdf [1] [78]:

$$P_\mu(I) = \frac{1}{\mu} \exp\left[-\frac{I}{\mu}\right] \ , \quad I \geq 0 \tag{4.62}$$

This noise is very strong since, according to the well-known property of exponential pdf, its standard deviation is equal to its average value μ. In practice, one often adds up several images of the same scene corresponding to different speckle realizations in order to reduce the noise variance. If L such "looks" are added together, then the intensity becomes a Gamma distributed random variable of order L and mean μ with the following pdf:

$$P_\mu(I) = \left(\frac{L}{\mu}\right)^L \frac{I^{L-1}}{\Gamma(L)} \exp\left(-\frac{L}{\mu}I\right), \quad I \geq 0 \tag{4.63}$$

The mean of this random variable is still equal to μ, but its variance is

$$VAR[I] = \frac{\mu^2}{L}$$

---

[1] There should be no confusion between the *exponential pdf*, which is a particular pdf defined in Eq. 4.62, and the *exponential family*, which is an ensemble gathering several types of pdf (including the exponential pdf !). These are standard denominations, and this is why we will use them, although they may be ambiguous.

The mean-to-standard deviation ratio, which can be considered as a signal-to-noise ratio, is equal to $\sqrt{L}$ and increases as more and more "looks" are added up (see Figure 4.10.b). It can be noticed that the exponential pdf is a Gamma pdf with order $L = 1$.

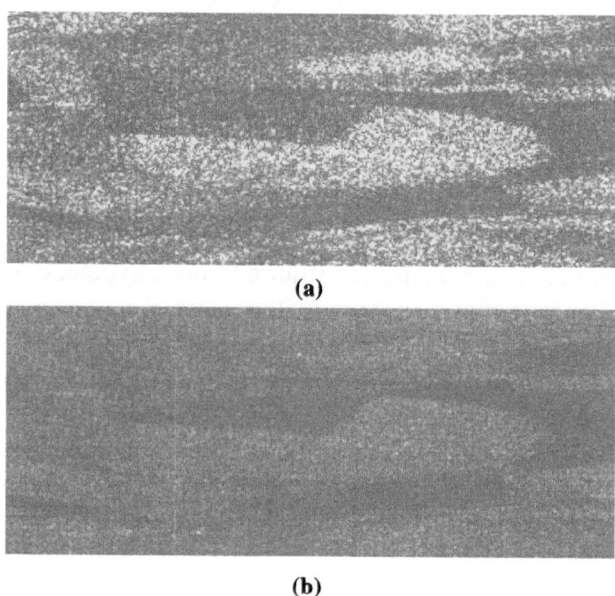

(a)

(b)

Figure 4.10.   **(a)** Mono-look SAR image of an agricultural region near Bourges, France, provided by the CNES and delivered by ESA. This image has been acquired by the satellite ERS1. It has been experimentally verified that the pdf of the noise in this image is close to an exponential pdf. **(b)** Result of adding up three mono-look images of the same region. The pdf of the noise is now Gamma with $L = 3$. We can observe that the variance of the fluctuations has decreased.

Speckle noise makes edge detection in SAR images a difficult task. It has been shown that classical differential edge detectors (i.e., based on the difference of averages computed on each side of a two-region sliding window) are not adapted to this type of image since their false alarm rate depends on the mean reflectivity [79]: They detect more spurious edges in brighter zones than in darker ones. We will show in this section that an efficient edge detector on SAR images can be designed by using the SIR model and the GLRT principle (this approach has been introduced in Ref. 66). We will also analyze how this detector can be used to perform edge location and we will characterize its precision for this application [80].

4.4.1   GLRT adapted to speckled images

Let us consider the SIR image model, where the masks $\mathbf{w}$ and $\bar{\mathbf{w}}$ are distributed as shown in the first image of Figure 4.11. According to what was said above about speckle noise statistics, we will represent the statistical distribution of the graylevels with a Gamma pdf. Applying the GLRT with this structure obviously allows one to perform the detection/localization of a vertical edge. Indeed, the two hypotheses of the statistical test correspond to:

- $H_{1,\boldsymbol{\tau}}$: An edge is present at pixel $\boldsymbol{\tau} = (\tau_x, \tau_y)$. The intensities in the two regions $\Delta_a^{\mathcal{T}}$ and $\Delta_b^{\mathcal{T}}$ therefore have Gamma pdf with different parameters.

- $H_{0,\boldsymbol{\tau}}$: No edge is present at pixel (τ_x, τ_y). The intensities within the two regions are therefore identically distributed with some Gamma pdf.

Similarly, the other windows represented in Figure 4.11 allow one to perform the detection of edges oriented at 22.5, 45, and 90 degrees from the vertical.

Figure 4.11. Directional windows for $\alpha = 0^{\circ}, 22.5^{\circ}, 45^{\circ}, 90^{\circ}$. The gray part represents the mask **w**, which defines the region $\Delta_a^{\mathcal{T}}$, and the black part its complementary $\bar{\mathbf{w}}$ which defines the region $\Delta_b^{\mathcal{T}}$.

The expression of the GLRT in this case is found in Eq. 4.42. If one assumes that the regions $\Delta_a^{\mathcal{T}}$ and $\Delta_b^{\mathcal{T}}$ have the same number of pixels N, in other words, $N = N_a = N_b = N_F/2$, it is easily shown that when the window **F** is centered on pixel (τ_x, τ_y), the GLRT can be written as:

$$r(\boldsymbol{\tau}) = 2\,N\,L\,\log\left[\frac{1}{2}\left(\sqrt{\widehat{\rho}(\boldsymbol{\tau})} + \sqrt{\frac{1}{\widehat{\rho}(\boldsymbol{\tau})}}\right)\right] \qquad (4.64)$$

where $\widehat{\rho}(\boldsymbol{\tau})$ represents the ratio between the estimates of the means in the two regions:

$$\widehat{\rho}(\boldsymbol{\tau}) = \frac{\widehat{\mu}_a(\boldsymbol{\tau})}{\widehat{\mu}_b(\boldsymbol{\tau})} \qquad (4.65)$$

with

$$\widehat{\mu}_u(\boldsymbol{\tau}) = \frac{1}{N_u}\sum_{(x,y)\in\Delta_u^{\mathcal{T}}} s(x,y) \quad \text{with } u = a \text{ or } b \qquad (4.66)$$

It is interesting to determine on which image parameters the performance of the GLRT depends. Speckle noise is a multiplicative noise, which means that the random variables representing the graylevel values of the scene **s** in regions

Δ_a^θ and Δ_b^θ can be written as:

$$
\begin{aligned}
s(x,y) &= \mu_a\, \bar{a}(x,y) \ \forall\ (x,y) \in \Delta_a^T \\
s(x,y) &= \mu_b\, \bar{b}(x,y) \ \forall\ (x,y) \in \Delta_b^T
\end{aligned}
\tag{4.67}
$$

where $\bar{a}(x,y)$ and $\bar{b}(x,y)$ are Gamma random variables with mean 1 and order L and μ_a and μ_b represent the true values of graylevel averages in regions a and b. Consequently, the ratio $\widehat{\rho}(\tau)$ can be expressed in the following way:

$$
\widehat{\rho}(\tau) = \rho\, \frac{\bar{\mu}_a(\tau)}{\bar{\mu}_b(\tau)}
\tag{4.68}
$$

where ρ represents the ratio of the true values of the averages in regions a and b:

$$
\rho = \frac{\mu_a}{\mu_b}
\tag{4.69}
$$

and $\bar{\mu}_a(\tau)$ and $\bar{\mu}_b(\tau)$ represent the average of $N/2$ random variables of type $\bar{a}(x,y)$ and $\bar{b}(x,y)$. They are thus independent of μ_a and μ_b. Consequently, it is clear that when N and L are fixed, the statistical distribution of $\widehat{\rho}(\tau)$, and thus the detection performance of the GLRT, depends only on the ratio ρ. In the following, we will call ρ the *contrast ratio* between the two regions.

It is interesting to note that the GLRT depends on the data only through the ratio $\widehat{\rho}(\tau)$ of the mean estimates $\widehat{\mu}_a(\tau)$ and $\widehat{\mu}_b(\tau)$, whereas usual edge detectors would rather depend on their difference. Indeed, a classical edge detector such as the Sobel detector would look like:

$$
d(\tau) = |\widehat{\mu}_a(\tau) - \widehat{\mu}_b(\tau)|
\tag{4.70}
$$

The detectors $r(\tau)$ and $d(\tau)$ have been applied to the synthetic Gamma-distributed image in Figure 4.12.a. The results, that is, the "edge strength maps" $d(\tau)$ and $r(\tau)$ obtained with the two detectors, are represented in Figures 4.12.b and 4.12.c. In these two images, the edge regions appear with higher values. However, in uniform regions (i.e., regions without edges), one can notice that the variance of the difference detector $d(\tau)$ is higher when the average intensity is higher (that is, in the right part of the image). On the contrary, the variance of $r(\tau)$ in a uniform region is independent of the average gray level of this region. Detectors which possess this property are called *Constant False Alarm Rate* (CFAR) since for the purpose of detection, a constant threshold applied to the plane $r(\tau)$ will provide a constant probability of false alarm in regions without edges, whatever the average values of the intensity in these regions [49]. Conversely, in the case of $d(\tau)$, the threshold required to obtain a prescribed probability of false alarm would depend on the average intensity of the region. In practice, the CFAR property is very important since it makes it possible to use a uniform threshold over the whole image to perform detection at a given false alarm rate. It is interesting to note that $d(\tau)$ would be CFAR if the noise in the image was additive (it would be the GLRT in the case of additive Gaussian noise, that is, with identical variances on the object and on the background). However, it is no longer CFAR in the presence of multiplicative noise, which clearly illustrates the importance of taking into account the statistical properties of the noise when designing image processing algorithms.

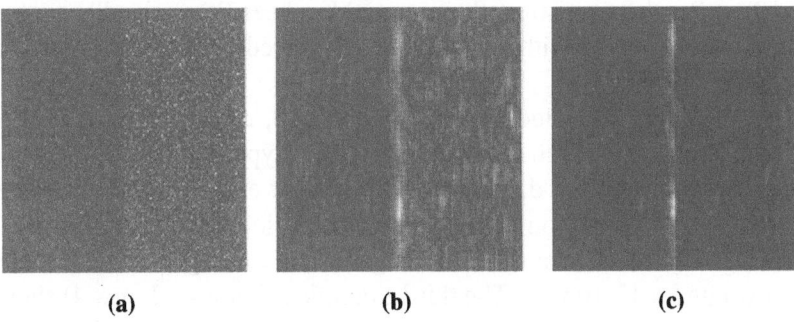

(a)   **(b)**   **(c)**

Figure 4.12. **(a)** Edge image in the presence of speckle, contrast $\rho = 3$. **(b)** $d(\tau)$ obtained with the difference detector in Eq. 4.70. **(c)** $r(\tau)$ obtained with the GLRT in Eq. 4.64.

The GLRT introduced above is valid for the detection of an edge with a given orientation. In practice, the orientations of the edges present in an image are unknown. A possible solution to this problem is to apply the GLRT with several differently oriented windows and to fuse the results delivered by each filter to get a unique edge strength map. Let $r(\tau, \alpha)$ be the edge strength map obtained after GLRT filtering with orientation α. One uses 8 different orientations (which is usually a good compromise) and the following fusion scheme has been retained to get the final edge strength map:

$$r_f(\tau) = \max_{\alpha} \; |r(\tau, \alpha)| \tag{4.71}$$

where $\alpha \in [0, 22.5, 45, 67.5, 90, 112.5, 135, 157.5]$ degrees. As described in Chapter 3, this approach is equivalent to considering α as a nuisance parameter and to estimating it on a discrete set. Note that even with this scheme, there are still edges whose orientation will not correspond to one of the windows.

The GLRT algorithm presented here may be used to determine edge location by taking the position of the maxima of the edge strength map $r(\tau)$. However, it should be kept in mind that the technique has been designed to perform a detection task and not edge location estimation. It is thus necessary to investigate its performance as an estimator of the edge position.

4.4.2   Bias on edge location

Careful analysis of the edge maps provided by the GLRT shows that the edge location can be biased, which means that, on average on speckle noise realizations, the estimated edge location is different from the true one [80]. Let us first give some examples of situations where edge location with the GLRT is biased. For this purpose, let us consider synthesized speckled edge images processed with the GLRT filter, using a vertical 10×10 pixel window (Figure 4.13). After the edge strength map $r(\tau)$ has been determined, a continuous edge line was extracted with an edge linking procedure, namely, a constrained watershed algorithm [81].

When the edge is an "ideal" step (Figure 4.13, row 1), the GLRT yields a correct location. On the other hand, for different types of "nonideal" edges, we observe a bias toward the darker side. In the first example, the step edge was tilted so that the orientation of the window is no longer adapted (Figure 4.13, row 2). In the second example, the edge is not a straight line but a sinuous frontier (Figure 4.13, row 3). The third example (Figure 4.13, row 4) shows the location bias when the image is blurred by a certain point spread function (psf).

One can notice that the occurence of the bias depends on whether the edge is "ideal" (the edge is a vertically oriented step) or not. When a "nonideal" edge is filtered, the estimation of the means nearby the frontier is somehow degraded, yielding a false location of the edge. We explore more precisely this phenomenon in the following.

Let us discuss the simple phenomenological model introduced in Ref. 80, which makes it possible to interpret the results observed in the previous section and to give an approximate expression of the bias.

If the window is shifted along the X axis (axis of equation $\tau_y = 0$) across the edge, then the position estimate $\widehat{\tau}_x$ is the value of x which maximizes the GLRT. Using 1-D notation, we have:

$$\widehat{\tau}_x = \arg \max_x [r(\tau_x)]$$

The bias δ is then defined as the expectation of the difference between the estimated edge position and the real edge position, which will be assumed to be $x = 0$ with no loss of generality. In this case:

$$\delta = \langle \widehat{\tau}_x \rangle$$

With this phenomenological model, the bias value is approximated by applying the GLRT to the speckle-free image, that is, to an image made of two domains of constant intensities μ_a and μ_b. As the value of x varies, we will call the two regions covered by the detection mask $\Delta_a^{\tau_x}$ and $\Delta_b^{\tau_x}$ and the averages of the graylevels in these regions $M_a(\tau_x)$ and $M_b(\tau_x)$. One can thus approximate the

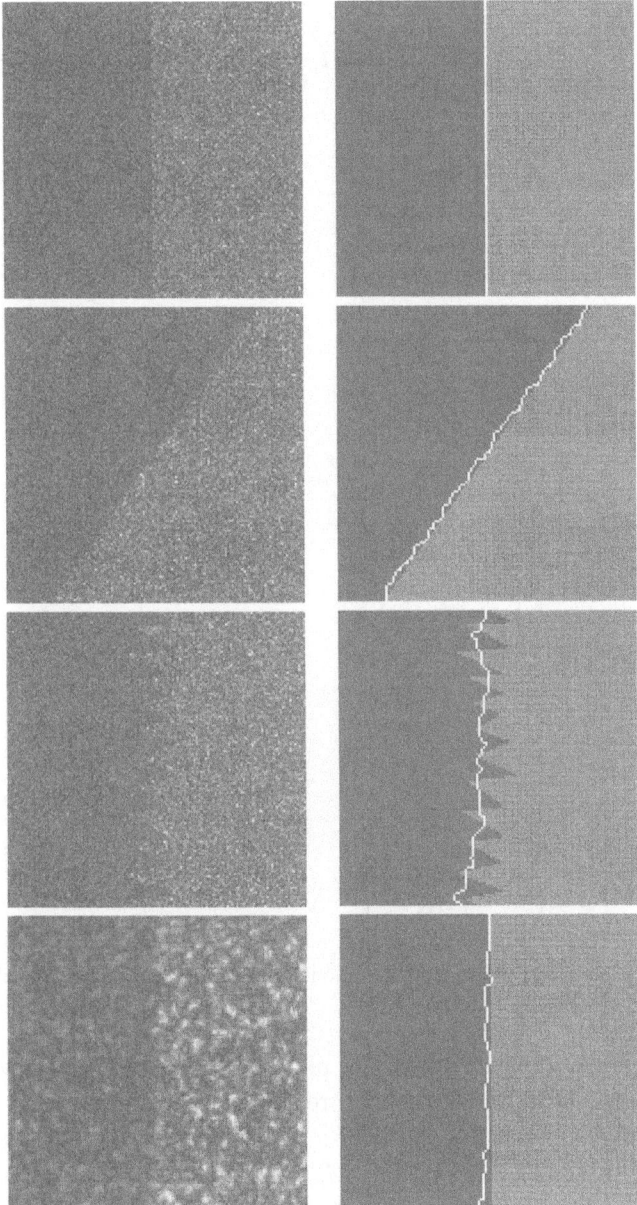

Figure 4.13.   Edge location for different situations. In all cases, the contrast is $\rho = 4$. First column: 128×128 speckled edges. Second column: Edge location performed by GLRT (10×10 vertical window) followed by an edge extraction step (watershed algorithm). The result is shown on the speckle-free image. First row: Ideal, straight, step edge. Second row: Edge tilted by 35 degrees. Third row: Sinuous edge. Fourth row: Blurred edge (circular kernel of diameter $d = 5$). ©2000 IEEE.

location bias in the following way:

$$\delta \simeq \arg\max_x \left\{ \log\left[\frac{1}{2}\left(\sqrt{q(\tau_x)} + \sqrt{\frac{1}{q(\tau_x)}} \right) \right] \right\} \qquad (4.72)$$

where

$$q(x,y) = \frac{M_a(\tau_x)}{M_b(\tau_x)} \qquad (4.73)$$

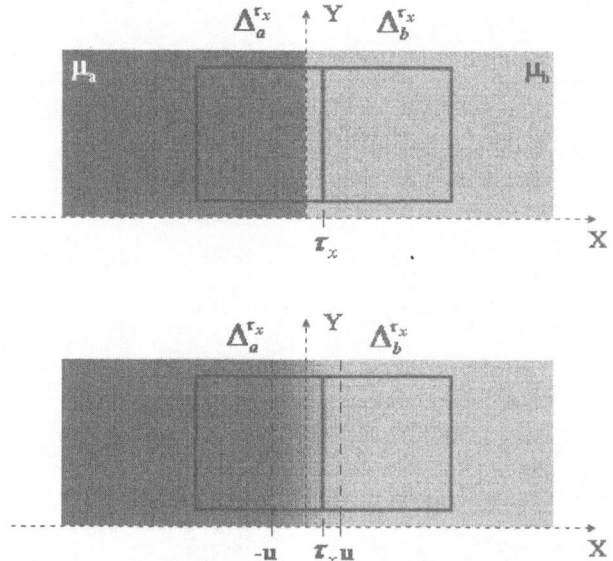

Figure 4.14.   Top: "Ideal" edge. Bottom: "Nonideal" edge with a "mixture zone" of length $2u$.
©2000 IEEE.

Let us analyze the behavior of the plots $M_a(\tau_x)$ and $M_b(\tau_x)$ with the help of Figures 4.14 and 4.15. In these figures, it has been set that $\mu_b > \mu_a$, and the phenomenological positive parameter u corresponds to the size of the "mixture zone." In the case of an "ideal" edge, $M_a(\tau_x)$ is constant ($M_a(\tau_x) = \mu_a$) for any $\tau_x \leq 0$ and increases linearly with τ_x (until it reaches μ_b). Similarly, $M_b(\tau_x)$ is constant ($M_b(\tau_x) = \mu_b$) for any $\tau_x \geq 0$ and decreases linearly (until it reaches μ_a) when τ_x decreases. Things are slightly different for a "nonideal" edge. $M_a(\tau_x)$ remains constant ($M_a(\tau_x) = \mu_a$) as long as $\Delta_a^{\tau_x}$ does not enter the zone containing a mixture of the two intensities μ_a and μ_b, i.e., for $\tau_x \leq -u$. Then $M_a(\tau_x)$ increases with τ_x toward μ_b but not in a linear manner. Similarly, $M_b(\tau_x)$ remains constant ($M_b(\tau_x) = \mu_b$) as long as $\tau_x \geq u$ and nonlinearly decreases toward μ_a when τ_x decreases. The behavior of $M_a(\tau_x)$ and $M_b(\tau_x)$

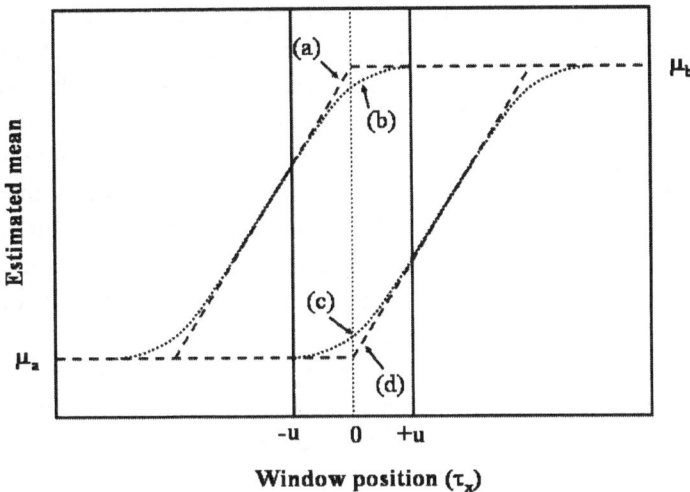

Figure 4.15. Variation of the estimated means as the window is shifted across the edge. **(a)** $M_b(\tau_x)$ for an "ideal" edge. **(b)** $M_b(\tau_x)$ for a "nonideal" edge. **(c)** $M_a(\tau_x)$ for a "nonideal" edge. **(d)** $M_a(\tau_x)$ for an "ideal" edge. The means are computed on the speckle-free images of Figure 4.14. ©2000 IEEE.

in the interval $\tau_x \in [-u, u]$ is not necessarily linear and quadratic expansions can be used to model it [80]:

$$
\begin{aligned}
M_a(\tau_x) &\simeq \mu_a + k(\tau_x + u)^2 \, \Theta(\tau_x + u) \\
M_b(\tau_x) &\simeq \mu_b - k(\tau_x - u)^2 \, \Theta(-\tau_x + u)
\end{aligned}
\tag{4.74}
$$

where k is a positive constant and Θ is the Heaviside function. Combining Eqs. 4.72 and 4.73 and expanding the calculation to the second order, one easily obtains the following expression:

$$
\delta = u \, \frac{1 - \rho}{1 + \rho} \quad \text{with} \quad \rho = \mu_a / \mu_b
\tag{4.75}
$$

which does not depend on k. Although it is very simple, this model permits us to describe three main facts:

- the bias is toward the darker side (see the sign of δ),

- the absolute value of the bias increases with the edge contrast ρ (approximately as $\left| \frac{\rho - 1}{\rho + 1} \right|$),

- the asymptotic value of the bias (as $\rho \to \infty$) is given by a phenomenological parameter u which is related to the size of the mixture zone within the sliding window.

The relevance of this model also lies in its generality since it allows the description of several different cases for which biased location occurs: One only needs to determine the size of the mixture zone. Thus, in the case of blurred edge (Figure 4.13, row 4), the mixture zone size is equal to the psf width d: $2u = d$. If the edge is tilted by an angle α with respect to the window (Figure 4.13, row 2), then it can easily be seen that the mixture zone has a size: $2u = L_y \tan \alpha$, where L_y is the height of the window. For a sinuous edge (Figure. 4.13, row 3), u is related to the variance of the edge position.

Simulations have been performed to test the reliability of this model [80]. For that purpose, a speckled edge blurred with a circular kernel of diameter d has been used (like the one presented on Figure 4.13, row 4). For a given set (ρ, d), 1000 realizations of the edge image are generated to estimate the bias. This estimation is compared to the bias measured on the speckle-free image and to the value given by expression 4.75. The results are shown in Figure 4.16. There is rather good agreement between the estimated bias and the measure on the speckle-free image: This shows that the phenomenological model in Eq. 4.75 is a good way of evaluating the bias. On the other hand, the expression of the bias given in Eq. 4.75 does not always fit very well with the estimations, especially for large d. This is due to the fact that this expression has been obtained with a second-order expansion. However, the general behavior of the curve remains well described.

In conclusion, the GLRT may lead to a biased location of an edge when it is not an "ideal" step edge of known orientation. One should not be surprised by such a result since the GLRT is basically a detection scheme and not an estimation technique. By use of a simple phenomenological model, it is possible to interpret the results for various cases where the bias is observed and to show very general results: The bias is toward the darker side of the edge, it increases with the edge contrast and is limited by an asymptotical value related to a certain phenomenological parameter u. In Chapter 5, we will see how the edge location bias can be removed by using a postprocessing step based on statistical actives contours.

4.5.   Conclusion

In this chapter, we have considered the SIR image model, which assumes that the image is composed of an object region and a background region whose pixel values are distributed with different pdf's. Despite its simplicity, this model can represent a wide spectrum of image processing tasks, such as target detection, target location, object segmentation, and object recognition. Image processing algorithms addressing these tasks are determined from this model by using the tools of decision and estimation theory described in Chapter 3. One of the advantages of the SIR image model is that it can be adapted to the statistics of the noise which actually perturbs the image. Moreover, if the pdf of this noise

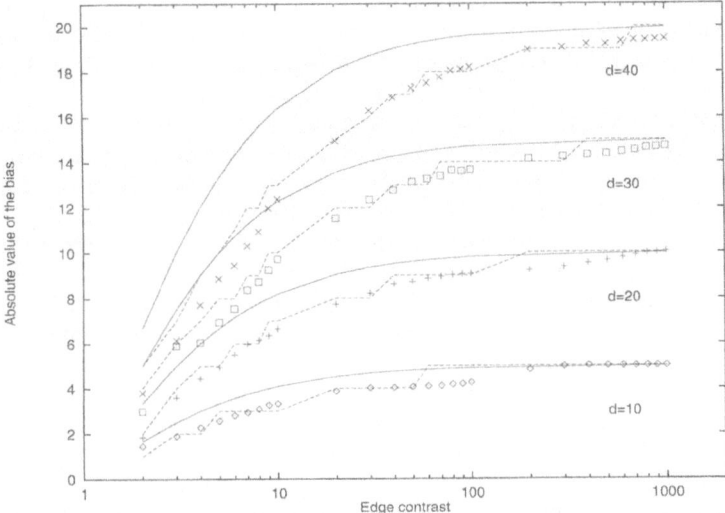

Figure 4.16. Bias of the edge position estimate as a function of the edge contrast ρ, for different sizes of correlation kernel d. For each d, three curves are plotted: bias estimated from 1000 realizations of the speckled image (points), bias measured on the speckle-free image (dashed curve), approximated expression of the bias (solid curve). The sliding window is 128×128. ©2000 IEEE.

belongs to the exponential family, the computational complexity of the obtained algorithms can be low. For example, target detection and location applications have the same computational complexity as linear correlation operations.

The usefulness of this technique for determining image processing algorithms has been illustrated on detection/location problems with different types of noises: Gaussian, Bernoulli (binary), and Gamma noise involved in SAR images. It has also been shown how the algorithms which are designed for spatially independent graylevels can be made efficient on correlated textures by means of a whitening preprocessing.

In the sequel of this book, the SIR image model will constitute the baseline from which image processing algorithms will be determined. For example, in the following chapter, we will focus on a very important domain of application – object segmentation – and we will show that efficient region-based active contours can be designed from this model.

APPENDIX 4.A: Basics of estimation theory

In this appendix, we briefly introduce the basic concepts and results of estimation theory which will be used throughout this book. More details can be found, for example, in Refs. 29 and 82.

The fundamental problem of estimation theory is modeled as follows. Let us consider a random variable X_λ, where the subscript λ represents the experimental outcome which gives rise to X_λ. Suppose that N realizations of this random variable are obtained and are gathered in a set denoted $\chi = \{x_1, ..., x_N\}$. This set will be called a *sample*.

The N realizations $x_1, \ldots, x_N$ that constitute the sample χ are fixed values, which can correspond for example to N measures of any physical phenomenon. On the other hand, if one considers all potential experimental outcomes, then the sample must be considered as a collection of N random variables. In this case, we will denote it $\chi_\lambda = \{X_{\lambda(1)}, \ldots, X_{\lambda(N)}\}$ where $\lambda = \lambda(1), \ldots, \lambda(N)$ is the series of experimental outcomes which gives rise to χ_λ. In other words, the sample χ can be considered as realization of χ_λ.

Let us now define some notions which will be useful in the following. First of all, any function $T[\chi] = t(x_1, \ldots, x_N)$ of the sample χ will be called a *statistics*. Although this term may seem ambiguous, we will use it since it is standard.

The pdf of X_λ is denoted $P_{\boldsymbol\mu}(x)$, where $\boldsymbol\mu$ represents the parameters of the pdf. In the following, in order to simplify the discussion, we will assume that $\boldsymbol\mu$ is a scalar parameter that we will denote μ. Similar results are obtained in the case of vectorial parameters (see Ref. 29 for more details).

One defines the *likelihood* $L(\chi|\mu)$ of the sample as the probability density of observing the realization χ of the sample χ_λ for a given value of μ:

$$L(\chi|\mu) = \prod_{i=1}^{N} P_\mu(x_i) \tag{4.A.1}$$

We can note that if $T[\chi]$ and $L(\chi|\mu)$ are defined on the sample realization χ, they are deterministic functions. On the other hand, if all potential experimental outcomes are considered, then $T[\chi_\lambda]$ and $L(\chi_\lambda|\mu)$ are random variables.

Bias and variance of an estimator

Given a sample χ and a pdf $P_\mu(x)$, statistical estimation consists in estimating the values of μ from the sample χ. Let us first consider a simple example. Suppose that:

$$P_\mu(x) = \frac{1}{\mu} \exp\left[-\frac{x}{\mu}\right] \tag{4.A.2}$$

It is well known that the mean and the variance of this pdf are:

$$
\begin{aligned}
<x> &= \mu \\
VAR[x] &= <x^2> - <x>^2 = \mu^2
\end{aligned}
$$

A simple way of estimating μ from a sample χ of dimension N is to approximate the ensemble average $<x>$ with the empirical mean of the sample. One thus obtains the following estimator of μ:

$$\hat{\mu}_1(\chi) = \frac{1}{N} \sum_{i=1}^{N} x_i \tag{4.A.3}$$

One could define another estimate of μ by approximating $VAR[x]$ with the empirical variance:

$$\hat{\mu}_2(\chi) = \sqrt{\frac{1}{N} \sum_{i=1}^{N} x_i^2 - \left(\frac{1}{N} \sum_{i=1}^{N} x_i\right)^2} \tag{4.A.4}$$

Generally speaking, an estimator $\hat{\mu}(\chi)$ will be a function of the sample χ. If all experimental outcomes are considered, it is a function $\hat{\mu}(\chi_\lambda)$ of χ_λ or, in other words, a *statistics* of the sample. It is thus a random variable.

It can be seen from Eqs. 4.A.3 and 4.A.4 that there are many possible estimators for a given parameter μ. One of the goals of estimation theory will be to determine the best estimator for a given statistical problem. In order to reach this goal, one first has to define a measure of the "quality" of an estimator.

Basically, a statistics $T[\chi_\lambda]$ will be a good estimator of the parameter μ if it is "close" to the true value μ_0 of μ. However, since $T[\chi_\lambda]$ is a random variable, this notion of "closeness" has to be defined on average. A standard but criticizable method consists in using the mean square error (MSE)[2]:

$$MSE\,(T[\chi_\lambda]) = \left\langle |T[\chi_\lambda] - \mu_0|^2 \right\rangle \tag{4.A.5}$$

In this equation, the mean $< . >$ is defined over the realizations of the sample χ_λ, that is:

$$
\begin{aligned}
\langle T[\chi_\lambda] \rangle &= \int T(\chi) L(\chi|\mu) d\chi \\
&= \int T(x_1, \ldots, x_N) L(x_1, \ldots, x_N|\mu)\, dx_1 \ldots dx_N \tag{4.A.6}
\end{aligned}
$$

The MSE can be expressed in a different way by defining the difference between the statistics and its mean:

$$\delta T[\chi_\lambda] = T[\chi_\lambda] - \langle T[\chi_\lambda] \rangle$$

and noticing that:

$$
\begin{aligned}
MSE\,(T[\chi_\lambda]) &= \left\langle |\delta T[\chi_\lambda] + (< T[\chi_\lambda] > - \mu_0)|^2 \right\rangle \\
&= \left\langle |\delta T[\chi_\lambda]|^2 \right\rangle + 2 \left\langle \delta T[\chi_\lambda] \right\rangle (\langle T[\chi_\lambda] \rangle - \mu_0) \\
&\quad + |< T[\chi_\lambda] > - \mu_0|^2
\end{aligned}
$$

The second term of the last equation is zero since by definition, $\langle \delta T[\chi_\lambda] \rangle = 0$. The MSE can thus be written in the following way:

$$MSE\,(T[\chi_\lambda]) = \sigma_T^2 + b_T^2 \tag{4.A.7}$$

where

- $b_T = \langle T[\chi_\lambda] \rangle - \mu_0$ is the *bias* of the estimator, that is, the difference between the average value of the statistics and the true value of the parameter.

- $\sigma_T^2 = \left\langle |\delta T[\chi_\lambda]|^2 \right\rangle$ is the variance of the estimator.

The MSE is thus the sum of the square of the bias and of the variance. Figure 4.A.1 represents the pdf of the statistics $T[\chi_\lambda]$ and a graphical illustration of the bias and of the variance.

Cramer-Rao Lower Bound

We will restrain our analysis to estimators with zero bias, which are called *unbiased estimators*. The quality of unbiased estimators is uniquely determined by their variance. A fundamental result of estimation theory specifies the minimal variance that any estimator can reach for a

---

[2]We discuss here the classical point of view of statistics. There exist other approaches such as for example the Bayesian one [51], which provide alternative solutions but which will not be discussed here for simplicity reasons.

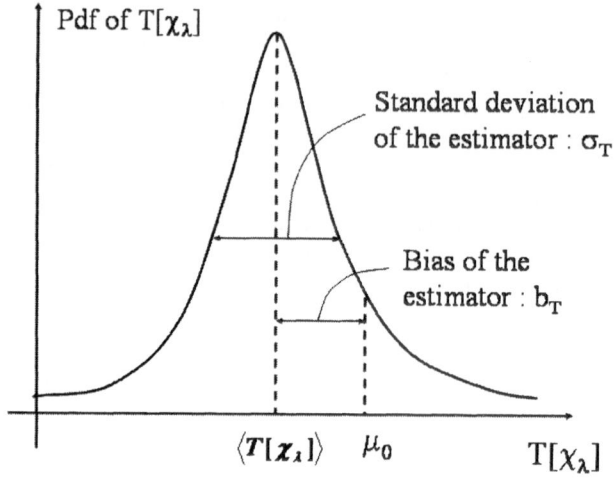

Figure 4.A.1.   Bias and variance of an estimator.

given statistical problem. More precisely, it states that if the domain of definition of x does not depend on μ, for a given value of the parameter μ, the variance $\sigma_T^2(\mu)$ of any unbiased estimator is such that:

$$\sigma_T^2(\mu) \geq \frac{1}{I_F(\mu)} \tag{4.A.8}$$

where I_F is the *Fisher information* defined as:

$$I_F(\mu) = -\left\langle \int \frac{\partial^2 \{\log[L(\chi_\lambda|\mu)]\}}{\partial \mu^2} \right\rangle \tag{4.A.9}$$

The bound on estimator variance defined in Eq. 4.A.8 is called the Cramer-Rao Lower Bound (CRLB).

Let us take the example of the pdf defined in Eq. 4.A.2. Its loglikelihood is equal to:

$$\log L(\chi|\mu) = -N \log \mu - \frac{1}{\mu} \sum_{i=1}^{N} x_i \tag{4.A.10}$$

and its second derivative with respect to parameter μ is:

$$\frac{\partial^2 \log[L(\chi|\mu)]}{\partial \mu^2} = \frac{N}{\mu^2} - \frac{2}{\mu^3} \sum_{i=1}^{N} x_i \tag{4.A.11}$$

The average value is:

$$\begin{aligned}
\left\langle \frac{\partial^2 \log[L(\chi|\mu)]}{\partial \mu^2} \right\rangle &= \frac{N}{\mu^2} - \frac{2}{\mu^3} \sum_{i=1}^{N} < x_i > \\
&= \frac{N}{\mu^2} - \frac{2}{\mu^3} N\mu = -\frac{N}{\mu^2}
\end{aligned} \tag{4.A.12}$$

The variance of any estimator of μ is thus superior to $\frac{\mu^2}{N}$.

The estimators which reach the CRLB are called *efficient*. We will discuss below that the problem of the determination of efficient estimators can be solved if the pdf of the sample belongs to the exponential family.

Exponential family and maximum-likelihood estimation

The exponential family of pdf (see Eq. 4.7) in the case of a scalar parameter μ is defined as follows:

$$P_\mu(x) = \exp\left[\alpha(\mu).\beta(x) + \phi(\mu) + \kappa(x)\right] \tag{4.A.13}$$

where $\beta(x)$ and $\kappa(x)$ are functions of x and $\alpha(\mu)$ is a function of μ. Examples of pdf which belong to the exponential family are given in Table 4.2.

If the pdf belongs to the exponential family, the likelihood of a sample χ has a simple expression since:

$$L(\chi|\mu) = \exp\left[\alpha(\mu)T[\chi] + N\phi(\mu) + F[\chi]\right] \tag{4.A.14}$$

with $T[\chi] = \sum_{i=1}^{N} \beta(x_i)$ and $F[\chi] = \sum_{i=1}^{N} \kappa(x)$. It is interesting to note that this likelihood can be written in the following form:

$$L(\chi \mid \mu) = g(T[\chi] \mid \mu) \, h(\chi) \tag{4.A.15}$$

where $g(x|\mu) = \exp\left[\alpha(\mu)T[\chi] + N\phi(\mu)\right]$ and $h(\chi) = \exp(F[\chi])$. When the likelihood can be decomposed in this way, $T[\chi]$ is called a *sufficient statistics*.

It can be shown that if there exists a sufficient statistics, then the ML estimate of the parameter μ will depend on the data χ only through the sufficient statistics. As seen in Section 3.2.2.1, the maximum likelihood estimate is the value of μ which maximizes the likelihood $L(\chi|\mu)$. It is thus the solution of the following equation:

$$\frac{\partial L(\chi \mid \mu)}{\partial \mu} = 0 \iff \alpha'(\mu)T[\chi] + N\phi'(\mu) = 0 \tag{4.A.16}$$

where $\alpha'(\mu) = \partial\alpha/\partial\mu$ and $\phi'(\mu) = \partial\phi/\partial\mu$. The ML estimate $\hat{\mu}_{ML}$ is thus obtained by inverting the following equation:

$$-\frac{\phi'(\hat{\mu}_{ML})}{\alpha'(\hat{\mu}_{ML})} = T[\chi]/N \tag{4.A.17}$$

It is clear from this relation that within the exponential family, the ML estimate $\hat{\mu}_{ML}$ depends on the sample values only through the sufficient statistics $T[\chi]$. Moreover, it also possesses the following important property:

Property:
If the following conditions are fulfilled:

- *The pdf $P_\mu(x)$ of the sample belongs to the exponential family,*
- *The domain of definition of x is independent of μ.*
- *The ML estimate $\hat{\mu}_{ML}$ of parameter μ is unbiased,*

then $\hat{\mu}_{ML}$ has minimum variance. Furthermore, if $\hat{\mu}_{ML}$ is proportional to the sufficient statistics $T[\chi]$, then $\hat{\mu}_{ML}$ is an efficient estimator, that is, it reaches the CRLB.

Let us take the example of the pdf defined in Eq. 4.A.2. It can be expressed in the following way:

$$P_\mu(x) = \exp\left[-x/\mu - \log\mu\right] \tag{4.A.18}$$

so that $\alpha(\mu) = -1/\mu$, $\beta(x) = x$ and $T[\chi] = \sum_{i=1}^{N} x_i$. From Eq. 4.A.17, the ML estimate is thus equal to $\widehat{\mu}_{ML} = T[\chi]/N$. It is easily seen that:

$$< \widehat{\mu}_{ML} > = \frac{1}{N} < x_i > = \mu$$

which means that the ML estimate is unbiased. Hence, according to the previous property, it must reach the CRLB. Indeed, it is easily shown that $\text{VAR}[\widehat{\mu}_{ML}] = \mu^2/N$, which is the CRLB for this pdf computed in Eq. 4.A.12.

In conclusion, within the exponential family, and if the domain of definition of x does not depend on the parameter μ, the ML estimate possesses three important properties:

- It depends on the data through a simple sufficient statistics $T[\chi] = \sum_{i=1}^{N} \beta(x_i)$.

- It is minimum variance if it is unbiased.

- It is efficient if it is unbiased and proportional to $T[\chi]$.

The second and third properties provide a strong justification to the use of the ML estimation of the nuisance parameters when the noise is modeled with a pdf belonging to the exponential family. The first property leads to computationally simple techniques of detection, location, and segmentation based on the SIR image model.

Chapter 5

STATISTICAL SNAKE-BASED SEGMENTATION ADAPTED TO DIFFERENT PHYSICAL NOISES

We now turn to another application of the approach introduced in the previous chapter for designing image processing algorithms adapted to different types of noise. We will address in the present chapter the problem of object segmentation, which consists in determining the contour of an object in an image. This issue well deserves a whole chapter, because of its importance in image processing. Indeed, image segmentation is often the first step in an image processing chain, and as such its importance is crucial for the global performance of a system. This is why image segmentation has been probably one of the most studied image processing problems. The literature about this topic is copious and a large number of different approaches have been proposed to solve different kinds of applications.

Consequently, when proposing a segmentation method, one first has to clearly delimit the problem which is addressed, the goals one wants to reach, and the type of images that are to be processed (presently, there exists no universal segmentation strategy valid for all kinds of images). In this chapter, we will thus limit ourselves to the segmentation of single objects in very noisy images. We will show how the SIR image model introduced in the previous chapter can be naturally generalized to yield segmentation algorithms by using the principle of active contours. As in the previous chapters, our two main goals will be to design algorithms which are adapted to the noise present in the image and which are potentially fast enough to be usable in real applications.

We will begin this chapter with a brief review of the state of the art in the domain of active contours in Section 5.1. This domain in itself is so huge that we cannot pretend to be exhaustive. Our goal will be simply to position the statistical approach with respect to the main existing ones. We will then explain in Section 5.2 how region-based active contours can be built from the SIR image model. We will show in particular that when the graylevel pdf belongs to the exponential family, the algorithm can be put in such a mathematical form that its computational complexity is fairly low. We then specialize this approach to the case where the contour is assumed to be polygonal in Section 5.3. We will conclude this chapter by addressing two practical applications of this algorithm. The first one concerns target tracking in video sequences (Section 5.4). In Section 5.5 we show how active contours can be used to refine edge location in SAR images and suppress the bias problem analyzed in Section 4.4.

5.1. Active contours

An important goal of computational vision and image processing is to automatically recover the shape of objects from various types of images. Over the years, many approaches have been developed to reach this goal. Let us consider the case of the segmentation of a single object in an image. A classical approach consists in detecting edges and then linking them in order to determine the shape of the object. However, this approach does not make use of the knowledge that the object is simply connected. On the other hand, deformable models which incorporate knowledge about the shape of the object are widely used. With these techniques, rather than expecting desirable properties such as continuity and smoothness of the contour to emerge from image data, those properties are imposed from the start. Deformable models encompass a variety of forms, principally active contours and deformable templates.

Active contours have become a topic of high interest since the work of Kass, Witkin, and Terzopoulos in 1988 [83]. An active contour, or *snake*, is a continuous curve which has the ability to evolve in order to match an object in the image (see Figure 5.1). This curve may be closed or not. It is initialized close to the searched contour, and it deforms according to the iterative optimization of a numerical criterion. This criterion is usually called "energy" since it was at the origin directly interpretable in terms of physical constraints such as rigidity, stiffness, and attraction to the image edges. Convergence of the snake toward the searched object can then be seen as the fulfillment of a mechanical stability condition.

5.1.1   Snake energy

We present here the original formulation of active contours by Kass *et al.* [83]. The active contour is defined as a parametric in-plane curve $\mathcal{C}$:

$$
\begin{aligned}
\mathcal{C}: \quad & [0,1] \;\rightarrow\; \mathcal{R}^2 \\
& s \;\rightarrow\; \mathbf{v}(s) = \begin{bmatrix} x(s) \\ y(s) \end{bmatrix}
\end{aligned}
\tag{5.1}
$$

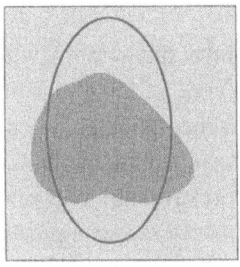 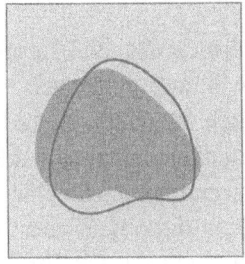 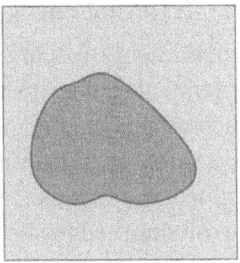

Figure 5.1.   Principle of active contour (snake) segmentation. Initial shape, convergence, and final result.

where s denotes the curvilinear abscissa and $\mathbf{v}(s)$ the point of coordinates $x(s), y(s)$ on the curve. The following energy is associated with the curve $\mathcal{C}$:

$$E(\mathcal{C}) = E_{int}(\mathcal{C}) + \gamma E_{ext}(\mathcal{C}) \tag{5.2}$$

where γ is a real-valued number. This energy is the weighted sum of two terms which represent two different types of constraints imposed on the active contour. By minimizing this weighted sum, one hopes to get a good compromise between these constraints.

5.1.1.1   Internal energy

The first term in Eq. 5.2 has the following expression:

$$E_{int}(\mathcal{C}) = \int_0^1 \alpha(s) \left| \frac{\partial \mathbf{v}}{\partial s} \right|^2 + \beta(s) \left| \frac{\partial^2 \mathbf{v}}{\partial s^2} \right|^2 \, ds \tag{5.3}$$

This energy is termed *internal* since it characterizes the shape of the contour independently of graylevels in the image. The internal energy has a simple physical interpretation. The coefficient $\alpha(s)$ represents the rigidity of the curve, that is, the cost which has to be paid to increase its length. The coefficient $\beta(s)$ represents the stiffness of the curve, that is, the cost to pay to "bend" it. If $\alpha(s)$ increases, the contour will have a tendency to regress, ultimately to a single point. If $\beta(s)$ increases, the contour becomes smoother. For simplicity's sake, these parameters are often taken constant. However, it may be interesting in some cases to let be them locally adaptive. For example, lowering $\beta(s)$ at some point will enable the contour to adapt to a sharp edge of the object.

Of course, optimizing only the internal energy defined in Eq. 5.3 will lead to a short and smooth contour, which will collapse to a single point. One thus needs to balance this "shrinking" effect of the internal energy with an *external energy* which depends on the information in the image.

5.1.1.2   External energy

External energy (second term in Eq. 5.2) represents the way in which the snake energy depends on the image graylevels. There is a great latitude in the choice of this energy, which must be adapted to the characteristics which make it possible to discriminate the object from the background. In the original formulation of the snake, the characteristic that is taken into account is the presence of sharp edges between the object and the background regions. Kass *et al.* thus propose the following expression for the snake energy:

$$E_{ext}(\mathcal{C}) = -\int_0^1 |\nabla I(\mathbf{v}(s))|^2 \, ds \qquad (5.4)$$

where $I(x, y)$ represents the image and ∇I represents the result of applying a gradient operator which enhances the edges present in the image. This energy is termed *external* since it represents the "force" imposed by the data to the snake. To minimize this energy, the snake will have to match as closely as possible the points on which the gradient is high, that is, where sharp edges are present.

5.1.1.3   Numerical aspects

From a numerical point of view, the segmentation problem consists in minimizing the following functional:

$$E(\mathcal{C}) = \int_0^1 \left[\alpha \left| \frac{\partial \mathbf{v}}{\partial s} \right|^2 + \beta \left| \frac{\partial^2 \mathbf{v}}{\partial s^2} \right|^2 - \gamma |\nabla I(\mathbf{v}(s))|^2 \right] ds \qquad (5.5)$$

where α, β, and γ are real-valued coefficients which balance the different contributions to the snake energy. For the sake of simplicity, we will suppose them invariant of the curvilinear abscissa s. If the contour is assumed to be closed, annulling the derivative of this functional with respect to the contour $\mathcal{C}$ leads to the following Euler-Lagrange equation [83]:

$$\alpha \, \kappa(s) \, \overrightarrow{\mathcal{N}} + \beta \frac{\partial^4 \mathbf{v}}{\partial s^4} - \gamma \nabla \left[|\nabla I(\mathbf{v}(s))|^2 \right] = 0 \qquad (5.6)$$

In this equation, we have used the fact that when a curve is parametrized by the curvilinear abscissa, one has the relation:

$$\frac{\partial^2 \mathbf{v}}{\partial s^2} = \kappa(s) \, \overrightarrow{\mathcal{N}}$$

where $\kappa(s)$ is the local curvature and $\overrightarrow{\mathcal{N}}$ the vector normal to the curve at abscissa s.

The usual approach to solve Eq. 5.6 is to perform the minimization of Eq. 5.5 with a steepest gradient descent method [84]. For this purpose, one assumes that the curve depends on time t and is denoted $\mathbf{v}(s, t)$. One iteratively solves the following equation [83]:

$$\frac{\partial \mathbf{v}(s, t)}{\partial t} = \alpha\kappa(s, t) + \beta\frac{\partial^4 \mathbf{v}(s, t)}{\partial s^4} - \gamma\nabla\left[|\nabla I(\mathbf{v}(s, t))|^2\right] \quad (5.7)$$

Starting with an initial shape $\mathbf{v}(s, 0)$ at time $t = 0$, one lets the dynamical system evolve until it stabilizes, which corresponds to $\frac{\partial \mathbf{v}(s,t)}{\partial t} = 0$ and thus to the solution of Eq. 5.6. In practice, Eq. 5.7 is discretized in time and space and solved numerically [83].

5.1.2 The limits of the classical snake

The classical snake model has been largely studied and utilized. It has been shown that the choice of the different parameters (weighting coefficients, time and space discretization steps) is difficult and critical to the performance of the snake, as well as the choice of the initial shape. Leymarie and Levine [85] have studied the influence of the parameter choice, the numerical resolution, and the stability of solutions in a precise application which consisted in tracking a biological cell moving in a plane.

More generally, the classical snake model can be classified in the category of edge-based segmentation approaches, since the image information which is used to drive the snake is the detected edges of the objects. This type of active contours is thus adapted to a class of problems in which the objects to segment form sharp edges with respect to the background. Moreover, they are sensitive to the choice of the initial shape, since the curve will not "evolve" if is not close enough to the object edges. Finally, this type of approach cannot segment objects which consist of several disconnected parts, since it assumes that the limits of the object are represented by a single closed curve. We now briefly discuss an alternative implementation of the edge-based snake which presents attractive characteristics in terms of topological properties and stability of numerical implementation.

5.1.3 Geodesic snakes

In order to address some limits of the classical snake, Caselles *et al.* have proposed the following evolution equation for a snake [86]:

$$\frac{\partial \mathbf{v}(s, t)}{\partial t} = g\left(|\nabla I[\mathbf{v}(s, t)]|^2\right)\kappa(s)\vec{\mathcal{N}} - \left(\nabla g\left(|\nabla I[\mathbf{v}(s, t)]|^2\right).\vec{\mathcal{N}}\right)\vec{\mathcal{N}} \quad (5.8)$$

where $g(x)$ is a decreasing function, such as $g(x) = \frac{1}{1+x^2}$. This is quite similar to the classical snake. The differences are that the stiffness term has disappeared and the rigidity term is multiplied by $g\left(|\nabla I[\mathbf{v}(s, t)]|^2\right)$. This modification means that at points where the gradient intensity is large, the snake "speed" will vanish. This is why $g(x)$ is usually called the *stop function*. Moreover, in the expression of the external energy (second term of Eq. 5.8), the gradient of the edge image is projected onto the normal direction to the curve, which seems reasonable since the evolution of the snake takes place in this direction. Caselles *et al.* have shown that there is an equivalence between Eq. 5.8 and the dynamical scheme built from a special case of the classical snake [86]. This type of active contours is called *geodesic*. The main advantage of this formulation is that the snake shape can be easily parametrized by the so-called *level sets*, which confers to the snake superior topological properties.

5.1.4   The level set implementation of snakes

Instead of being defined by its parametric representation $\mathbf{v}(s)$, the curve can be represented as the zero level set of a two-dimensional function $z = u(x, y) : [0 : a] \times [0 : b] \longrightarrow \mathcal{R}$ [87-89]. In other words, the curve consists of the set of points $\{(x, y) | u(x, y) = 0\}$. The main advantage of this implicit representation is to be able to represent any discontinuous topology of the contour while keeping a continuous topology for the surface $u(x, y)$ (see Figure 5.2). In particular, it is able to represent an object composed of several discontinuous parts. It can be shown that the evolution equation of the geodesic snake in Eq. 5.8 can be rewritten as a function of $u(x, y)$ in the following way [86]:

$$\frac{\partial u(x, y, t)}{\partial t} = g\left(|\nabla I(x, y)|^2\right) |\nabla u(x, y, t)| \, \kappa(x, y, t)$$
$$+ \nabla g\left(|\nabla I(x, y)|^2\right) \nabla u(x, y, t) \qquad (5.9)$$

where $\kappa(x, y, t)$ is the curvature of the contour, which can be expressed directly from u, as well as its normal [86]:

$$\vec{\mathcal{N}} = -\frac{\nabla u}{|\nabla u|} \quad \text{and} \quad \kappa = \text{div}\left(\frac{\nabla u}{|\nabla u|}\right) \qquad (5.10)$$

There exist robust numerical schemes to solve Eq. 5.9 [90]. Moreover, efficient implementation principles such as the "narrow band" approach permit faster convergence of this type of snake [89, 91].

It should be noted that the level set approach is a way of representing the snake which has very attractive topological properties, but does not allow one to determine which criterion has to be optimized for a given image model.

5.1.5   Region-based approaches

Contrary to edge-based methods, region-based approaches rely on the information contained in the graylevel structure of the regions of the image to perform segmentation. This is the approach we will use in the sequel of this book, and we briefly recall here pioneering works in this domain.

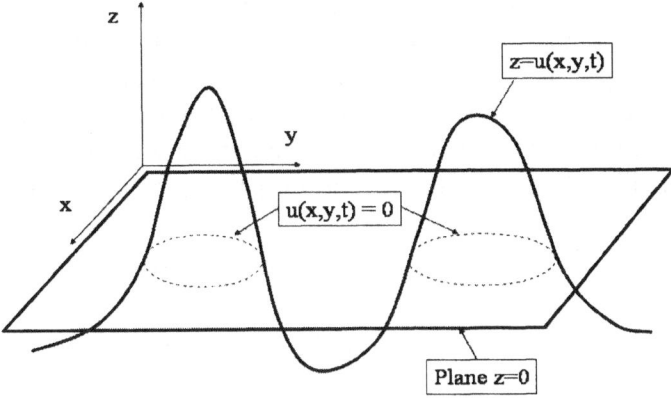

Figure 5.2. Principle of the level set implementation of active contours. The zero-level set of the continuous surface $z = u(x, y, t)$ implicitly defines a contour $u(x, y, t) = 0$ which can have a complex topology (here, it is divided into two parts).

One of the earliest works was performed by Ronfard [92]. He developed a model where the active contour is deformed by forces whose directions are normal to the contour and whose intensities stem from a measure of similarity between regions. The similarity between two regions R_1 and R_2 is evaluated with the *Ward distance* $D(R_1, R_2)$:

$$D(R_1, R_2) = \hat{\sigma}^2_{R_1 \cup R_2} - \hat{\sigma}^2_{R_1} - \hat{\sigma}^2_{R_2} \qquad (5.11)$$

where $\hat{\sigma}^2_R$ represents the empirical variance of the graylevels in region R. The intensity of the force F driving the active contour is defined as:

$$F = D(R_{in}, \delta R) - D(R_{out}, \delta R) \qquad (5.12)$$

where R_{in} and R_{out} represent the regions around the contour which lie inside and outside the object, and δR a smaller region in the neighborhood of the point at which the force is applied. According to the sign of the intensity, the force can thus dilate or retract the active contour. The evolution of the contour is driven by an iterative algorithm. The key point here is that the force does not depend on the gradient of the image but on the difference of the variances of the graylevels in regions located inside and outside the contour. However, only the pixels close to the contour are considered. Moreover, the expression of the force driving the snake, although reasonable, is not clearly justified.

Other approaches have thus consisted in embedding the region-based techniques in a more rigorous theoretical framework, and most of them have used statistical decision theory [61,93-96]. The basic hypothesis underlying this type of method is that the image is composed of two regions Δ_{in} and Δ_{out}

which represent the object of interest and the background. The graylevels of each region are supposed to have different statistical distributions $P_{in}(x|\boldsymbol{\mu}_{in})$ and $P_{out}(x|\boldsymbol{\mu}_{out})$ where $\boldsymbol{\mu}_{in}$ and $\boldsymbol{\mu}_{out}$ are the parameters of the pdf. These hypotheses are similar to the SIR model that has been defined in the previous chapter (see Eq. 4.1). The internal energy of the snake is defined as the opposite of the loglikelihood of the data, which is, in spatially continuous notation,

$$
\begin{aligned}
E_{int}(\mathcal{C}) \quad = \quad & - \int_{\Delta_{in}} \log P_{in}\left[s(x,y)|\boldsymbol{\mu}_{in}\right] dx dy \\
& - \int_{\Delta_{out}} \log P_{out}\left[s(x,y)|\boldsymbol{\mu}_{out}\right] dx dy
\end{aligned}
\tag{5.13}
$$

The optimization of this energy will lead to a contour which best separates the image into two regions whose graylevels have homogeneous statistical characteristics. One of the advantages of this type of internal energy over the classical edge-based snake is that it is more robust to the initial shape of the snake, which does not necessarily have to intersect the true boundaries to be able to converge. Moreover, in a noisy environment, edge extraction will lead to even noisier images and to numerous spurious contours. Since it is does not rely on edge extraction but on the graylevel characteristics in the whole regions Δ_{in} and Δ_{out}, the region-based snake is much more robust to noise.

An important problem in the expression of E_{int} in Eq. 5.13 is the choice of the pdf $P_{in}(x|\boldsymbol{\mu}_{in})$ and $P_{out}(x|\boldsymbol{\mu}_{out})$ and of their parameters $\boldsymbol{\mu}_{out}$ and $\boldsymbol{\mu}_{in}$. The general expression of the graylevel pdf may be inferred from the image formation process as was shown for the SIR image model in Section 4.1.2. The problem of the estimation of the pdf parameters is more delicate, and various approaches can be used. In Ref. 97, the regions to segment are assumed to have different means – after a preprocessing of the graylevels adapted to the textures of the objects – so that they form "peaks" in the histogram of the image. The peaks corresponding to each region are separated by assuming that the whole histogram is a mixture of Gaussians, and these Gaussians are used to represent the pdf of the graylevels in each region. This approach is limited to the case where the regions can be clearly identified on the histogram. A more general approach consists in estimating the pdf parameters from the graylevels in Δ_{in} and Δ_{out} with an Expectation-Minimization (EM) method which iterates the following steps [59,96,98]:

- Estimate $\boldsymbol{\mu}_{in}$ and $\boldsymbol{\mu}_{out}$ from a given shape of contour $\mathcal{C}$ defining regions Δ_{in} and Δ_{out}.

- Optimize $E_{int}(\mathcal{C})$ for a given value of $\boldsymbol{\mu}_{in}$ and $\boldsymbol{\mu}_{out}$ for a few iteration steps.

We will describe in Section 5.2.1 an efficient way of solving the problem of parameter estimation that avoids this two-step method.

Of course, the various types of internal energy terms and of contour representations that are used for the classical snake can also be applied to the region-based snake. For example, snakes based on B-splines and Fourier descriptors have been developed with success [59,60,93]

More recently, region-based snakes have been implemented with the level-set approach described in Section 5.1.3 in order to benefit from the advantages of this representation in terms of contour topology. For this purpose, Bayesian modelization of the graylevels and piecewise-constant image descriptions have been considered together with different types of graylevel descriptors [91, 99, 100]. The efficiency of these approaches has been demonstrated for the segmentation of objects in video sequences and medical images.

The main drawback of region-based approaches is that they are based on the hypothesis that the statistical properties of the object and background regions are homogeneous. When this hypothesis is too far from reality, techniques that rely only on the edges of the objects may be more efficient. Of course, region-based and edge-based energy terms can also be used together in order to benefit from their respective advantages [91].

5.2. The SIR Active Contour and its fast implementation

Region-based active contours can be designed using the statistical approach described in the previous chapters. As seen in Section 4.1, the SIR image model can be used to address image segmentation problems (see also Refs. 61 and 62). The observed scene is modeled as:

$$s(x,y) = a(x,y)w^{\theta}(x,y) + b(x,y)[1 - w^{\theta}(x,y)] \qquad (5.14)$$

where, as in Chapter 4, the pixel coordinates x and y belong to a discrete set. The image is assumed to be composed of two regions, the target and the background, and the target region is defined by a binary mask w^{θ}, where θ is the set of parameters which define the shape. Different shape descriptors can be used to define w^{θ}. If the contour is represented by a polygon with P nodes, θ is the $2P$-dimensional vector representing the positions (x, y) of all the nodes (see Figure 5.3). In the case of B-spline representation, the parameter vector θ is constituted of the control points [76] (see Figure 5.3). If Fourier descriptors are used [76], the parameter θ represents the Fourier coefficients for the different spatial frequencies. In Section 5.3, we will consider in more detail the polygonal description.

The purpose of segmentation is therefore to estimate the most likely shape parameters θ. As in the previous chapters, this is done by maximizing the log-likelihood of the scene s with respect to the parameters of interest θ, that is, $L(\mathbf{s}|\theta, \mu_a, \mu_b)$ (see Eq. 4.3). The nuisance parameters μ_a and μ_b can be estimated in the ML or MAP sense, or suppressed with the marginal Bayesian method.

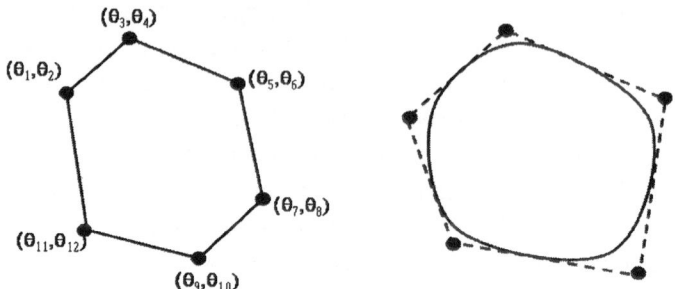

Figure 5.3.   Left: polygonal model of a shape, the nodes are black dots. Right: B-spline model of a shape, the control points are black dots.

5.2.1   Solution for exponential family laws

Let us consider that the pdf $P_{\boldsymbol{\mu}_a}(x)$ and $P_{\boldsymbol{\mu}_b}(x)$, which describe the graylevels in the target and background regions, belong to the exponential family. As seen in Section 4.1.2, if we choose to estimate the nuisance parameters with the ML or the MAP approach and to inject these estimates in the log-likelihood, or if we choose a marginal Bayesian approach, we obtain the function to be optimized with respect to $\boldsymbol{\theta}$ in the following form:

$$\ell(\mathbf{s}, \boldsymbol{\theta}) = -N_a(\boldsymbol{\theta})\, f_a\left(\frac{\mathbf{T}[\chi_a^{\boldsymbol{\theta}}]}{N_a(\boldsymbol{\theta})}\right) - N_b(\boldsymbol{\theta})\, f_b\left(\frac{\mathbf{T}[\chi_b^{\boldsymbol{\theta}}]}{N_b(\boldsymbol{\theta})}\right) + K \qquad (5.15)$$

where K is a constant which does not depend on $\boldsymbol{\theta}$ and:

$$\mathbf{T}[\chi_u] = \sum_{(x,y)\in\Delta_u^{\boldsymbol{\theta}}} \beta[s(x,y)] \qquad (5.16)$$

where $\beta(x)$ is defined in Eq. 4.7 and $u = a$ or b. We have denoted the likelihood $\ell(\mathbf{s}, \boldsymbol{\theta})$ in order to emphasize its dependence on the image data $\mathbf{s}$.

In order to respect a well-established tradition in the domain of active contours, we will consider that instead of maximizing $\ell(\mathbf{s}, \boldsymbol{\theta})$, we minimize an "energy function" which is equal to the opposite of the part of $\ell(\mathbf{s}, \boldsymbol{\theta})$ which depends on $\boldsymbol{\theta}$:

$$J(\mathbf{s}, \boldsymbol{\theta}) = -\ell(\mathbf{s}, \boldsymbol{\theta}) + K \qquad (5.17)$$

The parameter set $\boldsymbol{\theta}$ that minimizes $J(\mathbf{s}, \boldsymbol{\theta})$ then performs a maximum likelihood segmentation of the target in the scene. Within the active contour terminology, this criterion can consequently be regarded as an "external energy" acting on the snake (which is defined by $\boldsymbol{\theta}$) since the minimization of this term forces the contour to delimit the target. We describe below the expression of $J(\mathbf{s}, \boldsymbol{\theta})$ for different pdf belonging to the exponential family:

- **Bernoulli:**

$$J_{ber}(\mathbf{s}, \boldsymbol{\theta}) = N_a(\boldsymbol{\theta}) f[t_1^{(a)}(\boldsymbol{\theta})] + N_b(\boldsymbol{\theta}) f[t_1^{(b)}(\boldsymbol{\theta})] \qquad (5.18)$$

with

$$f(x) = -x \log(x) - (1 - x) \log(1 - x) \qquad (5.19)$$

- **Gaussian:**

$$J_{gau}(\mathbf{s}, \boldsymbol{\theta}) = \frac{N_a}{2}(\boldsymbol{\theta}) \log[\sigma_a^2(\boldsymbol{\theta})] + \frac{N_b}{2}(\boldsymbol{\theta}) \log[\sigma_b^2(\boldsymbol{\theta})] \qquad (5.20)$$

- **Gamma:**

$$J_{gam}(\mathbf{s}, \boldsymbol{\theta}) = N_a(\boldsymbol{\theta}) L \log[t_1^{(a)}(\boldsymbol{\theta})] + N_b(\boldsymbol{\theta}) L \log[t_1^{(b)}(\boldsymbol{\theta})] \qquad (5.21)$$

- **Rayleigh:**

$$J_{ray}(\mathbf{s}, \boldsymbol{\theta}) = N_a(\boldsymbol{\theta}) \log\left[t_2^{(a)}(\boldsymbol{\theta})\right] + N_b(\boldsymbol{\theta}) \log\left[t_2^{(b)}(\boldsymbol{\theta})\right] \qquad (5.22)$$

- **Poisson:**

$$J_{ray}(\mathbf{s}, \boldsymbol{\theta}) = N_a(\boldsymbol{\theta}) f[t_1^{(a)}(\boldsymbol{\theta})] + N_b(\boldsymbol{\theta}) f[t_1^{(b)}(\boldsymbol{\theta})] \qquad (5.23)$$

with

$$f(x) = -x \log(x) \qquad (5.24)$$

The following notations have been used in the previous equations:

$$t_k^{(u)}(\boldsymbol{\tau}) = \frac{1}{N_u} \sum_{(x,y) \in \Delta_u^{\boldsymbol{\tau}}} [s(x, y)]^k \qquad (5.25)$$

$$\sigma_u^2(\boldsymbol{\tau}) = t_2^{(u)}(\boldsymbol{\tau}) - [t_1^{(u)}(\boldsymbol{\tau})]^2 \qquad (5.26)$$

where $k = 1$ or 2 and $u = a, b$.

It is interesting to note that the criterion to optimize $J(\mathbf{s}, \boldsymbol{\theta})$ only depends on the contour parameters $\boldsymbol{\theta}$. This useful property is due to the fact that the considered graylevel statistics belong to the exponential family, in which the sufficient statistic has a very simple mathematical form (see Section 4.1.2.1).

In general, $J(\mathbf{s}, \boldsymbol{\theta})$ must be optimized iteratively. For this purpose, any optimization procedure can be used, and we will describe in Section 5.3.2 a very simple one adapted to polygonal shapes. Whatever the optimization procedure, for each deformation of the snake, that is, for each variation of θ (and thus of Δ_a^{θ} and Δ_b^{θ}), one has to recompute the criterion $J(\mathbf{s}, \theta)$. The computational complexity of this step is the key factor which determines the overall segmentation time. In the next section, we present a method to efficiently perform the computation of $J(\mathbf{s}, \theta)$ so as to drastically decrease the global segmentation time.

5.2.2   Implementation of a fast statistic calculation

In this section, we will show that one can transform the double summation calculation of the statistics $\mathbf{T}(\chi_a)$ over the domain Δ_a^{θ} into a one-dimensional summation along the boundary δ_a^{θ} of Δ_a^{θ} and thus one can obtain an efficient optimization of the segmentation algorithm. In order to realize this transformation, one has to perform a simple precomputation before beginning the segmentation process. Then, during snake convergence, the heaviest task involved in the criterion calculation will be a one-dimensional summation. An efficient optimization is thus obtained [62].

5.2.2.1   Definition of the interior of the contour

Since the implementation of the snake requires summations over the surface of the object, it is necessary to specify how the interior of the object is defined [62, 101, 102]. The choice of this definition has important implications in the implementation of the snake and this is why the present section is devoted to this question.

Let us first define some notions that will be used in the following.

- The image plane is composed of points on a discrete grid whose coordinates are vectors of integer values and thus belong to $\mathbb{Z}^2$. The points of the contour belong to this grid (Figure 5.4.a). They form the *discrete polygonal contour* δ, which is a subset of $\mathbb{Z}^2$.

- From δ can be defined the *continuous lace* $\tilde{\delta}$ which is the curve of $\mathbb{R}^2$ joining the successive points of δ (Figure 5.4.b).

- Int $\left[\tilde{\delta}\right]$ will denote the strict interior of $\tilde{\delta}$ which is a surface of $\mathbb{R}^2$ (Figure 5.4.c).

The most straightforward way of defining the interior Δ of the contour consists in taking into account the pixels which are strictly inside $\tilde{\delta}$ plus the pixels on δ. This interior is defined formally by the following relation (see Figure 5.5.a):

$$\Delta = \left(\mathbb{Z}^2 \cap \text{Int}\left[\tilde{\delta}\right]\right) \cup \delta \qquad (5.27)$$

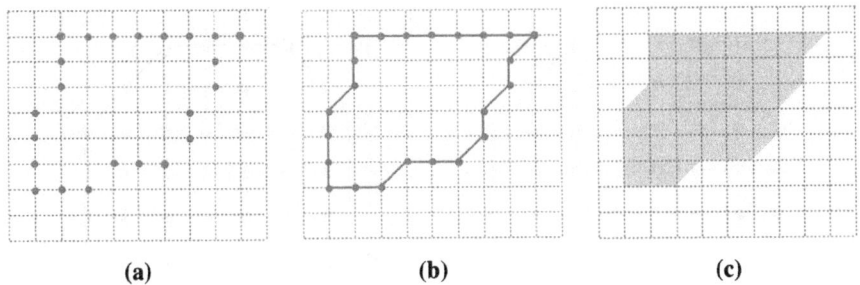

(a) (b) (c)

Figure 5.4. Notations used to define the interior of a discrete contour. (a) Discrete contour δ. (b) Corresponding continuous lace $\tilde{\delta}$. (c) Strict interior of the continuous lace Int $\left[\tilde{\delta}\right]$.

However, this convention presents some drawbacks in terms of implementation of the snake [62, 101]. First, it introduces some difficulties in the implementation of the fast algorithm that will be described in the next section. Second, it cannot be generalized to the partition of an image with an active "grid," as will be seen in Section 6.3. Indeed, it introduces a dissymmetry between two contiguous regions since the contour cannot simultaneously belong to each of them.

Another convention can be used to suppress this ambiguity on the pixels belonging to the discrete contour [101]. Let us consider again the object interior Δ delimited by the discrete contour δ and the corresponding continuous lace $\tilde{\delta}$. It is easily shown that the lace translated of a vector $(1/2, 1/4)$ has no intersection with the discrete grid $\mathbb{Z}^2$ [103]. Δ can thus be defined without ambiguity as the set of pixels belonging to the strict interior of the translated lace. Formally, this definition is written as:

$$\Delta = \mathbb{Z}^2 \cap \text{Int} \left[\mathcal{T}_{(\frac{1}{2},\frac{1}{4})} \left(\tilde{\delta} \right) \right] \tag{5.28}$$

where $\mathcal{T}_{(\frac{1}{2},\frac{1}{4})}$ represents the operation of translation along $(1/2, 1/4)$. This convention is equivalent to considering that the discrete contour is no longer located on the discrete grid $\mathbb{Z}^2$ but on a dual grid. The representation of this dual grid is obtained by translating the direct grid of $(1/2, 1/4)$. This convention differs from the straightforward one by the fact that all the points of the discrete contour δ do not belong to the interior Δ (see Figure 5.5.b).

5.2.2.2 Principle of fast algorithm

Equation 5.15 shows that the heaviest task involved in computing $J(\mathbf{s}, \boldsymbol{\theta})$ is to determine $\mathbf{T}(\chi_a)$. All the other operations are pointwise. In particular, $\mathbf{T}(\chi_b) = \mathbf{T}(\mathbf{s}) - \mathbf{T}(\chi_a)$ where $\mathbf{T}(\mathbf{s})$ is the sum, over all the pixels of the image, of $\beta[s(x, y)]$ ($\mathbf{T}(\mathbf{s})$ can be computed once for all before beginning

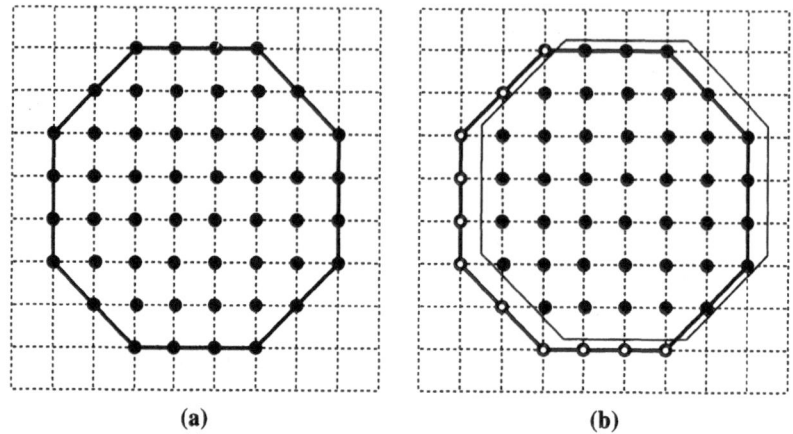

(a)   (b)

Figure 5.5. Comparison of the two conventions for defining the interior of a discrete contour (the interior pixels are represented in black). **(a)** Straightforward convention (Eq. 5.27). **(b)** Convention defined in Eq. 5.28.

the segmentation process). The computation of $\mathbf{T}(\chi_a)$, which is a double summation of $\beta[s(x,y)]$ over the domain Δ_a^{θ}, is time consuming. In order to obtain a fast algorithm, one has to find a method to calculate $\mathbf{T}(\chi_a)$ rapidly since it has to be recomputed at each iteration of the snake optimization process. The problem is thus to rapidly evaluate the sum:

$$\sum_{(x,y)\in\Delta_a^{\theta}} s^k(x,y) \text{ , for } k = 1,2 \tag{5.29}$$

Let us first consider the continuous case. The theorem of Green-Ostrogradsky transforms the integral of a scalar function over a surface Δ into a curvilinear integral over the contour of this surface:

$$\int\int_{\Delta} \text{div}\left(\vec{V}\right) \, dxdy = \int_{\delta} \vec{V}.\vec{n} \, dt \tag{5.30}$$

where $\vec{V}$ is a vectorial function, div the divergence operator, and $\vec{n}$ the normal to the surface at a point of curvilinear abscissa t on the contour δ (Figure 5.6). We can thus write:

$$\int\int_{(x,y)\in\Delta} s^k(x,y) \, dxdy = \int_{(x,y)\in\delta} V_k(x,y) \, \vec{e_x}.\vec{n} \, dt \tag{5.31}$$

where $\vec{e_x}$ is the unitary vector along axis x, and:

$$V_k(x,y) = \int_0^x s^k(u,y) \, du \tag{5.32}$$

This result can be generalized to the discrete case to establish the equivalence between the sum over a region and the sum over its contour [62]. For this purpose, one makes use of the discrete equivalent of (5.32):

$$F_k(x, y) = \sum_{t=0}^{x} s^k(t, y) \tag{5.33}$$

which represents the image "integrated along the lines." Let us give a more rigorous explanation of this algorithm by assuming that the region Δ is convex. This property insures that each line of ordinate $y_{min} \leq y \leq y_{max}$ cuts the contour δ in two points – which may possibly be identical – denoted $x_{min}(y)$ and $x_{max}(y)$ (Figure 5.6). The sum (5.29) can then be rewritten in the form:

$$\sum_{(x,y)\in\Delta} s^k(x, y) = \sum_{y=y_{min}}^{y_{max}} \sum_{x=x_{min}(y)}^{x_{max}(y)} s^k(x, y) \tag{5.34}$$

$$= \sum_{y=y_{min}}^{y_{max}} F_k[x_{max}(y), y] - F_k[x_{min}(y), y] \tag{5.35}$$

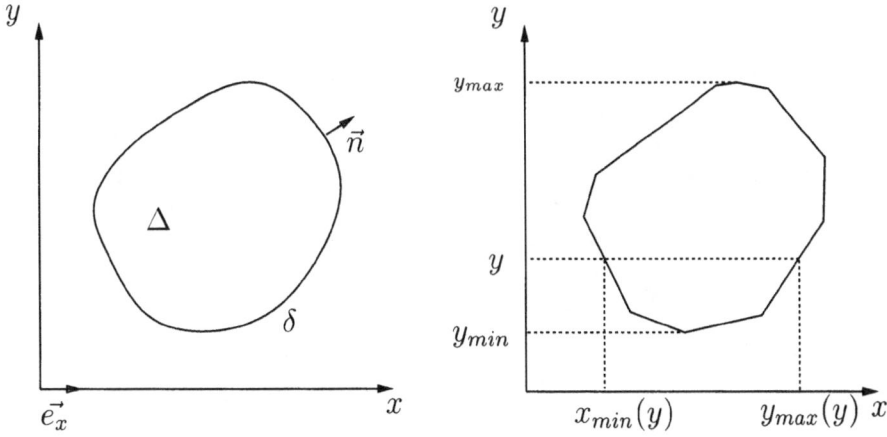

Figure 5.6. Transposition of the Green-Ostrogradsky theorem from the continuous to the discrete case.

The expression (5.35) finally appears as a linear combination of values of $F_k(x, y)$ evaluated at all the pixels of the contour:

$$\sum_{(x,y)\in\Delta} s^k(x, y) = \sum_{(x,y)\in\delta} c(x, y) F_k(x, y) \tag{5.36}$$

where $c(x, y)$ is a coefficient which can take the values 1, -1, or 0. It can be shown [62] that the relation (5.36) is in fact valid whether the region is convex or not. The main problem then consists in properly defining the coding $c(x, y)$ of the contour pixels, which is the aim of the next section.

To conclude, let us note that (5.36) can be compared to the relation obtained in the continuous case (Eq. 5.31). The coefficient $c(x, y)$ can be interpreted as the equivalent, in discrete space, of the term $\vec{e_x}.\vec{n}\, dt$ in the contour integral of Eq. 5.31.

5.2.2.3   Coding of the contour

The fast computation algorithm described in the previous section requires the coding of the contour pixels in order to be able to use Eq. 5.36. Figure 5.7 illustrates the principle of this coding.

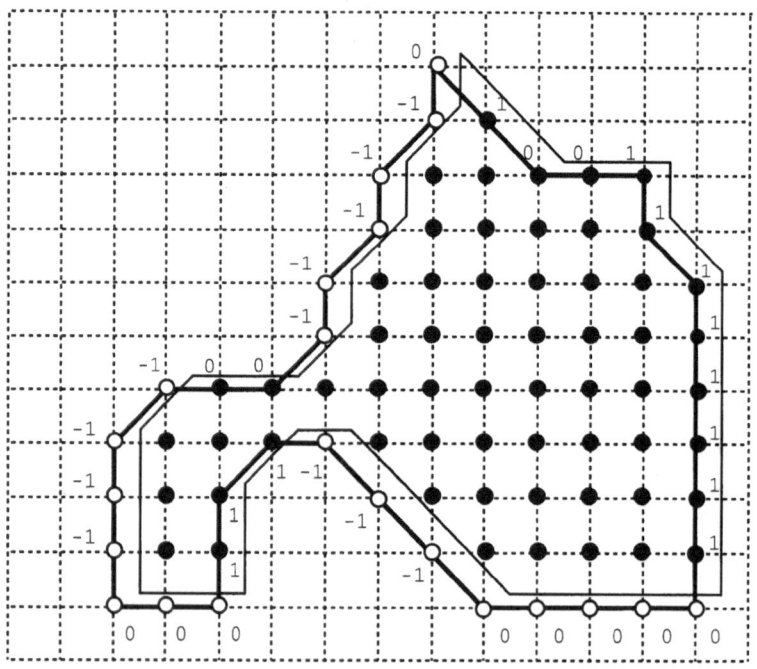

Figure 5.7. Example of coding $c(x, y)$ of the pixels of a discrete contour. $\tilde{\delta}$ is represented as the bold line and its translated version $\mathcal{T}_{(\frac{1}{2}, \frac{1}{4})}(\tilde{\delta})$ the thin line. The pixels belonging to the interior are represented in black.

In order to attribute a code to a pixel of the contour, let us consider the image row by row. On a given row, one identifies the *connex horizontal section*, that is, the connex sets of pixels belonging to the interior of the contour and with the same ordinate y. The pixel at the extreme right side of the horizontal section must belong to the contour. One attributes to it the code $c = 1$. Symmetrically, the pixel located immediately at the left of the pixel at the extreme left side of the horizontal section belongs to the contour – but not to the interior of the object, according to the convention described in Section 5.2.2.1. One affixes to this pixel the code $c = -1$. The code $c = 0$ is attributed to all the other pixels of the contour.

This coding agrees with the convention described in Section 5.2.2.1 to define the interior of a closed contour. For example, the pixels with code $c = -1$ have no contribution to the computed sum since they do not belong to the interior of the contour.

However, this coding rule is quite computation intensive. Indeed, it requires the knowledge of the pixels belonging to the interior of the contour in order to be able to determine the connex horizontal sections. This means that one would need to handle surfaces, which is precisely what was to be avoided by using the transform 5.36. Fortunately, it is possible to determine the code of a pixel locally, knowing simply the positions of its two neighbors on the contour [101]. Let us choose the positive trigonometric sense to go along the contour. With this choice, the interior of the surface always lies on the "left side" of the contour. There exist $8 \times 8 = 64$ possible configurations of three consecutive pixels on the contour, which are encoded by using a Freeman code (see Figure 5.8). The coding corresponding to each of these configurations is represented in Table 5.1.

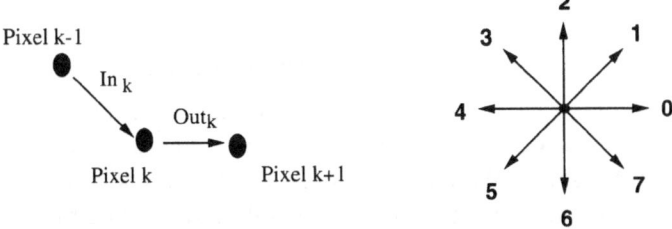

Figure 5.8.   Notation used to describe a configuration of three consecutive pixels on the contour (see Table 5.1). If one goes along the contour in the positive trigonometric sense, In_k and Out_k are the vectors formed by pixel k with its two neighbors: the previous one $(k - 1)$ and the next one $(k + 1)$. The directions of these vectors are encoded with the Freeman code (graph on the right side).

To conclude, the fast algorithm consists in computing at the beginning the images $F_k(x, y)$ with $k = 1, 2$ (Eq. 5.33) and using them to rapidly evaluate the new value of the criterion after each deformation of the contour. The evaluation requires one to go along the contour and to estimate at each pixel the value $c(x, y)F_k(x, y)$. The code $c(x, y)$ is directly determined along the contour by using Table 5.1. It is important to note that accelerating the algorithm in this

$\text{In}_k \setminus \text{Out}_k$	0	1	2	3	4	5	6	7
0	0	0	0	0	0	-1	-1	-1
1	1	1	1	1	1	0	0	0
2	1	1	1	1	1	0	0	0
3	1	1	1	1	1	0	0	0
4	0	0	0	0	0	-1	-1	-1
5	0	0	0	0	0	-1	-1	-1
6	0	0	0	0	0	-1	-1	-1
7	0	0	0	0	0	-1	-1	-1

Table 5.1. Coding table for the fast algorithm. When one goes along the discrete contour in the positive trigonometric sense, this table gives the coding $c(x, y)$ of a pixel (x, y) (Eq. 5.36), knowing the relative positions of the previous and of the next pixel on the contour (see Figure 5.8).

way does not imply any approximation in the computation of the statistics, and thus of the value of the criterion, which is determined in a rigorously exact way. However, this coding is valid only if the shape is simply connected, that is, if it is not composed of several disjoint pieces. Thus, during the snake convergence, it is of prime importance to forbid all intersections between parts of the snake. A solution to this problem exists for polygonal shapes [62] and is described in Appendix 5.A.

5.3.   Application to polygonal active contour

The region-based snake algorithm described in the previous section can be implemented with any contour description method, such as polygons, B-splines, Fourier descriptors, etc. We will focus in this section on polygonal description. This case is particularly interesting for several reasons. The first one is, obviously, the simplicity of its implementation. Moreover, a polygonal description is well adapted to describe manufactured objects (car, plane, etc.) and can also correctly segment smooth shapes, like organs in medical images, provided that a sufficient number of segments and an appropriate regularizing energy term are used. Another advantage of this description is the fact that, when a node of the polygon is moved, the deformation of the shape is local and concerns only the two segments which contain this node (Figure 5.9). Thus, the summation over δ_a^θ can be reduced to a summation only over the pixels which belong to these two segments, which further accelerates the algorithm.

We address in the next sections two important issues for the implementation of the polygonal snake: the regularization term and the optimization procedure. We then give some examples of application to images perturbed by different types of physical noises.

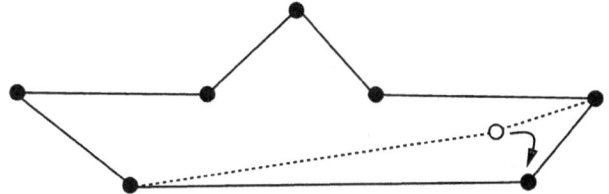

Figure 5.9. For a polygonal description, when a node of the polygon is moved, the deformation is local since it concerns only the two segments which contain this node. Then, in order to reduce the computation time of the segmentation algorithm, one can reduce the summation along the boundary to a summation only over the pixels which belong to these two segments. ©1999 IEEE.

5.3.1   Regularization of the contour

The criterion $J(\mathbf{s}, \boldsymbol{\theta})$ defines the "external energy" of the snake, that is, the way it depends on the data. However, as explained in Section 5.1, it is often necessary to define an "internal energy" for the snake. This energy term does not depends on the data, but only on the current shape of the snake. It is used to impose constraints on the snake, in general to enforce smoothness of the contour. In the framework of statistical decision theory, such a regularizing term is very naturally introduced as a prior distribution of the shape parameter vector $\boldsymbol{\theta}$. Different types of priors can be used with various mathematical expressions (see Ref. 104 and references therein for details). Our purpose here is not to study precisely the influence of the choice of the prior, but to show that such a term can be easily introduced. We will thus consider in the following a simple regularization energy analogous to the elastic energy employed in the classical snake [83, 104]:

$$P(\boldsymbol{\theta}) = A \, \exp\left\{-\frac{1}{2\varphi^2} U_{in}(\boldsymbol{\theta})\right\} \tag{5.37}$$

where

$$U_{in}(\boldsymbol{\theta}) = \sum_{i=1}^{N_{node}} d_i^{\,2} \tag{5.38}$$

and where d_i is the distance between the node number i (N_i) and the center of the segment defined by the nodes N_{i-1} and N_{i+1}. N_{node} is the number of nodes of the polygon, A a normalization constant, and φ a parameter of the prior. It is interesting to note that this prior penalizes situations in which nodes are not aligned.

In order to take into account the prior $P(\boldsymbol{\theta})$, one chooses the shape parameter $\boldsymbol{\theta}$ which maximizes the posterior distribution $P(\mathbf{s}|\boldsymbol{\theta})P(\boldsymbol{\theta})$, which corresponds to a MAP approach. $P(\mathbf{s}|\boldsymbol{\theta})$ is the likelihood function of the SIR model defined in Eq. 5.15. The maximization of the MAP function thus leads to the minimization of the following criterion:

$$J'(\mathbf{s}, \lambda, \boldsymbol{\theta}) \; = \; (1 - \lambda) \, J(\mathbf{s}, \boldsymbol{\theta}) \; + \; \lambda \, U_{in}(\boldsymbol{\theta}) \qquad (5.39)$$

where $J(\mathbf{s}, \boldsymbol{\theta})$ is the optimal criterion defined in Section 5.2.1 and λ a parameter which balances the internal and the external energies.

5.3.2   Minimization procedure

Once the criterion $J'(\mathbf{s}, \lambda, \boldsymbol{\theta})$ has been specified on the basis of the statistical properties of the scene (external energy) and of the required properties of the contour itself (internal energy), one has to determine the parameter $\widehat{\boldsymbol{\theta}}$ which minimizes it. This minimization is difficult since $\boldsymbol{\theta}$ lies in a high dimensional space and $J'(\mathbf{s}, \lambda, \boldsymbol{\theta})$ usually has many local minima. There exist many algorithms which are designed to find the global minimum in such problems, such as simulated annealing [105]. These algorithms could be useful in our case, however, they tend to be very time consuming. We will thus consider here simpler optimization techniques which are faster but may find local minima. Another reason for this choice is that when segmenting an object in a complex scene, it is likely that the global minimum of $J'(\mathbf{s}, \lambda, \boldsymbol{\theta})$ will not correspond to the correct segmentation of the object of interest, if many other objects are present in the scene. On the other hand, if the initial shape of the snake is close enough to the object, the local minimum will correspond to the correct segmentation. In cases where the initial shape is further from the true one, we will show later how a simple multiresolution scheme makes it possible to skip over the first spurious local minima.

We will consider two different minimization approaches. The first one is a stochastic optimization algorithm, which can be considered as a simulated annealing at zero temperature, or a Monte Carlo method (see Figure 5.10). It consists of randomly choosing a node of the snake and moving it in a random way inside a square-shaped neighborhood. This move is accepted if it decreases the criterion $J'(\mathbf{s}, \lambda, \boldsymbol{\theta})$ and refused otherwise. The parameters of this algorithm are the maximum amplitude of the deformations and the number of iterations (number of tested node moves).

The second algorithm is deterministic (see Figure 5.11). For each node of the snake, one searches exhaustively, within a given range, the position of the node which minimizes the criterion. This process is iterated until no node moves. This method has the advantage of yielding a very simple stopping criterion. However, it appears to be more sensitive to snake initialization than the stochastic algorithm.

During the optimization process, it may happen that the snake gets trapped in a local minimum corresponding to a wrong segmentation. This problem gets more serious as the initial snake shape is further from the true object. In order to

- *Step 1*   Initialization. One chooses a vector of parameters $\boldsymbol{\theta}$ which describes the initial contour.

- *Step 2*   Test a local deformation of the snake:

  - Randomly choose one of the k nodes of the snake (each node has the same probability of being chosen).

  - Move the chosen node from its initial position (x, y) to its new position $(x + \delta_x, y + \delta_y)$ where δ_x and δ_y are randomly chosen, uniformly in the range $\{-a, \ldots, 0, \ldots, a\}$. The new parameter set corresponding to this deformation is $\boldsymbol{\theta}'$.

- *Step 3*   Compute the new criterion value $J'(\mathbf{s}, \boldsymbol{\theta}', \lambda)$.

- *Step 4*

  - Validate the deformation if it decreases the value of the criterion $J'(\mathbf{s}, \boldsymbol{\theta}', \lambda) < J'(\mathbf{s}, \boldsymbol{\theta}, \lambda)$.

  - Do not take it into account otherwise.

- *Step 5*   Return to step 2.
                    This process is iterated N_{iter} times.

Figure 5.10.   Stochastic minimization algorithm.

solve the problem, a simple multiresolution method can be implemented [61], which consists in progressively increasing the number of nodes (see Figure 5.12) and decreasing the amplitude a of the potential deformations. In other words, the segmentation starts with a polygonal contour having a limited number of nodes (4 or 5), a large value of a and without regularization ($\lambda = 0$). A rough segmentation, with a reduced number of nodes, can thus be obtained. A certain number of nodes are then added to the obtained contour and this modified contour is used as the initial shape for a second segmentation process with a smaller value of a, and so on. In practice, the final segmentation can be obtained after two or three steps. An important characteristic of the algoriihm is that the number of nodes added depends on the resolution of the contour which is needed on the final image [106]. A possible solution consists of adding n new nodes between two existing nodes if their distance D is greater than a given value d. n is the integer part of D/d and d can thus be viewed as the resolution of the final snake.

- *Step 1*   Initialization. One chooses a vector of parameters $\boldsymbol{\theta}$ which describes the initial contour.

- *Step 2*   Go through all nodes.

  - For each node of the polygon (k varies from 1 to N_{nodes}).

    * For each value of (δ_x, δ_y) belonging in
      $\{(-a, -a), (-a, 0), (-a, a), (0, -a), (0, a), (a, -a), (a, 0), (a, a)\}$
      · Move node k from its initial position (x, y) to position $(x + \delta_x, y + \delta_y)$. The parameter value corresponding to this deformation is $\boldsymbol{\theta}_k$.
      · Compute the criterion value $J'(\mathbf{s}, \boldsymbol{\theta}_k, \lambda)$.

  - Validate the deformation which decreases more $J'(\boldsymbol{\theta}_k, \lambda)$.

  - Do not take the deformations into account otherwise.

- *Step 3*   Return to step 2 as long as at least one deformation has been validated.

Figure 5.11.   Deterministic algorithm.

5.3.3   Discussion

Let us discuss the main characteristics of the snake approach introduced in this section:

- Since it is based on a statistical model of image perturbations, it can be adapted to different types of imaging systems.

- Since it is a region-based approach, it can be robust to noise and to snake initialization.

- It does not require the knowledge of the pdf parameters of the image. These parameters are reestimated at each iteration of the snake. Contrary to some other approaches, the snake evolution and the parameter estimation are not performed in separate steps, but simultaneously. This ensures that all the available image information is used at each snake iteration.

- If the pdf describing the graylevels of the scene belong to the exponential family, the snake can be efficiently implemented so as to provide a fast segmentation method. Moreover, polygonal modeling of the shape permits further acceleration of the algorithm, since only the two segments connected to the moved node have to be considered for target and background statistics update at each iteration.

Of course, this approach has also some drawbacks, the main of which are:

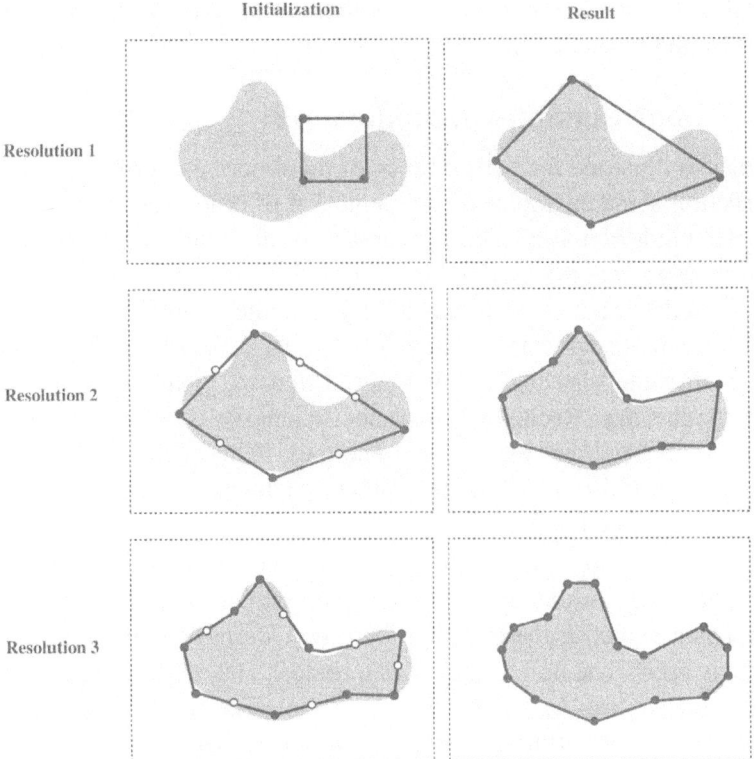

Figure 5.12. Multiresolution technique used to improve the convergence of the snake. Segmentation is performed in three steps, which correspond to increasing resolutions. After each step, the number of nodes is increased in order to obtain a finer estimation of the object shape.

- The topology is limited to a single simply connected object. We will see in Section 6.3 how the SIR model can be generalized to address multiregion segmentation applications. If the topology of the object changes during time, e.g., it separates in two parts, handling such topological change is in principle possible, but the solution is not as efficient as that proposed by level set implementations (see Section 5.1.4).

- The number of nodes of the polygonal contour must be fixed by the user. Since the optimal number of nodes necessary to correctly segment the object is not known beforehand, this often leads to using an excessive number. We will see in Section 6.2 how an optimal number of nodes can be determined by using information theoretic concepts.

- The pixels are supposed statistically independent, which means that the model does not account for spatially correlated textures. If such textures are present in the image, we will show in the next section that, as in the case of detection (see Section 4.2.1), a whitening preprocessing can be used to improve the results.

5.3.4   Some examples of application

In order to illustrate the performance of the described approach, we present in this section some examples of segmentation of objects in images perturbed by different kinds of noises. The segmentations are obtained with the stochastic minimization procedure described in Section 5.3.2 in a three-step process. The number of nodes is increased progressively (the internode distance is $d = 20$ at the end of the first segmentation step and $d = 10$ at the end of the second one) and the coefficient λ is equal to 0 in the first step, to 0.1 in the second one, and to 0.2 in the last one. Recall that λ is a coefficient which allows one to balance between the internal and the external energies. The internal energy imposes smoothness constraints and penalizes situations in which one node moves away from others (see Section 5.3.1).

In Figure 5.13, we consider images perturbed by three different types of noises. The image in the first row is an echographic image of a kidney. The noise pdf is assumed to be Rayleigh and the external energy $J_{ray}(\mathbf{s}, \boldsymbol{\theta})$ (see Eq. 5.22) is used. The initial shape has 5 nodes. The image in the second row is binary and we use $J_{ber}(\mathbf{s}, \boldsymbol{\theta})$ (see Eq. 5.18). The third image is synthetic and represents a car perturbed with Poisson noise: The snake energy adapted to this pdf, $J_{poi}(\mathbf{s}, \boldsymbol{\theta})$ (see Eq. 5.23), has thus been used. For these two images, the snake is initialized with a 4-node polygon. We show in the right column of Figure 5.13 the result of the segmentation of these three images.

Let us now consider the images in Figure 5.14. We have seen in Section 4.2.1 that such images are not well represented by a white (i.e., spatially uncorrelated) random field. A whitening preprocessing has thus been applied, which approximately transforms such correlated scenes into images which are well modeled with white Gaussian random fields. In other words, instead of $\mathbf{s}$, the segmentation is achieved on a preprocessed image $\mathbf{z}$ with a quasi white spectrum ($\mathbf{z}$ is defined as the inverse Fourier transform of $\dfrac{\tilde{s}(k_x, k_y)}{\sqrt{\epsilon + |\tilde{s}(k_x, k_y)|^2}}$, where ϵ is a small positive real number). This preprocessing has been applied to the scenes in Figure 5.14, the graylevels of each scene have been rescaled so that they lie between 0 and 1, and ϵ has been set to 10^{-6}. Whitened versions of the central parts of the two last scenes in Figure 5.14 are shown in Figure 5.15. Finally, the snake with the external energy adapted to Gaussian pdf $J_{gau}(\mathbf{s}, \boldsymbol{\theta})$ (see Eq. 5.20) has been applied to these whitened images. The results of the segmentations are shown in the right column of Figure 5.14.

Finally, in order to illustrate the advantage of using a region-based approach on these images rather than an edge-based technique, we show in Figure 5.16

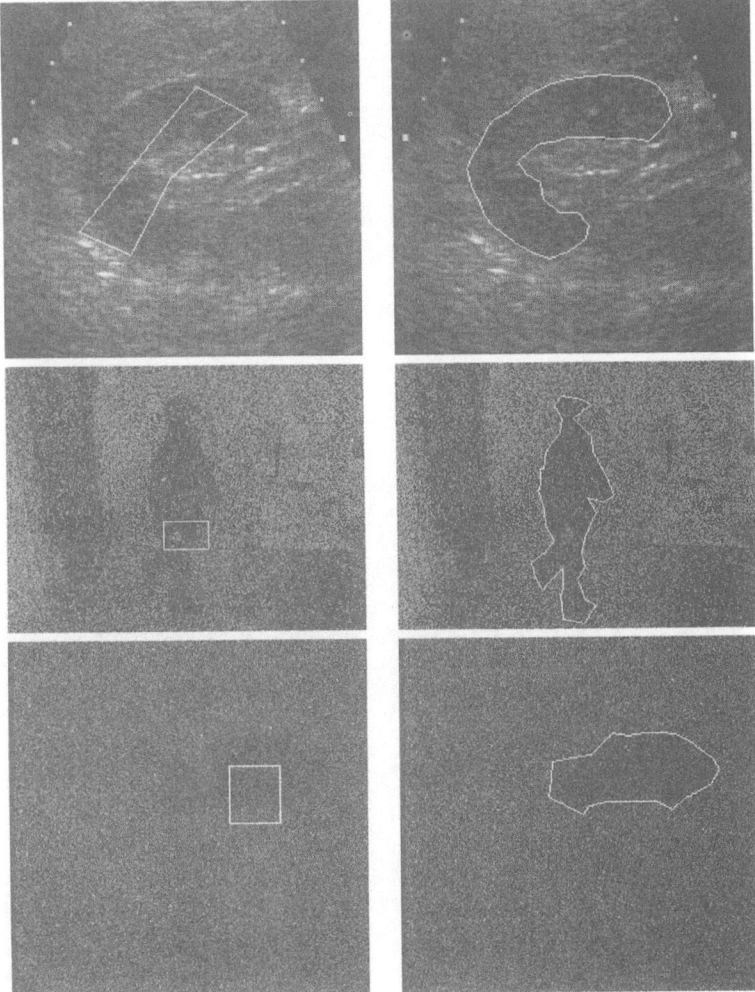

Figure 5.13. Left column: initialization of the snake in different images. Right column: final state of the snake after optimization of $J'_{ray}(\mathbf{s}, \lambda, \boldsymbol{\theta})$ for the first image, which is an echographic image of a kidney, $J'_{ber}(\mathbf{s}, \lambda, \boldsymbol{\theta})$ for the second image which is binary, and $J'_{poi}(\mathbf{s}, \lambda, \boldsymbol{\theta})$ for the last image which is perturbed with Poisson noise. ©1999 IEEE.

edge-enhanced versions of the airplane image of Figure 5.14 and the Poisson noise-perturbed image of Figure 5.13. The Canny-Deriche filter has been used for edge enhancement [107]. It is clear that the object borders have no sufficient contrast, compared to the background clutter, for an edge-based technique to yield a satisfactory result.

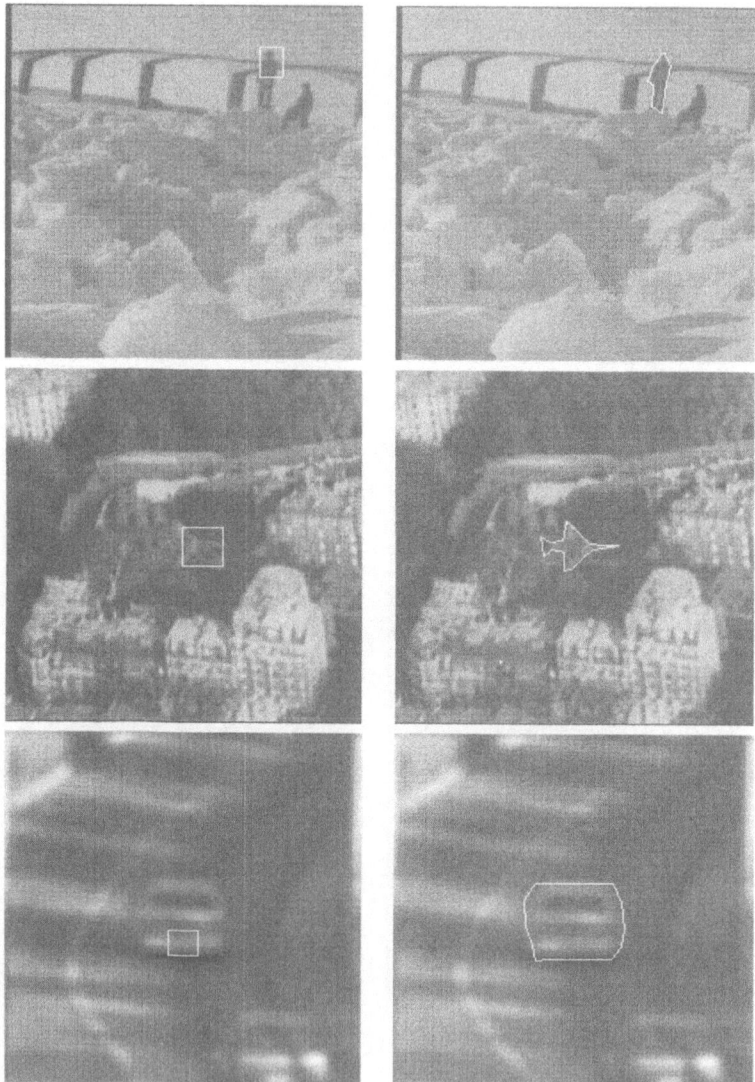

Figure 5.14.   Left column: initialization of the snake for different images. Right column: final state of the snake after optimization of $J'_{gau}(\mathbf{s}, \lambda, \boldsymbol{\theta})$ on the whitened version of the input image. ©1999 IEEE.

The images processed above have a dimension of 256×256 pixels. The total computation time for all of them is less than 200 ms on a 700-MHz PC. With the conventional algorithm implementation (the approach which consists in computing the sufficient statistics with summations over the region $\Delta_a^{\boldsymbol{\theta}}$),

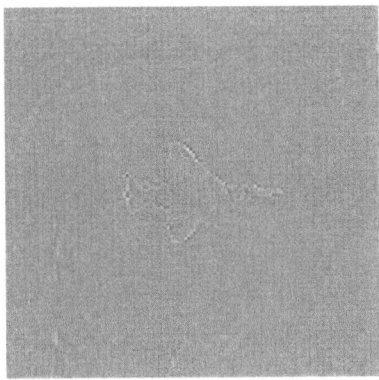

Figure 5.15. Left to right: whitened version of the central parts of the second and third images of Figure 5.14. ©1999 IEEE.

the computation time is larger than 80 s (for each image). This huge gain in computation time (a factor of almost 400) is possible because the graylevel pdf have been assumed to belong to the exponential family. It is with this assumption that the summations over the region Δ_a^{θ} can be converted into summations over its contour δ_a^{θ}.

Of course, there exist image formation processes in which this assumption is not realistic. We will see an example of this situation when we study polarimetric imaging in Chapter 7. We will then show that rather than considering the true ML-based external energy, which would be time consuming, it is preferable to apply a transformation which makes the pdf of the transformed image similar to the Gaussian pdf. Then, the fast implementation of the snake algorithm $J_{gau}(\mathbf{s}, \boldsymbol{\theta})$ (see Eq. 5.18) can be applied efficiently to the transformed image.

In the following sections, further examples of applications of the polygonal statistical snake method to object tracking in video sequences and to edge estimation refinement in SAR images are presented.

5.4.   Applications to tracking in video sequences

We now illustrate how the location and snake-based segmentation approaches based on the SIR image model can be used jointly to achieve target tracking on image sequences, even if the shape of the target changes during the sequence. This process is illustrated in Figure 5.17. Assume that on the image acquired at time $t - 1$, the target has been segmented using the described snake method (Figure 5.17.a). This segmentation produces a binary reference shape which can be used to locate the target in the image acquired at time t (Figure 5.17.b), for example with the GLRT algorithm. This is possible since the GLRT or other correlation-based location algorithms are robust to limited deformation of the

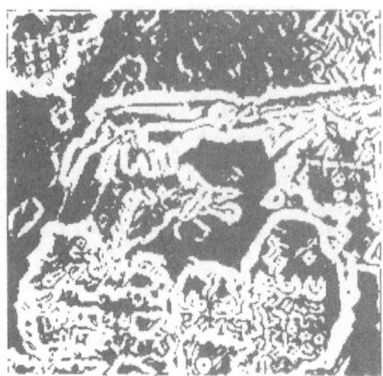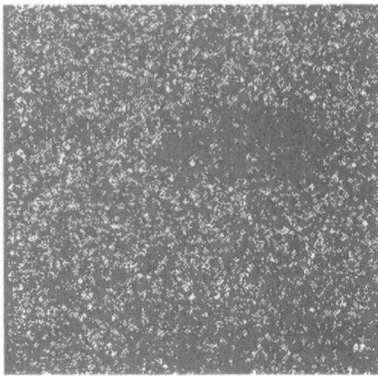

Figure 5.16. Left to right: results of processing the airplane images of Figure 5.14 and the Poisson noise-perturbed image of Figure 5.13 with a Canny-Deriche filter. ©1999 IEEE.

target with respect to the reference object. They can thus locate the target even if its shape has slightly changed compared to the previous frame. Note that in principle, the GLRT is equivalent to the snake algorithm with the parameter θ restrained to a global translation of the target.

When the object has been roughly located in the new frame, one can apply a global registering, which consists in jointly optimizing the orientation and the position of the reference so as to best fit the object (Figure 5.17.c). Note that this step consists in considering a parameter θ limited to translations and to rotations, but the range of considered translations is smaller here than during the location step. After global registering, one obtains a shape which is close to the object and thus provides a good initialization for snake segmentation (Figure 5.17.d). This process is repeated until the end of the sequence. In summary, using jointly location and snake algorithms consists in first determining the object location (which corresponds to very constrained variation of the shape), then in estimating its attitude and finally its precise shape with the snake algorithm. In many instances, the shape variations from one image to the next are small and only few snake iterations are needed to converge to the new shape.

In order to illustrate this approach, we will address in the following sections the problem of tracking objects in video sequences. We will successively consider the case where the camera is fixed or only slightly vibrating, and the case where it is moving.

5.4.1   Fixed camera

As an example of case where the camera is not moving during the acquisition of a video sequence, let us first consider tracking vehicles on a highway, which can be illustrative of a traffic surveillance application. This example will also

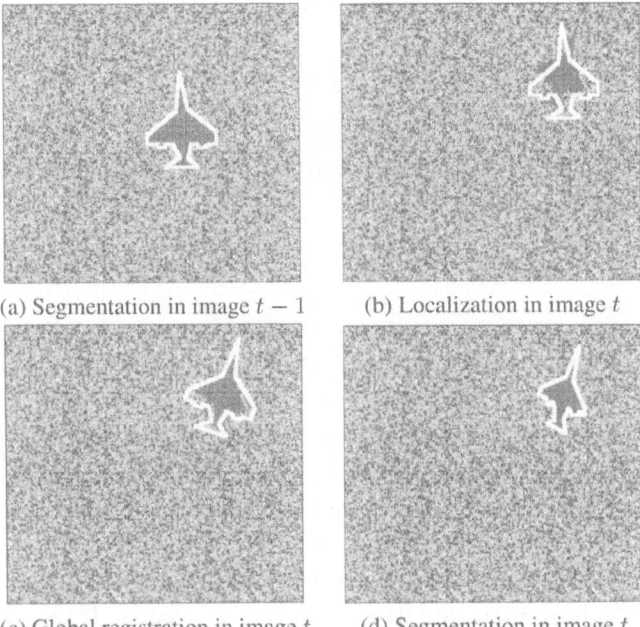

(a) Segmentation in image $t-1$ (b) Localization in image t

(c) Global registration in image t (d) Segmentation in image t

Figure 5.17. Principle of an automatic tracking system based on joint utilization of segmentation, location, and global registration algorithms. On image $t-1$, the object is segmented (a). The segmented shape is used to locate the target at time t (b) . The reference shape is rotated and finely translated so as to best fit the object in the new image (c) and provide a good shape initialization to the segmentation algorithm, which refines the estimation of the new shape (d). These four steps are iterated along the sequence.

give us the opportunity to introduce some preprocessing techniques which are often necessary to obtain correct location and segmentation results in practical applications.

A typical image is shown in Figure 5.18.b. We can note that in this scene, the natural background is very complex and nonhomogeneous. This can induce problems on the snake segmentation algorithm, which is based on the simple hypothesis of homogeneous object and background regions. However, in the application which is considered here, the camera observing the scene is fixed and the background does not move. A reference image of the road without vehicle (see Figure 5.18.a) can thus be acquired and subtracted to the subsequent frames. However, in traffic surveillance applications, the camera is submitted to vibrations and thus always moves a little from one frame to the other. Consequently, the frames have to be registered before subtraction. We will assume that the vibrations of the camera are small enough so that the movements of the images can be considered as translations. The translation vector τ is estimated by simple correlation of the reference image $r(x, y)$ with the current image $s(x, y)$:

$$\hat{\tau} = \arg \max_{\tau} \sum_{x,y} s(x, y)\, r(x + \tau_x, y + \tau_y) \qquad (5.40)$$

The image $s(x, y)$ is then subtracted to the registered reference to yield the following difference image:

$$d(x, y) = s(x, y) - r(x + \hat{\tau}_x, y + \hat{\tau}_y) \qquad (5.41)$$

An example of image $d(x, y)$ is represented in Figure 5.18.c. It must be noted that this subtraction approach can be applied only if the background graylevels in the current frame are similar to those in the reference image. This may not be the case, for example when outdoor scene illumination varies during the day. In this case, one must periodically refresh the reference image, or define it as the moving average of images taken over a long period, so as to average out the effects of rapid events such as driving cars, etc.

One can see in Figure 5.18.c that most details of the background have disappeared and that the vehicle and its shadow appear clearly. If the segmentation algorithm was applied to this image, it would segment the vehicle and also its shadow. To get rid of the latter, one can use the whitening preprocessing introduced in Section 4.2.1. However, this processing requires computing the Fourier transform of the image, which may be too time consuming for some applications. We can notice that the whitening operation defined in Eq. 4.30 consists in convolving the image with a "whitening filter," which is adaptive in the sense that it depends of the image itself, and which has an infinite impulse response (IIR). A solution would consist in approximating it with a nonadaptive filter having finite impulse response (FIR). It can be shown [108] that the effect of the whitening filter can be well approximated by the following 3×3 FIR filter:

$$\mathbf{h}_\gamma = \frac{1}{8} \begin{bmatrix} \gamma - 1 & \gamma - 1 & \gamma - 1 \\ \gamma - 1 & 8 & \gamma - 1 \\ \gamma - 1 & \gamma - 1 & \gamma - 1 \end{bmatrix} \qquad (5.42)$$

When $\gamma \simeq 0$, $\mathbf{h}_\gamma$ is close to a Laplacian filter, which is high-pass. Such a filter will thus suppress uniform areas such as shadows. In the following applications, γ will be fixed empirically to 0.1. The result of the whitening operation appears in Figure 5.18.d.

Figure 5.19 represents three images belonging to a traffic sequence. The first row represents the first image of the sequence, for which no *a priori* knowledge of the shape of the vehicle is taken into account. The snake is thus initialized as a square partially overlapping the vehicle and the multiresolution snake segmentation scheme described in the previous section is used. The energy criterion is $J'_{gau}(\mathbf{s}, \lambda, \boldsymbol{\theta})$, which is adapted to Gaussian pdf and contains a shape

Figure 5.18.   **(a)** Reference image $r(x, y)$. **(b)** First image of the sequence. **(c)** Difference image $d(x, y)$. **(d)** Whitened difference image.

regularization term. Segmentation is performed in three steps and the number of nodes is increased so that the internode distance is set to $d = 20$ pixels at the beginning of the second step and to $d = 10$ at the beginning of the third one. The regularisation coefficient λ is set to 0 in the first step (no regularization), 0.1 in the second and 0.2 in the third. In the first and second steps, the stochastic optimization algorithm is used with a deformation amplitude $a = 3$

and numbers of iterations respectively equal to 500 and 1000. In the third step, the deterministic algorithm is used with a deformation amplitude $a = 2$. We can see the result in the first row of Figure 5.19. In the subsequent frames, the object is located and the shape is initialized as a rectangle centered on this position. The shape is then segmented with the deterministic algorithm using a deformation amplitude $a = 2$. The results obtained on some subsequent images of the sequence are shown in Figure 5.19. On all the images of the sequence, the computation time corresponding to the segmentation step was less than 60 ms on a 700-MHz PC.

Another example is shown in Figure 5.20, where a walking person is tracked. We can note that due to the walking movement, the apparent shape of the person changes during the sequence. Moreover, the background is complex and we have also subtracted the current frame to a reference frame after registration. The tracking algorithm is similar to that used for tracking cars on the highway, the only difference being that on frame t, the initial shape of the snake is taken as the final shape obtained at $t - 1$.

5.4.2   Moving camera

In practice, there are cases where the background of the scene is also evolving from frame to frame. This is in particular the case if the camera observing the scene is moving. In such cases, it is not always possible to register the background so as to substract it out from the scene. Basic solutions to get rid of a complex background are to limit the region of interest to a small part of the image around the target and to apply a whitening preprocessing. A further possibility is to use shape consistency, which amounts to assuming that the shape of the object does not change a lot from one frame to the next. Applying these techniques, it may be possible to track objects with low contrast on very complex backgrounds.

Introduction of shape consistency in the statistical snake scheme can be done very naturally by considering a prior density on the current shape of the object which depends on the previous one. Let us denote $\boldsymbol{\theta}^t = \{(x_k^t, y_k^t) \mid k \in [1, N_{nodes}]\}$ the shape parameters (i.e., the positions of the nodes) on the image acquired at time t. We assume that the statistical distribution of $\boldsymbol{\theta}^t$ depends on the value of $\boldsymbol{\theta}^{t-1}$ in the following way:

$$P\left(\boldsymbol{\theta}^t | \boldsymbol{\theta}^{t-1}\right) = A \, \exp\left\{-\frac{1}{2\xi^2} \sum_k^{N_{nodes}} \left[(x_k^t - x_k^{t-1})^2 + (y_k^t - y_k^{t-1})^2\right]\right\}$$

$$(5.43)$$

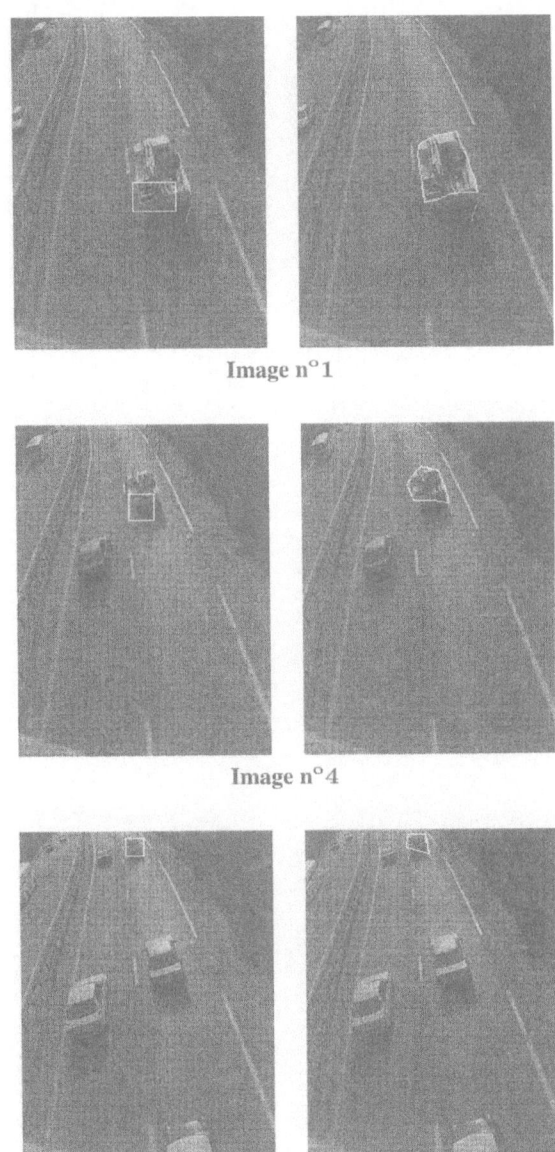

Image n°1

Image n°4

Image n°9

Figure 5.19. Example of tracking on an image sequence. The sequence is composed of 9 frames, and frames 1, 4 and 9 are shown. Left column: image with initial snake. This initial snake is a square approximately centered on the object. Right column: image with snake after convergence. The GLRT and the snake are both adapted to Gaussian pdf. ©1999 SPIE.

Image n°1

Image n°4

Image n°9

Figure 5.20. Example of tracking on an image sequence. The sequence is composed of 9 frames, and frames 1, 4 and 9 are shown. Left column: image with initial snake. This initial snake is the shape segmented in the previous image translated to the estimated object location obtained with the GLRT. For image $n°1$, the initial shape is a square. Right column: image with snake after convergence. The GLRT and the snake are both adapted to Gaussian pdf. ©1999 SPIE.

This distribution is used as a regularizing prior for the snake. The optimized criterion is thus:

$$J''(\mathbf{s}, \eta, \boldsymbol{\theta}) = (1 - \eta) J_{gau}(\mathbf{s}, \boldsymbol{\theta}) + \eta \sum_{k}^{N_{nodes}} \left[(x_k^t - x_k^{t-1})^2 + (y_k^t - y_k^{t-1})^2 \right]$$

(5.44)

η is a scalar coefficient which balances the two terms. In the following example, the first image in the sequence is segmented with the method used in the traffic sequence (see Figure 5.19). In the subsequent frames, a region of interest is extracted around the target position estimated in the previous image, the Laplacian whitening filter of Eq. 5.42 is applied and the shape is segmented by optimizing $J''(\mathbf{s}, \eta, \boldsymbol{\theta})$ with the deterministic convergence algorithm (see Figure 5.11) and a deformation amplitude $a = 2$. The tracking results on a particular sequence are shown in Figure 5.21. The images are synthetic and represent an airplane flying over a urban area. Images 1, 4 and 24 of the sequence are shown and it can be seen that the airplane is correctly tracked even when it appears with a very low contrast over the background.

5.5. Application to accuracy improvement of edge location

In Section 4.4, we have described an edge detector adapted to SAR images perturbed with Gamma noise. We have shown that it is quite efficient in detecting the presence of edges, but it provides edge location with a bias. This limitation can be overcome with a deformable contour that precisely matches the searched edges. For this purpose, one can use a region-based active contour working on the original image (and not on the edge strength map) to refine the accuracy of edge location [101].

In the presence of noise with Gamma pdf, the expression of the criterion to optimize for snake segmentation is (see Eq. 5.21):

$$J(\mathbf{s}, \boldsymbol{\theta}) = N_a(\boldsymbol{\theta}) L \, \log \widehat{\mu_a}(\boldsymbol{\theta}) + N_b(\boldsymbol{\theta}) L \, \log \widehat{\mu_b}(\boldsymbol{\theta}) \qquad (5.45)$$

where $\widehat{\mu_a}(\boldsymbol{\theta})$ and $\widehat{\mu_b}(\boldsymbol{\theta})$ are the estimates of the graylevel averages in each region.

We present some segmentation examples obtained with active contours on SAR images, where this technique is used for refining the estimation of the object borders. For a given scene, edges are first located by application of the GLRT-based edge detector and then extracted with the watershed algorithm [81] in order to provide continuous borders. The result is then "sampled" (i.e., approximated by a polygon) to initialize the active contour: one goes along the contour and sets a new node each time a given distance t has been covered. As a result, a polygonal contour with C/t equally-spaced nodes (where C is the length of the contour) is obtained. The statistical active contour is then used to refine the edge location. Figure 5.22 illustrates that the segmentation accuracy

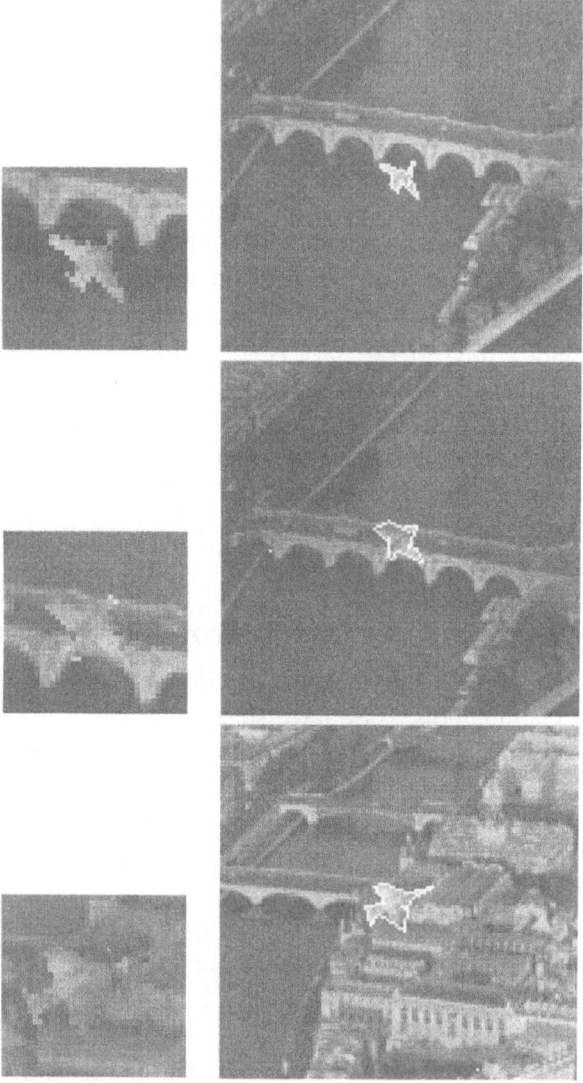

Figure 5.21. Example of tracking an airplane flying above a city (synthetic images). The sequence consists of 32 images and images 1, 4 and 24 are shown here. First column: region of interest in the image. Second column: Segmented shape after convergence of the snake.

can be significantly improved. This accuracy can be quantified by computing the number of misclassified pixels (NMP), which is the number of pixels which belong to the target region but are outside the estimated contour, plus the number of pixels which belong to the background but which lie inside the estimated contour. On this example, the NMP decreases from 421 in Figure 5.22.b to 147 in Figure 5.22.d

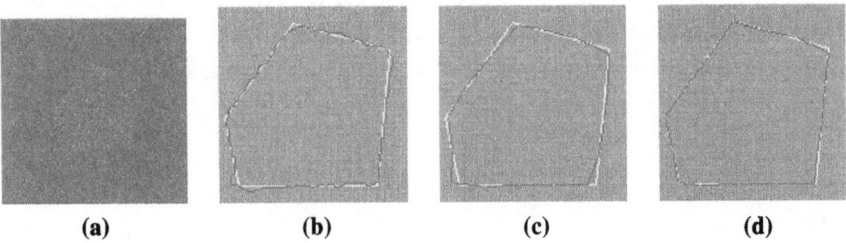

(a)   (b)   (c)   (d)

Figure 5.22. **(a)** Synthetic monolook speckled image. The contrast is $\rho = 1.7$ (see Eq. 4.69). **(b)** Result of GLRT edge detection on (a) with 8 differently-oriented 18×18 windows and edge extraction performed with the watershed algorithm. **(c)** Initial active contour (16 node polygon) obtained by sampling (b). **(d)** Final contour estimation given by the active contour ($\lambda = 0.02$). The estimated edges (black) are superimposed on the true ones (white). ©2001 IEEE.

It is easy to adapt this technique in order to precisely locate any edge in the scene. For this purpose, we consider the domain defined by a window centered on the detected edge. Within this window, the two-region SIR model is assumed to be valid, in other words, a unique edge is separating two homogeneous regions. It is therefore possible to refine the edge position within the window. Figure 5.23 illustrates such a case. In this example, a sinuous edge separates two domains and the strong radiometric contrast yields a bias when the segmentation is performed with the GLRT. The active contour permits, here again, to eliminate this bias and achieve better edge location since the NMP passes from 247 (fig. 5.23.b) to 133 (fig. 5.23.d).

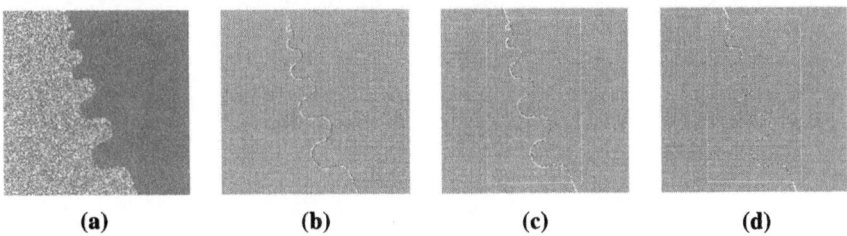

(a)   (b)   (c)   (d)

Figure 5.23. **(a)** Synthetic monolook speckled image. The contrast is $\rho = 7$. **(b)** Result of GLRT edge detection on (a) with 8 differently-oriented 10×10 windows and edge extraction performed with the watershed algorithm. **(c)** Initial active contour (56-node polygon) obtained by sampling (b). **(d)** Final contour estimation given by the active contour ($\lambda = 0.01$). The edge location is refined inside the window defined by the bright-gray rectangle. The estimated edges (black) are superimposed on the real ones (white). ©2001 IEEE.

This approach has also been applied to an intensity monolook SAR image acquired by the satellite ERS-1. Two extracts of this image are processed using the same parameters. On Figure 5.24, a lake appears with a high contrast. The estimate of contour location provided by the GLRT edge detector is thus biased and the active contour permits one to refine the edge location. On Figure 5.25, the aim was to regularize the frontier of a field which appears with a much lower contrast.

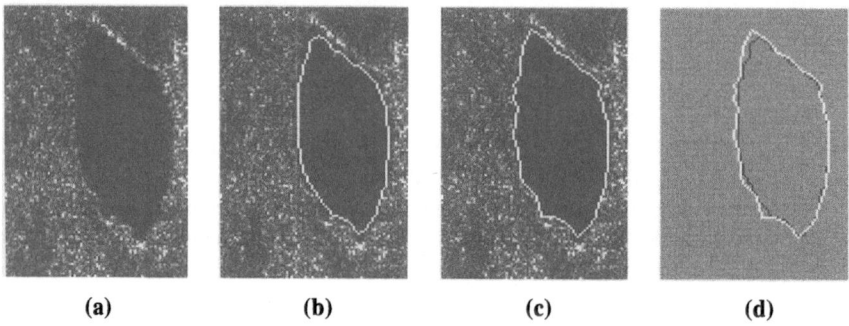

(a)   (b)   (c)   (d)

Figure 5.24. **(a)** Extract of an ERS-1 monolook SAR image of a region in Ukraine, provided by the CNES and delivered by ESA. **(b)** GLRT filtering of (a) with 8 differently-oriented 10×10 windows. Edge extraction is performed with the watershed algorithm. **(c)** Final contour estimation given by the active contour (47-node polygon, $\lambda = 0.20$). The initialisation is obtained by sampling the contour (b) every 4 pixels. **(d)** Comparison of contours in (b) (black) and (c) (white). ©2001 IEEE.

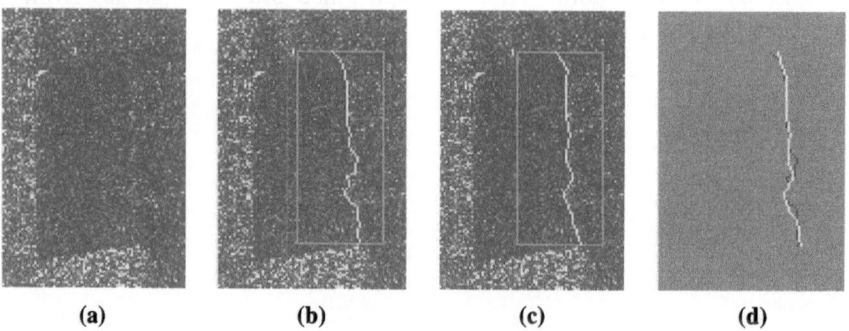

(a)   (b)   (c)   (d)

Figure 5.25. **(a)** Extract of an ERS-1 monolook SAR image of a region in Ukraine, provided by the CNES and delivered by ESA. **(b)** GLRT filtering of (a) with 8 differently-oriented 10×10 windows. Edge extraction is performed with the watershed algorithm. **(c)** Final contour estimation given by the active contour (26 node polygon, $\lambda = 0.20$). The initialization is obtained by sampling the contour (b) every 4 pixels. **(d)** Comparison of contours in (b) (black) and (c) (white). ©2001 IEEE.

These examples show that region-based active contours can be efficient for refining edge location obtained with a coarser, but faster, detection method such as the GLRT. However, the active contour technique used here is adapted to images consisting of two regions whereas many images, including the SAR images of agricultural regions considered here, consist in a patchwork of regions with different radiometric properties. In order to solve the more general problem of the partition of an image composed of several regions, a method based on a statistical active "grid" will be described in Section 6.3.

APPENDIX 5.A: Crossing tests

In this section, one describes a fast algorithm which tests and prevents intersections between the segments of the polygonal active contour during its convergence.

First, one can note that when the node N_{n_0} of the polygonal contour is moved, the deformation of the polygon concerns only the two segments which contain this node: $\overline{N_{n_0-1} \, N_{n_0}}$ and $\overline{N_{n_0} \, N_{n_0+1}}$.

Thus, when the node N_{n_0} is moved, one only has to check possible intersections of segments $\overline{N_{n_0-1} \, N_{n_0}}$ and $\overline{N_{n_0} \, N_{n_0+1}}$ with the other segments of the polygon. One can note that there is no test to compute between adjacent segments (since these segments have a common node). Then, if the polygon has k nodes, this represents $2(k-3)$ crossing-tests. As the crossing-tests are time-consuming, it is necessary to achieve them only for segments which are likely to intersect. This point is described in the next section.

It is important to note that it is assumed that none of the segments intersect in the initial polygonal contour.

Preselection

Let us first present the preselection algorithm to analyze rapidly if the segment $\overline{N_{n_0} \, N_{n_0+1}}$ can cross other segments of the polygon. This algorithm is rapid since it only involves tests on the integer coordinates of the nodes of the polygon.

First, the ex-inscribed rectangle of the segment $\overline{N_{n_0} \, N_{n_0+1}}$ (i.e., rectangle having the diagonal $\overline{N_{n_0} \, N_{n_0+1}}$) is determined. This rectangle shares the image in 9 regions which are E, NE, N, NW, W, SW, S, SE and C (Figure 5.A.1). Moreover, four sets are defined: $E_0 = \{SE, E, NE\}$, $E_1 = \{NE, N, NW\}$, $E_2 = \{NW, W, SW\}$ and $E_3 = \{SW, S, SE\}$.

Then, one has simply to visit each node (anti-clockwise) and determine to which region it belongs. If two consecutive nodes N_i and N_{i+1} belong to two regions of a same set E_i, then no intersection is possible with segment $\overline{N_{n_0} \, N_{n_0+1}}$. However, if two consecutive nodes belong to two regions of two different sets, the segment $\overline{N_n \, N_{n+1}}$ is likely to cross the segment $\overline{N_{n_0} \, N_{n_0+1}}$ and the crossing-test, described below, must be run.

Crossing test

Now, let us present the method used to determine if the segment $\overline{N_{n_0} \, N_{n_0+1}}$ intersects with the segment $\overline{N_n \, N_{n+1}}$.

Let us introduce four angles $\alpha_1 = \widehat{N_{n_0+1} N_{n_0} N_n}$, $\alpha_2 = \widehat{N_{n_0+1} N_{n_0} N_{n+1}}$, $\alpha_3 = \widehat{N_{n+1} N_n N_{n_0}}$, and $\alpha_4 = \widehat{N_{n+1} N_n N_{n_0+1}}$ (Figure 5.A.1).

The test is based on the determination of the signs of these angles. Two cases must be considered. In the first case, the four nodes (N_{n_0}, N_{n_0+1}, N_n, and N_{n+1}) are not aligned. Then, if α_1 and α_2 have opposite signs and if α_3 and α_4 have opposite signs, the two segments

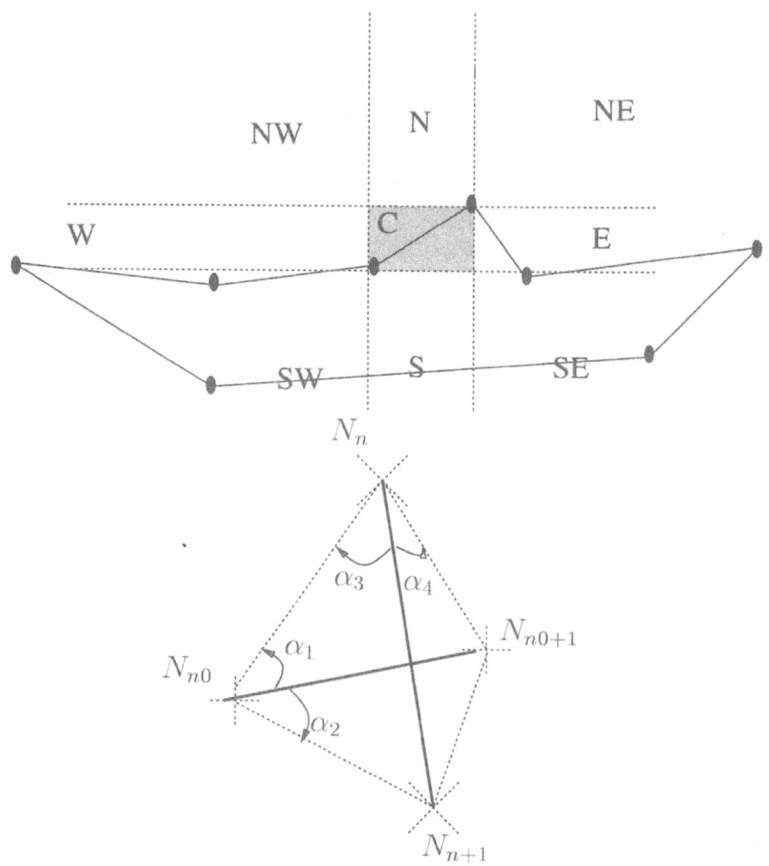

Figure 5.A.1.   Upper graph: notations used for the preselection of a segment which can cross the segments in area C. Lower graph: notations used for the crossing tests. ©1999 IEEE.

$\overline{N_{n_0}\,N_{n_0+1}}$ and $\overline{N_n\,N_{n+1}}$ necessarily intersect. In the second case, the four nodes are aligned. As the nodes passed the previous preselection, the two considered segments necessarily intersect in this case.

In order to compute the sign of angle α_i, one can calculate the sign S_i of the z-component of the related vectors cross-product. $S_1 = \mathrm{sgn}(\overrightarrow{N_{n_0}\,N_{n_0+1}} \wedge \overrightarrow{N_{n_0}\,N_n})$,
$S_2 = \mathrm{sgn}(\overrightarrow{N_{n_0}\,N_{n_0+1}} \wedge \overrightarrow{N_{n_0}\,N_{n+1}}), S_3 = \mathrm{sgn}(\overrightarrow{N_n\,N_{n+1}} \wedge \overrightarrow{N_n\,N_{n_0}})$,
and $S_4 = \mathrm{sgn}(\overrightarrow{N_n\,N_{n+1}} \wedge \overrightarrow{N_n\,N_{n_0+1}})$. sgn is a function equal to 1 if its argument is strictly positive, equal to 0 if its argument is null, and equal to -1 otherwise. To summarize, S_i is equal to 1 and -1 when the angle is, respectively, positive and negative and S_i is equal to 0 when the angle is null or equal to π. Thus, if $S_1 \neq S_2$ and $S_3 \neq S_4$ or if $S_1 = S_3 = 0$, the crossing test is true (the two segments $\overline{N_{n_0}\,N_{n_0+1}}$ and $\overline{N_n\,N_{n+1}}$ intersect), and deformation of the snake must not be allowed.

Chapter 6

SOME DEVELOPMENTS OF THE POLYGONAL STATISTICAL SNAKE AND THEIR APPLICATIONS

We have shown in the previous chapter that the polygonal statistical snake is an efficient method to segment objects in very noisy backgrounds, without prior knowledge of the statistical parameters of the target and of the background. We present in this chapter further improvements and generalizations of this technique, while keeping reasonable computation times.

The first described generalization comes from the fact that the polygonal statistical snake has been until now only applied to scalar images. However, there exist sensors which produce multichannel images in which each pixel value is defined by a vector instead of a single graylevel. This is the case for example if the image of the same scene is acquired with several wavelengths. One can obtain color, multispectral, or hyperspectral images depending on the number of different wavelengths that are used. Polarimetric images which represent the polarimetric state of the light reflected by each point of a scene (see Chapter 7) are another example of multichannel images. Another way of generating multichannel images is to observe the same scene at different dates. This method is frequently employed in radar imaging in order to decrease the variance of speckle noise. The problem of segmenting multidate SAR images will be addressed in Section 6.1 as an example of the generalization of the polygonal statistical snake to multichannel images.

The idea for the second proposed improvement comes from the remark that if the object of interest is a polygon with k nodes, the estimated contour should also have exactly k nodes. However, when using the simple multiresolution method introduced in Section 5.3.2, the segmentation quality can be good but the final number of nodes of the polygon is in general larger than the true value of k. Determining the precise number of nodes of the contour is equivalent to estimating the *order* of the model which describes the shape of interest. Model order estimation is a very important topic in signal processing and statistics, and many approaches have been considered to solve this problem [109-111]. We will describe in Section 6.2 a solution to the problem of estimation of the number of nodes based on the Minimum Description Length principle [112].

The third discussed generalization addresses the fact that the polygonal statistical snake is limited by construction to the segmentation of a single object over a background, that is, to the partition of the image into two regions with homogeneous statistics. A generalization of the snake to scenes which do not correspond to this model, because the background and/or the object regions cannot be realistically considered homogeneous, will be presented in Section 6.3. For this purpose, a *polygonal statistical active grid* which is a generalization of the snake to a patchwork of homogeneous regions will be described.

6.1. Generalization of the statistical snake to multichannel images

In this section, we generalize the statistical snake to multichannel images with statistically independent channels [113]. Let us consider a set $S = \{s^{(k)} | k \in [1, K]\}$ of K images. The intensity at pixel (x, y) for the channel k is denoted $s^{(k)}(x, y)$. We intend to segment a target appearing in each of the K channels which are assumed to be independent of each other. The case of statistically dependent channels will be addressed in Section 7.6 in the context of polarimetric images. For each channel k, the intensity of the target (resp. the background) is supposed to have a pdf $P_{\mu_u^{(k)}}(x)$, with $u = a, b$. Making use of the hypothesis of statistical independence of the channels, the loglikelihood of the image is:

$$\mathcal{L}(S \mid \boldsymbol{\theta}, \boldsymbol{\mu}_a, \boldsymbol{\mu}_b) = \sum_{k=1}^{K} \mathcal{L}^{(k)} \left[s^{(k)} \mid \boldsymbol{\theta}, \mu_a^{(k)}, \mu_b^{(k)} \right] \tag{6.1}$$

with

$$\mathcal{L}^{(k)} \left[s^{(k)} \mid \boldsymbol{\theta}, \mu_a^{(k)}, \mu_b^{(k)} \right] = \sum_{(x,y) \in \Delta_a^{\boldsymbol{\theta}}} \log P_{\mu_a^{(k)}}[s^{(k)}(x, y)] +$$

$$\sum_{(x,y) \in \Delta_b^{\boldsymbol{\theta}}} \log P_{\mu_b^{(k)}}[s^{(k)}(x, y)] \tag{6.2}$$

where the regions $\Delta_u^{\boldsymbol{\theta}}$ with $u = a, b$ are defined in Section 4.1.1 and $\boldsymbol{\mu}_u = \{\mu_a^{(k)} \mid k \in [1, K]\}$. In other words, thanks to the hypothesis of channel independence, the global loglikelihood is simply the sum of the likelihoods in each channel. If the nuisance parameters are estimated in the ML sense, one obtains:

$$\widehat{\mu_u^{(k)}}(\boldsymbol{\theta}) = \arg \max_{\mu_u^{(k)}} \left\{ \mathcal{L}^{(k)} \left[s^{(k)} \mid \boldsymbol{\theta}, \mu_a^{(k)}, \mu_b^{(k)} \right] \right\} \tag{6.3}$$

with $u = a, b$ and $\widehat{\mu_u^{(k)}}(\boldsymbol{\theta})$ is thus simply the ML estimate obtained from the data in channel k. The pseudo-loglikelihood is thus:

$$
\ell(\mathbf{S}, \boldsymbol{\theta}) = \sum_{k=1}^{K} \mathcal{L}^{(k)} \left[\mathbf{s}^{(k)} \mid \boldsymbol{\theta}, \widehat{\mu_a^{(k)}}(\boldsymbol{\theta}), \widehat{\mu_b^{(k)}}(\boldsymbol{\theta}) \right]
$$

$$
= \sum_{k=1}^{K} \ell^{(k)} \left(\mathbf{s}^{(k)}, \boldsymbol{\theta} \right) \tag{6.4}
$$

and thus the multichannel statistical snake criterion, which is equal to $-\ell(\mathbf{s}, \boldsymbol{\theta})$ (see Eq. 5.17), is:

$$
J^{multi}(\mathbf{S}, \boldsymbol{\theta}) = \sum_{k=1}^{K} J^{(k)}(\mathbf{s}^{(k)}, \boldsymbol{\theta}) \tag{6.5}
$$

where $J^{(k)}(\mathbf{s}^{(k)}, \boldsymbol{\theta}) = -\ell^{(k)}(\mathbf{s}^{(k)}, \boldsymbol{\theta})$. This equation shows that if the channels are independent, the snake criterion on the multichannel image is simply the sum of the criteria computed on each channel separately.

Let us take the example of SAR imaging [113]. Synthetic Aperture Radar (SAR) sensors easily provide multidate images when they are on board satellites since they periodically fly over the same ground area. In each channel k corresponding to a different date, $P_{\mu_u^{(k)}}(x)$ is a Gamma pdf of mean $\mu_u^{(k)}$ with $u = a, b$. The orders L of the Gamma pdf are assumed to be the same whatever the channel and the region (target or background). According to Eqs. 5.21 and 6.5, the snake criterion is:

$$
J_{gam}^{multi}(\mathbf{S}, \boldsymbol{\theta}) = N_a(\boldsymbol{\theta}) \sum_{k=1}^{K} L \log \left[\widehat{\mu_a^{(k)}}(\boldsymbol{\theta}) \right] + N_b(\boldsymbol{\theta}) \sum_{k=1}^{K} L \log \left[\widehat{\mu_b^{(k)}}(\boldsymbol{\theta}) \right]
$$

$$
\tag{6.6}
$$

Let us illustrate this snake approach on the real-world SAR multidate image represented in Figure 6.1. It is a set of three intensity monolook SAR images acquired by the satellite ERS-1 over an agricultural zone in Bourges, France. The aim is to segment a field whose contrast varies greatly from one channel to the other. The active contour initialization is shown in Figure 6.1.d. The optimization of the criterion is performed as described in Section 5.3.2. A polygonal active contour is used and its number of nodes k is progressively increased during the convergence process. Once the contour with k nodes has converged, new nodes are added so that the distance between two adjacent nodes remains below a given threshold d and this new contour is made to converge again. Three steps have been used here, with a maximal amplitude deformation $a = 7$. The internode distance has been set to $d = 20$ after the first step, to $d = 10$ after the second. No regularization term has been used in the two first steps, but a slight regularization has been used in the last one, with $\lambda = 0.1$. The performance of the multichannel method (Figure 6.1.e) is compared with the single channel method applied to the average of all channels (Figure 6.1.f). Obviously, the multichannel method leads to a better segmentation of the field. In particular, the left part of the object, which has a weak contrast in the averaged image, is not correctly segmented with the single channel snake whereas it is successfully processed by the multichannel method.

6.2.   MDL-based statistical snake

We describe in this section a method to estimate efficiently and rapidly the sufficient number of nodes needed to describe the shape of an object [59, 114]. This approach will thus lead to a segmentation procedure without free parameter. It will also be shown that, for polygonal objects, this technique leads to better segmentation results than using a regularization strategy based on the smoothness of the contour and that it is still robust if the object's shape is not a simple polygon. This method is based on applying the Minimun Description Length (MDL) principle [59,112] to the polygonal statistical snake.

6.2.1   The MDL principle

The MDL principle [112] has been proposed to solve the problem of order estimation in parametric statistical models, which is a crucial issue in signal and image processing [110, 115]. Indeed, it is well known that the best models are seldom those which use the largest number of parameters [66]: Beyond a certain level of complexity, adding new parameters leads to a performance decrease. Estimating the optimal number of model parameters has thus long been a major concern in signal processing. The main problem is that when trying to estimate the model order from a single realization of the data, the more parameters are used, the better the fit with the data. One thus has to associate a cost to the increase of model complexity.

For this purpose, it is possible to use Rissanen's stochastic complexity, which is an estimate of the averaged complexity of the data, taking into account the description of the data itself and the encoding of the model parameters. Basically, one defines a class of models $\mathcal{M} = \{\mathcal{M}_k, k \in [1, K]\}$, where each model $\mathcal{M}_k$ is defined by a parameter vector $\boldsymbol{\theta}^k$ whose size may vary with k. Let us denote the available data sample χ. The stochastic complexity is the code length $\mathcal{C}(\chi|\mathcal{M}_k)$ necessary to describe the data. It is divided into two parts: the length of encoding the data knowing the model parameters $\mathcal{C}(\chi|\boldsymbol{\theta}^k)$ and the length of encoding the model parameters $\mathcal{C}(\boldsymbol{\theta}^k)$ themselves. One thus has:

$$\mathcal{C}(\chi|\mathcal{M}_k) = \mathcal{C}(\chi|\boldsymbol{\theta}^k) + \mathcal{C}(\boldsymbol{\theta}^k) \tag{6.7}$$

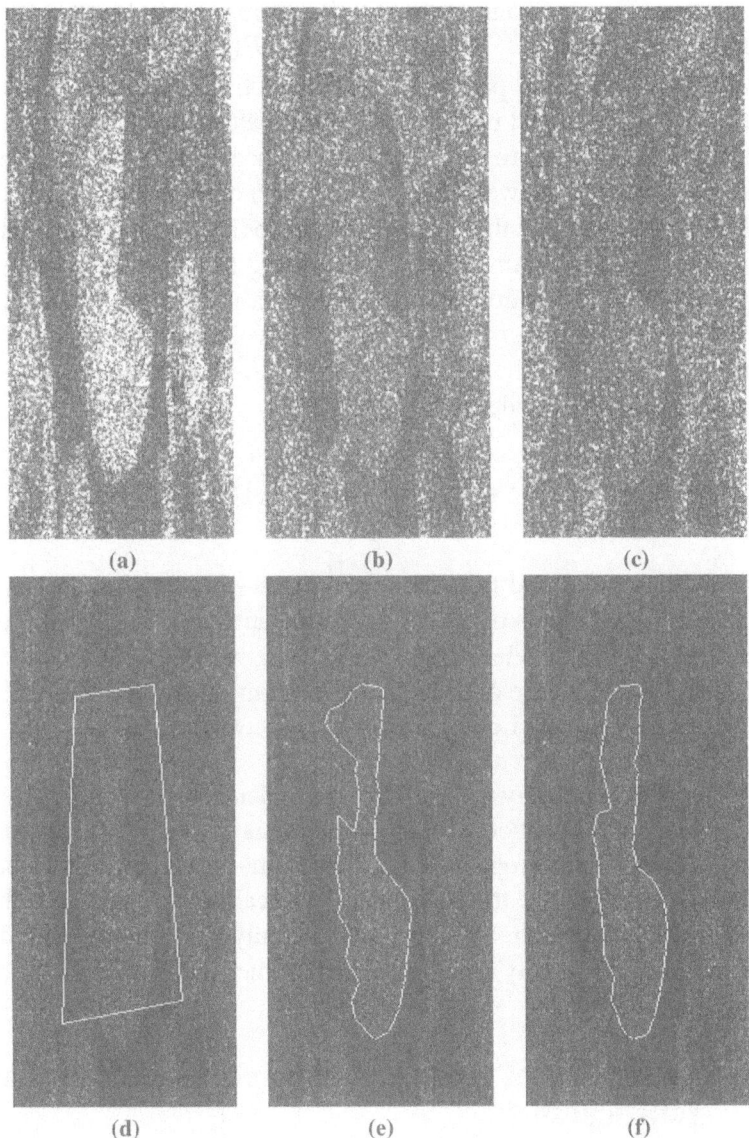

Figure 6.1. Three-date intensity monolook SAR images of an agricultural region in Bourges, provided by the CNES and delivered by ESA. (**a**) Date 1 (30/04/93). (**b**) Date 2 (04/06/93). (**c**) Date 3 (09/07/93). (**d**) Initialization of the contour. (**e**) Result obtained by segmenting the average of the three channels with the single channel snake. (**f**) Result obtained with the multichannel snake method. The results are presented on the sum of the three channels. ©1999 OSA.

If the data is described by random variables having a statistical distribution $P(\chi|\boldsymbol{\theta}^k)$, according to Shannon theory, the average code length of the data is its entropy: $C(\chi|\boldsymbol{\theta}^k) = -\int \log P(\chi|\boldsymbol{\theta}^k) \log P(\chi|\boldsymbol{\theta}^k)d\chi$ [116]. The code length $C(\boldsymbol{\theta}^k)$ of the model parameters depends on their nature. Some subtle issues arise when the model parameters are real-valued, since their code length is infinite. One thus has to consider them truncated to a certain precision, and perform a trade-off between the coding length of the parameters necessary to reach this precision and the increase of the coding length of the data term $C(\chi|\boldsymbol{\theta}^k)$ due to a poor parameter precision [112]. In our case, the parameters of interest will be the positions of the nodes of a polygon on a discrete grid. These are discrete parameters and we will thus avoid these issues.

Whatever the nature of the model parameters, the MDL principle consists in choosing the model with index $\hat{k}$ according to:

$$\hat{k} = \arg\min_{k} \{C(\chi|\mathcal{M}_k)\} \tag{6.8}$$

This principle has profound implications in terms of estimation theory, which are described in Rissanen's book [112]. On a practical point of view, its main advantage is to provide an elegant and very general way of weighting the goodness of fit to the data on the one hand, and the complexity of the model on the other. The common yardstick to measure these two quantities is the coding length expressed in bits.

The MDL principle is very well adapted to problems formulated in a Bayesian framework since the likelihood, which is the basis of such methods, is an approximation of the data entropy, and thus of the data-linked part of the stochastic complexity $C(\chi|\boldsymbol{\theta}^k)$. Since the polygonal statistical snake is based on the likelihood, half of the job is already done and one only needs to estimate the part of the stochastic complexity which depends on the model parameters, that is, $C(\boldsymbol{\theta}^k)$.

6.2.2 Application of the MDL principle to the polygonal statistical snake

As stated above, the statistical snake described in Chapter 5 is well adapted to the application of the MDL principle, since the external energy which is minimized during the iterative process is directly linked to the likelihood of the data. This principle has been applied with success by Figueiredo *et al.* [59] for estimating the number of control points of a B-spline to model the contours of smooth objects. This approach consists in adding the control points one by one. When considering the $(k+1)th$ point, a new convergence is performed, starting from the shape which has been obtained with k points, in order to refine the shape estimation. The MDL is computed from the new value of the likelihood plus a term depending on the number of nodes. Finally, the value of k which minimizes this MDL function is chosen as the optimal order of the model.

However, we will show in the following that in the case of the polygonal statistical snake, this approach is not directly applicable. The main reason is that the polygonal shapes that we are considering are not smooth and can have sharp cusps. The multiresolution convergence method which consists in adding nodes one by one does not allow one to find the absolute minimum of the MDL criterion for such shapes. We will describe in this section how to apply the MDL principle to the polygonal statistical snake [114].

As stated above, the MDL principle consists in minimizing the number of bits, or code length, required to describe the image at hand. In the SIR model, the image is composed of three parts: the object region, the background region, and the contour. The code length $\mathcal{C}_T(\mathbf{s}|\boldsymbol{\theta}^k, k)$ is thus the sum of three terms: the length $\mathcal{C}_a(\mathbf{s}|\boldsymbol{\theta}^k)$ of the description of the graylevels of the object region, the length $\mathcal{C}_b(\mathbf{s}|\boldsymbol{\theta}^k)$ of the description of the graylevels of the background region, and the length $\mathcal{C}(\boldsymbol{\theta}^k)$ of the description of the polygonal contour consisting of k nodes. One thus has the following relation:

$$\mathcal{C}_T(\mathbf{s}|\boldsymbol{\theta}^k, k) = \mathcal{C}_a(\mathbf{s}|\boldsymbol{\theta}^k) + \mathcal{C}_b(\mathbf{s}|\boldsymbol{\theta}^k) + \mathcal{C}(\boldsymbol{\theta}^k) \qquad (6.9)$$

Information theory makes it possible to determine an approximation of these three terms [116, 117].

Let us begin with a very simple approximation of the term $\mathcal{C}(\boldsymbol{\theta}^k)$. Each node of the polygon $\boldsymbol{\theta}^k$ can be located at one of the N pixels of the image. There are thus approximatively N^k possibilities to encode the k nodes of the polygon $\boldsymbol{\theta}^k$ and we will consider, as a first approximation, that this value corresponds to the number of possible polygons with k nodes. If one considers that each polygon is equiprobable, the number of bits necessary to encode one of them is then of the order of $\log_2(N^k)$ (where $\log_2$ corresponds to the logarithm in base 2).

Let us now determine an approximation of the terms $\mathcal{C}_a(\mathbf{s}|\boldsymbol{\theta}^k)$ and $\mathcal{C}_b(\mathbf{s}|\boldsymbol{\theta}^k)$. According to Shannon theory, the minimal average number of bits necessary to encode a set of realizations of random variables corresponds to the entropy of the probability law from which they are drawn [116]. In the case of the snake, the minimal average number of bits necessary to encode the $N_a(\boldsymbol{\theta}^k)$ graylevels of the object region, which are distributed with the pdf $P_{\boldsymbol{\mu}_a}(x)$, is thus equal to the product of $N_a(\boldsymbol{\theta}^k)$ with the entropy S_a of $P_{\boldsymbol{\mu}_a}(x)$, and thus $\mathcal{C}_a(\mathbf{s}|\boldsymbol{\theta}^k) \simeq N_a(\boldsymbol{\theta}^k) S_a$. The entropy S_a is classically given by the following relation [116, 117]:

$$S_a = -\int P_{\boldsymbol{\mu}_a}(x) \log_2 \left[P_{\boldsymbol{\mu}_a}(x) \right] dx - \log_2(q)$$

where q represents the quantification step. Since the contribution of the term $\log_2(q)$ only consists in adding a constant value to the total description length, it will not be considered in the following. Similarly, the code length of the background region is equal to $C_b(\mathbf{s}|\boldsymbol{\theta}^k) \simeq N_b(\boldsymbol{\theta}^k) \, S_b$. The total description length of the image is thus equal to:

$$C_T(\mathbf{s}|\boldsymbol{\theta}^k, k) = N_a(\boldsymbol{\theta}^k) \, S_a + N_b(\boldsymbol{\theta}^k) \, S_b + k \, \log_2(N) \qquad (6.10)$$

If one replaces ensemble averaging with empirical averaging in the expression of the entropies, one obtains approximate values $\widehat{S}_u(\boldsymbol{\theta}^k)$ of the entropies:

$$N_u(\boldsymbol{\theta}^k) \, \widehat{S}_u(\boldsymbol{\theta}^k) = - \sum_{(x,y)\in\Delta_u^{\boldsymbol{\theta}^k}} \log_2 \left(P_{\boldsymbol{\mu}_u}[s(x,y)] \right) \qquad (6.11)$$

where $u = a$ or b. It is then easy to notice that $-N_a(\boldsymbol{\theta}^k)\, \widehat{S}_a(\boldsymbol{\theta}^k) - N_b(\boldsymbol{\theta}^k)\, \widehat{S}_b(\boldsymbol{\theta}^k)$ represents the loglikelihood (in base 2) $l_2[\mathbf{s}|\boldsymbol{\theta}^k, \boldsymbol{\mu}_a, \boldsymbol{\mu}_b]$ of the hypothesis that the polygonal contour $\boldsymbol{\theta}^k$ corresponds to the silhouette of the object present in the scene $\mathbf{s}$. In the case where the pdf parameters $\boldsymbol{\mu}_a$ and $\boldsymbol{\mu}_b$ are unknown, one will consider a second approximation which consists in replacing their true values with their maximum likelihood estimates. One will thus consider that the pseudo-loglikelihood $\ell_2[\mathbf{s}|\boldsymbol{\theta}^k, \widehat{\mu}_a(\mathbf{s}, \boldsymbol{\theta}^k), \widehat{\mu}_b(\mathbf{s}, \boldsymbol{\theta}^k)]$ constitutes an approximation of $-N_a(\boldsymbol{\theta}^k)\, \widehat{S}_a(\boldsymbol{\theta}^k) - N_b(\boldsymbol{\theta}^k)\, \widehat{S}_b(\boldsymbol{\theta}^k)$. This pseudo-loglikelihood precisely corresponds to the opposite of the criterion $J(\mathbf{s}, \boldsymbol{\theta}^k)$ which is minimized in the polygonal statistical active snake (see Section 5.2.1).

In our case, the MDL principle thus consists in minimizing a criterion denoted $J^{MDL}(\mathbf{s}, \boldsymbol{\theta}^k, k)$ which corresponds to the length C_T of the image description and which can be written in the following way (note that the base-2 logarithms have been replaced with natural logarithms, without loss of generality):

$$J^{MDL}(\mathbf{s}, \boldsymbol{\theta}^k, k) = C_T(\mathbf{s}|\boldsymbol{\theta}^k, k) = J(\mathbf{s}, \boldsymbol{\theta}^k) + k \log(N) \qquad (6.12)$$

Thus in practice, in order to form the MDL criterion adapted to a given type of noise, one simply has to consider the energy criterion adapted to this noise, as described in Section 5.2.1, and add to it the term $k \log(N)$ representative of the description length of the contour. The polygonal contour $\widehat{\boldsymbol{\theta}^k}$ which minimizes the criterion J^{MDL} corresponds to the contour which approximates the object in the sense of the stochastic description of the images [112]. It will thus be called in the following *MDL contour estimate*.

The minimization of the criterion J^{MDL} is a difficult problem since the optimization has to be performed both with respect to the contour parameters $\boldsymbol{\theta}^k$ for a given value of k and to the number of nodes k of the contour.

6.2.3   Two-step optimization process

With an optimization process where the resolution of the polygon is increased progressively, the segmentation quality is satisfying only when the number of nodes which define the contour is large enough. Consequently, the MDL approach cannot be implemented in general by only increasing the number of nodes and stopping the process when the MDL criterion increases. A possible alternative consists in implementing a two-step optimization process of the MDL criterion.

With this approach, during the first step of the process, the object is segmented by progressively increasing the number of nodes with the multiresolution strategy described in Section 5.3.2. Considering a small value of the parameter d_f which represents the internode distance at the end of the process, one obtains a final contour with a large number of nodes denoted k_{max}.

As an illustration of this first step, let us consider the synthetic image represented in Figure 6.2. Figure 6.2.a represents the shape of the object and Figure 6.2.b the noisy image built from this object perturbed by Gamma noise. The evolution of the criterion $J^{MDL}(\mathbf{s}, \hat{\boldsymbol{\theta}}^k, k)$ evaluated during the first step of the optimization process is represented in Figure 6.3. The values obtained for five increasing values of d_f are represented with symbols $(+)$ (continuous line). Figure 6.2.c presents the contour obtained after the first phase of the process when $d_f = 2$ pixels and when the initial contour is that represented in Figure 6.2.b. This contour has $k_{max} = 192$ nodes. One can see that all the details of the shape have been correctly segmented, but the number of nodes is manifestly excessive, which leads to a noisy contour estimate (no regularization term has been used in this experiment). On the other hand, during this first step, the result of segmenting the image with a 10-node polygon is shown in Figure 6.2.d and it is clearly seen that the result is not satisfying.

During the second step of the process, one reduces the complexity of the contour obtained after the first step by progressively removing the "less useful" nodes. The criterion $J^{MDL}(\mathbf{s}, \hat{\boldsymbol{\theta}}^k, k)$ is evaluated each time a node is removed, which makes it possible to estimate the optimal number of nodes in the MDL sense. More precisely, this second step, which corresponds to a pruning process, is performed in the following way:

- Suppose that the initial contour is composed of k nodes.

- For each node of the contour taken successively, one computes the value of the likelihood-based part of the MDL criterion $J(\mathbf{s}, \boldsymbol{\theta}^{k-1})$ obtained when this node is removed.

- Among all the nodes, the one which leads to the smallest value of the criterion $J(\mathbf{s}, \boldsymbol{\theta}^{k-1})$ is suppressed.

- The positions of the remaining nodes are refined by applying the technique of the polygonal statistical snake (minimization of $J(\mathbf{s}, \boldsymbol{\theta}^{k-1})$ with respect to $\boldsymbol{\theta}^{k-1}$).

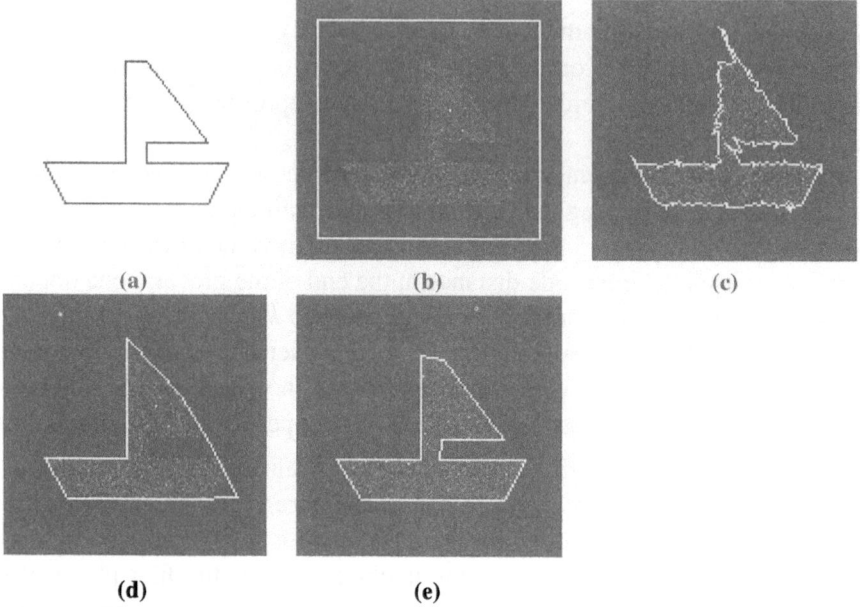

Figure 6.2.   Illustration of the two-step optimization procedure on a synthetic image. **(a)** Polygonal shape defined by 10 nodes. **(b)** Synthetic image of size 128×128 pixels generated from the shape represented in (a); the graylevel pdf is Gamma of order 1 with mean equal to 1 in the background region and to 4 in the object region. The initial contour (starting point of the optimization process) is represented in white. **(c)** Polygonal contour obtained after the first step of the process with $d_f = 2$ pixels; it consists of 192 nodes. **(d)** MDL contour obtained during the increasing resolution optimization process when the number of nodes is fixed to the true value ($k = 10$). **(e)** MDL estimate $\widehat{\boldsymbol{\theta}^k}$ obtained after the second step of the process; it consists of 10 nodes, which corresponds to the true number of nodes of the shape.

- The value of the criterion $J^{MDL}(\mathbf{s}, \boldsymbol{\theta}^{k-1}, k-1)$ associated with this polygonal contour with $k - 1$ nodes is computed.

This operation is repeated until the number of remaining nodes is equal to a predefined value k_{inf} (typically $k_{inf} = 4$). The value of k which minimizes the criterion J^{MDL} corresponds to the optimal estimation of the number of nodes in the MDL sense and the contour associated with this value of k corresponds to the MDL estimate of the object shape $\widehat{\boldsymbol{\theta}^k}$. Note the difference between $\hat{\boldsymbol{\theta}}^k$, which is the contour which minimizes J^{MDL} for a given value of k, and $\widehat{\boldsymbol{\theta}^k}$, which is the MDL contour estimate, that is, which minimizes J^{MDL} with respect to k.

The evolution of $J^{MDL}(\mathbf{s}, \hat{\boldsymbol{\theta}}^k, k)$ during the second step has been represented in Figure 6.3 with a dotted line. It is seen that during this second step,

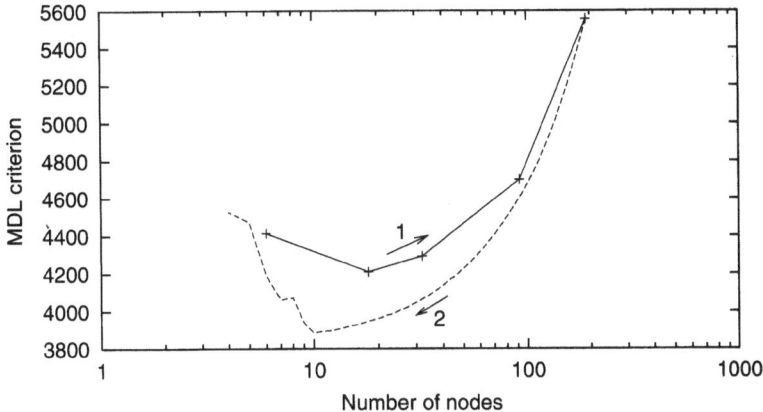

Figure 6.3. Evolution of the criterion $J^{MDL}(\mathbf{s}, \hat{\boldsymbol{\theta}}^k, k)$ as a function of k in the case of the synthetic image presented in Figure 6.2.b. Continuous line: increasing resolution step; dotted line: node removing step.

$J^{MDL}(\mathbf{s}, \hat{\boldsymbol{\theta}}^k, k)$ has a well-defined global minimum when $k = 10$. In this example, the MDL estimate of the contour thus corresponds to the true number of nodes. The segmentation result obtained for this number of nodes after the second step is shown in Figure 6.2.e and we can observe that it corresponds to a very good estimation of the shape of the object.

We have also represented in Figure 6.4 the evolution of the segmentation quality (in terms of the number of misclassified pixels, NMP) as a function of the number of nodes k. One can see that the contour which leads to the lowest NMP is the one obtained when the number of nodes is equal to 10. Moreover, this value of the NMP is lower than that obtained during the first step of the optimization process. This implementation of the MDL principle thus constitutes an efficient estimate of the object shape. Its efficiency will be illustrated in the next section with other objects and other types of noises.

6.2.4   Results obtained with different types of noises

Let us consider the image in Figure 6.5.a which has been built from a polygonal shape consisting of 15 nodes. The graylevels are distributed with Gaussian pdf with zero mean and different standard deviations ($\sigma = 1$ in the object region and 4 in the background region). The results obtained in this image are presented in Figure 6.5. In particular, Figure 6.5.c represents the contour which minimizes the MDL criterion during the first step (increasing resolution) of the process. This contour consists of 46 nodes and one can notice that it approximates only very coarsely the silhouette of the object. Figure 6.5.d represents

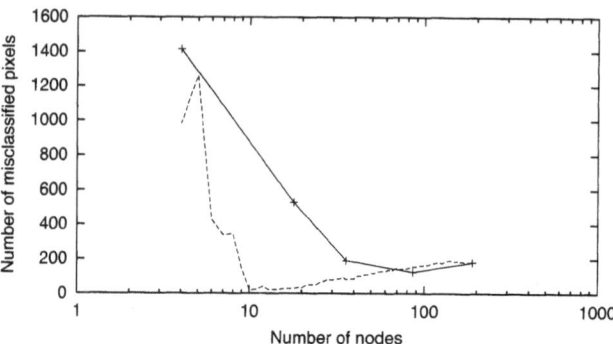

Figure 6.4. Evaluation of segmentation quality (NMP) obtained with the two-step optimization process on the synthetic image represented in Figure 6.2.b. The values of the NMP are plotted as a function of k, in the continuous line for the first step of the process and in the dotted line for the second step (i.e., the pruning process).

the contour obtained with $d_f = 2$ pixels at the end of the first step. These two results have been obtained from the same initial contour which is drawn with black lines on each image. Finally, Figure 6.5.e represents the contour estimate obtained when the number of nodes corresponds to the minimum value of the MDL criterion during the pruning process. This contour estimate consists of 15 nodes, which is the true number of nodes of the shape from which the image was built. One can also notice that the segmentation quality is satisfying. Finally, the evolution of the MDL criterion $J^{MDL}(\mathbf{s}, \hat{\boldsymbol{\theta}}^k, k)$ during the two steps of the optimization process is represented in Figure 6.6.a.

The third example is also a synthetic image of size 128×128 pixels, but in which the object cannot be described with a simple polygon (see Figure 6.7.a). The graylevels of the image are Poisson random variables with mean 1 on the object and 2 on the background. The obtained segmentation results are shown in Figure 6.7. The contour estimate obtained with the two-step method is much better, which can also be verified by computing the NMP. The evolution of the criterion $J^{MDL}(\mathbf{s}, \hat{\boldsymbol{\theta}}^k, k)$ during the two-step optimization procedure is represented in Figure 6.6.b. Contrary to what was observed on the previous images, the obtained minimum is not sharp, which is due to the fact that the object to segment is not a simple polygon and there is no exact value of k associated with its shape.

Let us now illustrate the two-step optimization process on two real-world images of a hand acquired with a digital camera. The snake energy adapted to Gaussian noise is used. The obtained results are represented in Figures 6.8.a and 6.8.b. The contour estimates consist of respectively 48 and 93 nodes. Figures 6.8.c and 6.8.d represent the magnified version of the two windows F1 and F2

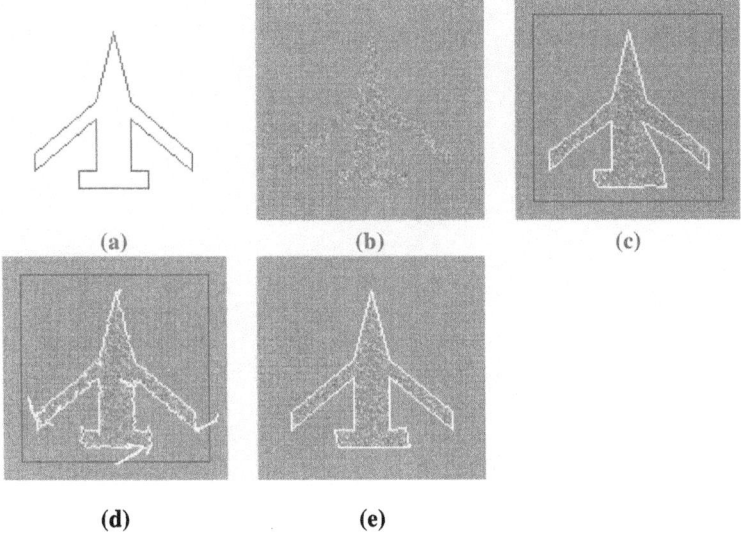

Figure 6.5. Application of the two-step optimization process on an image of size 128 × 128 pixels with Gaussian graylevel statistics of zero mean and $\sigma = 1$ on the object region and $\sigma = 4$ on the background region. **(a)** Shape from which the image has been built ($k = 15$). **(b)** Noisy image. **(c)** Contour estimate which minimizes the MDL criterion during the first step (increasing resolution) of the optimization process ($k = 46$). **(d)** Contour estimate obtained at the end of the first step with $d_f = 2$ pixels ($k_{max} = 254$). **(e)**: MDL contour estimate obtained after the second step (pruning process) ($k = 15$).

marked in Figure 6.8.b. They illustrate the distribution of the nodes on the MDL contour estimate and we can observe that there is a higher concentration of nodes in regions where the curvature of the contour is large. On the other hand, in regions where the contour is straight, only a few nodes are necessary.

We show in Figure 6.9 two examples of segmentation of more complex shapes. In the first row, we consider a polygonal object with very sharp cusps. In the second row, the scene is composed of two objects appearing on a textured background. In the latter example, one can observe on the star-shaped object that the nodes concentrate in the cusps, where more nodes are required to correctly define the local curvature.

As a final example of the potential applications of the described method, Figure 6.10 displays some segmentation results of different types of tools. The segmented shapes could be used for example in a tool recognition system.

In conclusion, it is important to recall that the segmentation method described in this section makes it possible to segment objects with quite complex shapes without having to specify the number of nodes that is used: All the segmentation results which have been shown start with a rectangular initial contour. The

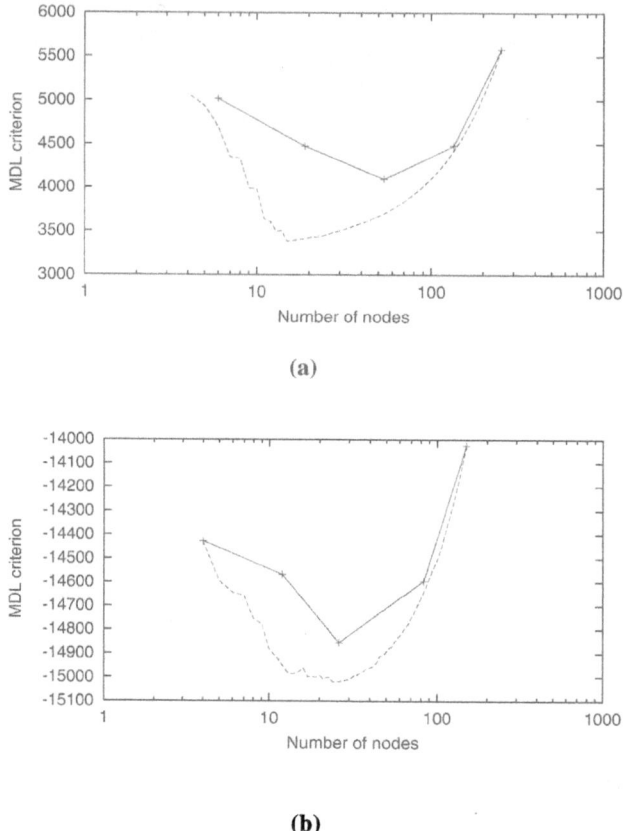

Figure 6.6. Evolution of the criterion $J^{MDL}(\mathbf{s}, \hat{\boldsymbol{\theta}}^k, k)$ as a function of k with the two-step optimization process during the first step (continuous line) and the second step, that is, the pruning process (dotted line). **(a)** On the image in Figure 6.5.b. **(b)** On the image in Figure 6.7.b.

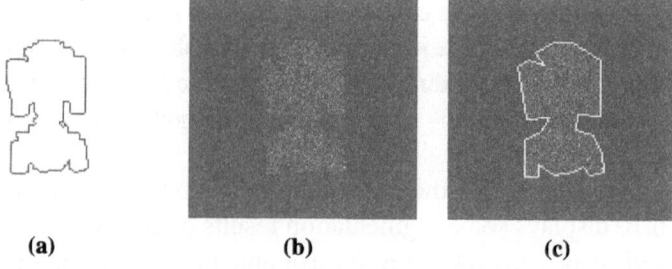

Figure 6.7. Application of the two-step optimization process on an image of size 128×128 with Poisson noise. **(a)** Silhouette of object. **(b)** Scene. **(c)** MDL estimate after the second step ($k = 24$).

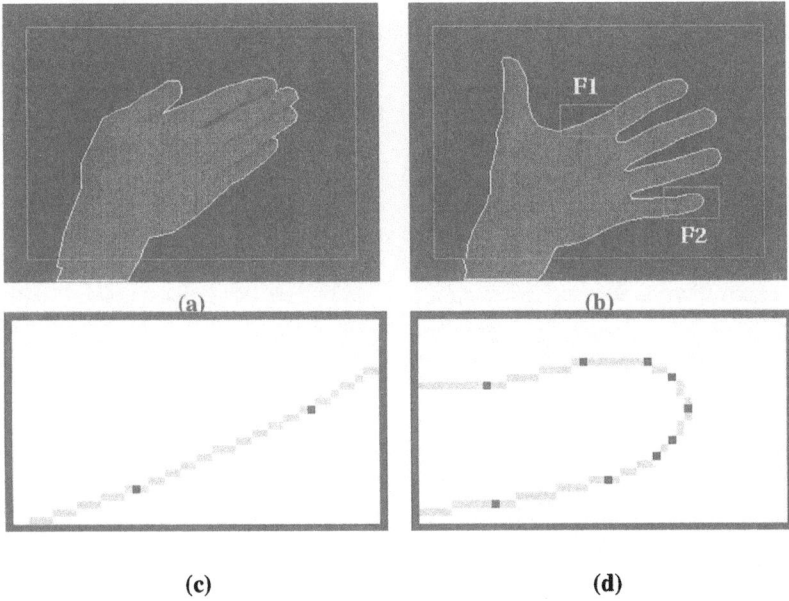

Figure 6.8.   MDL contour estimates (in white) obtained with the two-step optimization process with $d_f = 2$ pixels. The initial contour is represented in gray. (**a**) Contour estimate obtained on the first image of the hand. (**b**) Contour estimate obtained on the second image of the hand. (**c**) Magnification of window F1 (Figure 6.8.b); the nodes are represented in black. (**d**) Magnification of window F2 (Figure 6.8.b).

correct segmentation is thus obtained without requiring the user to specify *ad hoc* parameters since no regularization term is used in the optimized criterion.

6.2.5   Quantitative evaluation of the segmentation performance

We will in this section evaluate more accurately the performance of the MDL-based snake and compare its performance with a classical segmentation approach where a simple regularization method is applied and the number of nodes is not estimated. We will then comment on the approximation that is used to estimate the entropy of the contour.

6.2.5.1   Comparison with simple regularization methods

In this section, we compare in a simple example the MDL shape estimate and estimates obtained with classical regularization methods when the object of interest is a polygon. We also analyze the results obtained when the object is circular.

As seen in Section 5.3.1, the regularization of a contour estimate is classically obtained by adding an "internal energy" term U_{int} to the data-dependent

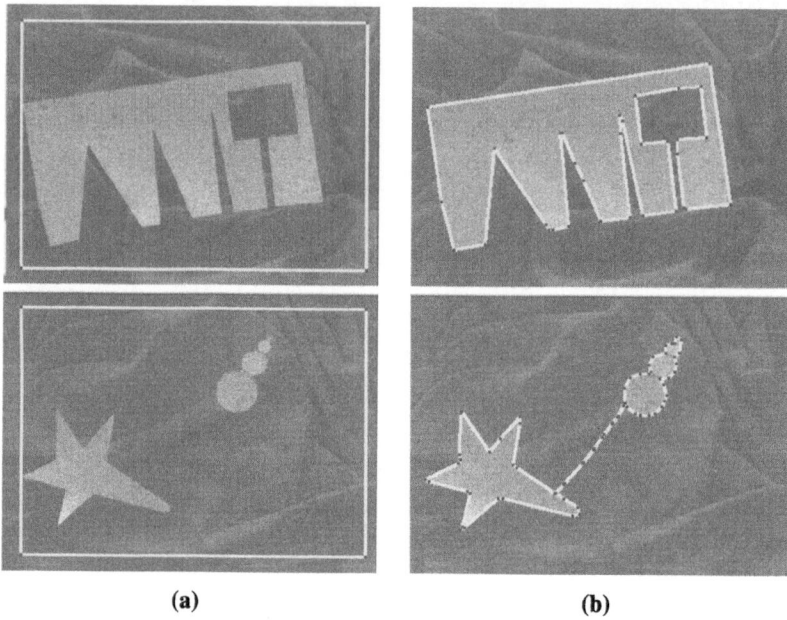

(a)   (b)

Figure 6.9.   Segmentation of real images acquired with a CCD camera. **(a)** Image and initial contour.   **(b)** Contour estimate obtained after the second step.   The nodes of the polygonal contours are represented in black.

"external" energy. We will consider in this section the internal energy term used in Chapter 5 (see Eq. 5.38). This energy term is added to the likelihood to provide the following snake criterion:

$$J'(\mathbf{s}, \boldsymbol{\theta}^k, \lambda) = (1 - \lambda)\, J(\mathbf{s}, \boldsymbol{\theta}^k) + \lambda\, U_{int} \qquad (6.13)$$

where $J(\mathbf{s}, \boldsymbol{\theta}^k)$ is defined in Eq. 6.12 and λ is the weighting term which sets the compromise between the regularization level and the fit to the data. The main problem of classical regularization methods lies in the choice of the value of λ. The optimal value of λ will depend on the expression of the chosen regularization term U_{int} as well as on the considered image. In order to consider the best possible result that can be obtained with this regularized approach, the value λ^{opt} of λ which minimizes the NMP has been used.

Let us consider the scene in Figure 6.11.a which represents an object perturbed with a Gamma noise of contrast ratio $\rho = 3$. The MDL-based method (Figure 6.11.b) and the regularized snake (Figure 6.11.c) with the value of λ leading to the best NMP have been applied to this scene. We can see that the MDL contour estimate is not only smoother, but also more accurate.

Let us now consider the circular shape represented in Figure 6.12.a. This is an unfavorable case since the polygonal description is not adapted to this kind

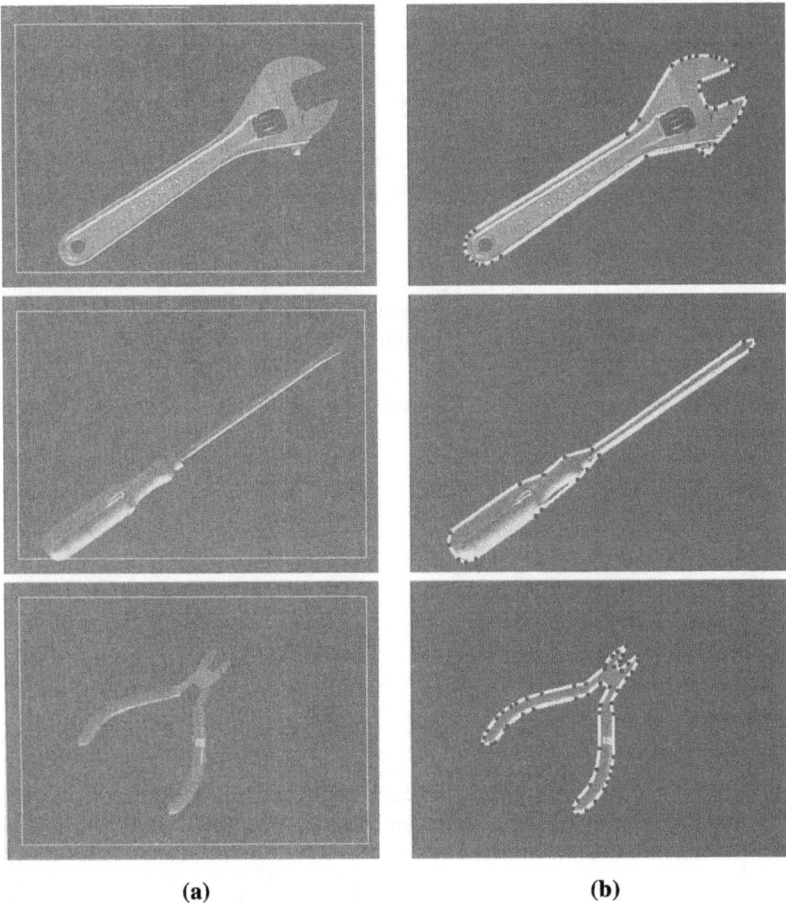

Figure 6.10. Segmentation of different types of tools on images acquired with a CCD camera. (a) Image and initial contour. (b) Contour estimate obtained after the second step. The nodes of the polygonal contours are represented in black.

of shape. Moreover, the regularized snake is more adapted to the segmentation of such smooth shapes. The two contours represented in Figures 6.12.c and 6.12.d have been obtained by applying to the image in Figure 6.12.b the same experimental procedure as previously. The MDL estimate is a polygon with 11 nodes whereas the regularized estimate is more curved but also noisier. The obtained NMP for both contour estimates are similar.

6.2.5.2 A note about the entropy of the contour

In order to establish the expression of the MDL-based snake, the entropy of the contour has been defined with a very simple approximation, assuming that

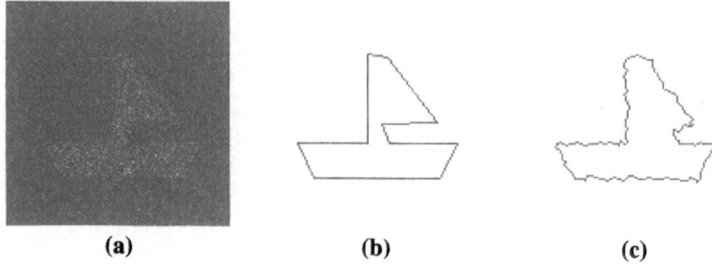

(a)   **(b)**   **(c)**

Figure 6.11. Comparison of the segmentation quality on a low-contrast image with the MDL-based and regularized snakes. **(a)** Image built from the object in Figure 6.2.b, with Gamma noise of order $L = 1$ and contrast $\rho = 3$. **(b)** MDL contour estimate with two-step optimization approach. **(c)** Contour obtained with the regularized snake when λ is optimized on the number of misclassified pixels.

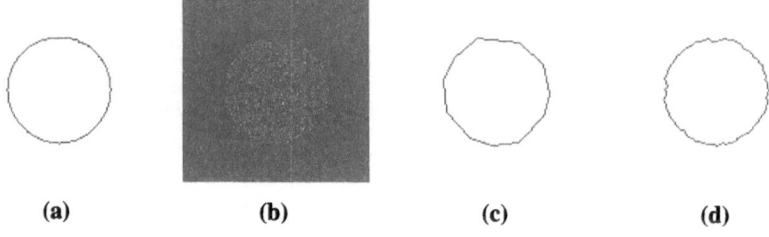

(a)   **(b)**   **(c)**   **(d)**

Figure 6.12. Results obtained on a circular object with the same procedure as that used in Figure 6.11. **(a)** Silhouette of the considered circular object. **(b)** Image with contrast $\rho = 3$. **(c)** MDL contour estimate. **(d)** Contour obtained with $\lambda = 0.2$.

each of the k nodes of the polygon could lie on one of the N pixels of the image. However, it is clear that a lot of these configurations are unsatisfying, since they do not define a contour without crossings. As seen in Appendix 5.A, the absence of crossings at each step of the snake optimization procedure is imposed in order to apply the two described algorithm accelerations. In practice, the number of possible positions is lower than N, and the number of possible configurations of the polygon lower than N^k. In this section, we will consider that the contour entropy has the value kC, where the coefficient $C < \log(N)$. The influence of the value of C on segmentation results will be illustrated in two very simple simulations. First, the two-step MDL segmentation procedure with different values of C is applied to the two synthetic images represented in Figures 6.2.b (Gamma noise) and 6.7.b (Poisson noise). We have represented in Figure 6.13 the evolution of the criterion $J^{MDL}(\mathbf{s}, \hat{\boldsymbol{\theta}}^k, k)$ during the second step of the optimization process for different values of C inferior to $\log(N) = 9.70$ (where $N = 128 \times 128 = 16384$ is the number of pixels of the image).

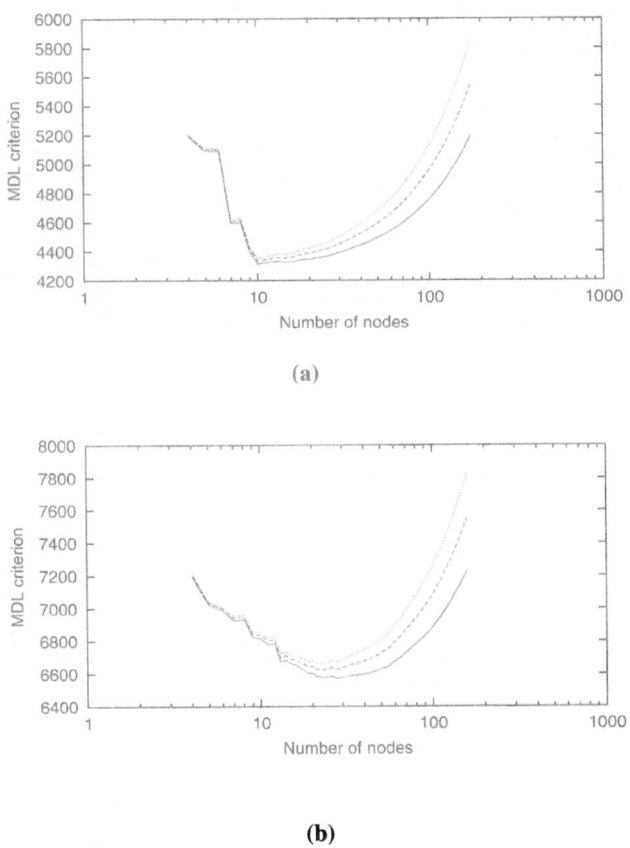

(a)

(b)

Figure 6.13.   Evolution of $J^{MDL}(s, \hat{\theta}^k, k)$ as a function of k, for different values of C. (**a**): on image in Figure 6.2.b (polygonal contour, $k = 10$). (**b**): on image in Figure 6.7.b (non-polygonal contour). _____: $C = 6$; _ _ _ _: $C = 8$; _____: $C = \log(16384) = 9.70$.

In Figure 6.13, we can see that in the case of the polygonal contour, the choice of the value of C has little influence on the estimated number of nodes, since it is still equal to the true value $k = 10$ for all the considered values of C. On the other hand, in the case of the nonpolygonal object, the estimated number of nodes slowly increases as C decreases: One obtains $k = 24$ when $C = \log(N) = 9.70$, $k = 25$ when $C = 8$, and $k = 26$ when $C = 6$. This can be easily interpreted: The larger C, the more penalized the contours with a large number of nodes, and the MDL contour estimate will be defined by fewer nodes. However, one can notice that in Figure 6.14, it is difficult to see the difference between the MDL contour estimates obtained with the three different values of C.

In conclusion, the approximation which consists in considering that the contour entropy is equal to $k \log(N)$ seems to be a satisfactory approximation in the cases considered here.

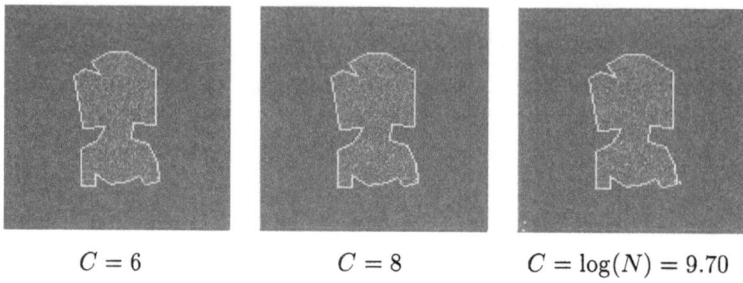

| $C = 6$ | $C = 8$ | $C = \log(N) = 9.70$ |

Figure 6.14.   MDL contours obtained on the images in Figure 6.7.b with the two-step optimization procedure and three different values of C.

6.3.   Statistical active grid and application to SAR image segmentation

Polygonal statistical snakes introduced in Chapter 5 are limited to the segmentation of one single object over a background, that is, to the partition of the image into two regions with homogeneous statistics. However, it is possible to generalize this approach to images where the background and/or the object regions cannot be realistically considered homogeneous [101].

In this section, we describe this generalization of the concept of the polygonal statistical snake to a *polygonal statistical active grid* which is able to segment images made of several (more than two) homogeneous regions [101]. In principle, this grid is just a natural extension of the SIR image model to a multiregion partition of the image. However, in practice, the optimization problem involved when simultaneously determining the number of regions and their frontiers is difficult. We will thus consider that the number of regions as well as their approximate positions are known *a priori*. The aim is then to refine and regularize an initial rough segmentation by using a deformable partition that will be called an *active grid* in the following. In Section 6.3.1, we describe the mathematical formulation of the problem and the solution which is obtained in the case where the pdf of the intensities belong to the exponential family. In Section 6.3.2, the practical implementation of the method is discussed and it is shown that a fast algorithm analogous to the one described in Section 5.2.2 can be applied. Finally, some segmentation results with computing times are presented in Section 6.3.3.

6.3.1   Statistical active grid

We present in this section how to generalize the statistical snake to a statistical active grid in terms of shape modeling and of optimization procedure.

6.3.1.1   Multiregion SIR image model

Let us consider a scene $\mathbf{s} = \{s(x, y)\}$. This scene is modeled as a tessellation of R statistically independent and simply-connected regions. Note that the estimation of R will not be addressed and that this parameter will be assumed known *a priori*. In each region Δ_r ($r \in \{1, 2, ..., R\}$), we assume that the pixel intensities are realizations of independent and identically distributed random variables with a pdf of parameter vectors $\boldsymbol{\mu}_r$. Let $\mathbf{w}^{\boldsymbol{\theta}} = \{w^{\boldsymbol{\theta}}(x, y)\}$ be an R-valued function that denotes a partition of the image into R regions $\Delta_r^{\boldsymbol{\theta}}$, so that $w^{\boldsymbol{\theta}}(x, y) = r$ if and only if (x, y) belongs to region $\Delta_r^{\boldsymbol{\theta}}$.

The shape parameter $\boldsymbol{\theta}$ can be estimated in the maximum likelihood (ML) sense. For a given partition $\mathbf{w}^{\boldsymbol{\theta}}$ of the scene, the loglikelihood can be written:

$$\log P[\mathbf{s}|\boldsymbol{\theta}, \boldsymbol{\mu}_1, \boldsymbol{\mu}_2, ..., \boldsymbol{\mu}_R] = \sum_{r=1}^{R} \log P(\chi_r^{\boldsymbol{\theta}}|\boldsymbol{\mu}_r) \tag{6.14}$$

where $\chi_r^{\boldsymbol{\theta}} = \{s(x, y)|(x, y) \in \Delta_r^{\boldsymbol{\theta}}\}$. Following the approach introduced in Section 5.2.1, we consider that the pdf of the pixel intensities belong to the exponential family. For simplicity reasons, we will only consider the case where the nuisance parameters $\boldsymbol{\mu}_r$ are estimated in the ML sense. In this case, maximizing the pseudo-likelihood is equivalent to minimizing the following criterion:

$$J(\mathbf{s}, \boldsymbol{\theta}) = \sum_{r=1}^{R} N_r(\boldsymbol{\theta}) \, f \left[\frac{T(\chi_r^{\boldsymbol{\theta}})}{N_r(\boldsymbol{\theta})} \right] \tag{6.15}$$

In Eq. 6.15, $N_r(\boldsymbol{\theta})$ is the number of pixels in $\Delta_r^{\boldsymbol{\theta}}$, $f(x)$ is a function depending on the graylevel pdf, and $T(\chi_r^{\boldsymbol{\theta}})$ is the sufficient statistic for this pdf computed on the sample $\chi_r^{\boldsymbol{\theta}}$ (see Eq. 4.15).

The partition $\mathbf{w}^{\boldsymbol{\theta}}$ which minimizes $J(\mathbf{s}, \boldsymbol{\theta})$ then performs the maximum likelihood segmentation of the scene s relative to a certain graylevel pdf. The active grid is therefore deformed to minimize this criterion.

6.3.1.2   Optimization procedure

As in the case of the polygonal snake, the partition $\mathbf{w}^{\boldsymbol{\theta}}$ will be induced by a polygonal grid, that is, a set of nodes linked by segments to define the boundaries of polygonal regions. The shape parameter vector $\boldsymbol{\theta}$ is then the list of the nodes of this polygonal grid. The grid can be simply deformed by moving the nodes and the minimization of the criterion in Eq. 6.15 is then performed by iterating the same process as in the polygonal snake (see Section 5.3.2):

- randomly choose and randomly move one of the grid nodes,

- accept the move if it has lowered the criterion; cancel it otherwise,

until a stopping condition is met.

Like all approaches based on deformable models, the active grid suffers from the classical difficulty of sensitivity to the initialization. To reduce this sensitivity, a multiresolution scheme analogous to the one described in Section 5.3.2 can be implemented. It consists in increasing the number of nodes during the optimization process. In a first step, the grid contains few nodes and the convergence yields a rough result. This result is used as an initialization of a second step in which the number of nodes is increased to refine the result. In general, the segmentation is achieved in two or three steps. The addition of nodes is performed as follows: One goes along each segment of the grid and adds new nodes on the segment so that the distance between two consecutive nodes does not exceed a given threshold d, which can then be viewed as the resolution of the grid.

6.3.2   Implementation issues

The practical implementation of the active grid is similar to that of the polygonal snake. However, since the topology of the grid is more complex, it requires some modifications in terms of topology description and of implementation of the fast convergence procedure.

6.3.2.1   Topology of the grid

The active grid includes P nodes and R polygonal regions. This grid is described by two data structures:

- One structure contains the spatial coordinates of the P nodes. This structure changes during the convergence.

- The other structure is relative to the grid topology, i.e., the relationship between nodes and regions. This structure remains invariant during the convergence.

The topology of the grid is represented by an oriented, valued graph (Figure 6.15). To each node in the grid corresponds a vertex in the graph. When two nodes M and M' are linked by a segment, the valuation of the corresponding arc $\overrightarrow{MM'}$ is the label of the region on the left side of the edge $\overrightarrow{MM'}$. This graph gives access to two types of information. Given a region, one can determine the list of nodes that define its polygonal boundary. Given a node, one can determine the different regions to which it belongs.

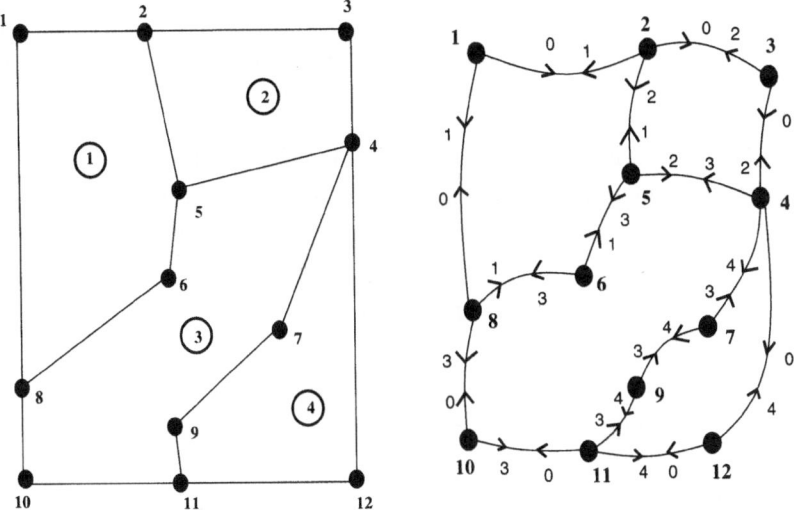

Figure 6.15. Example of grid with 12 nodes and 4 regions (left) and its topology graph representation (right). ©2001 Elsevier Science.

6.3.2.2 Fast algorithm

The basic step of the iterative optimization algorithm is to move one of the nodes and compute the criterion for the new state of the grid. When a node is moved, the deformation of the grid is local: Only the regions to which the node belongs are affected. Thus, computing the new value of the criterion (Eq. 6.15) only requires the calculation of the new values of $N_r(\theta)$ and $T(\chi_r^\theta)$ for each region Δ_r^θ affected by the move. The involved calculations are 2-D summations over Δ_r^θ of functions $\beta[s(x,y)]$ like $\sum_{(x,y)\in\Delta_r^\theta}[1]$, $\sum_{(x,y)\in\Delta_r^\theta}[s(x,y)]$, or $\sum_{(x,y)\in\Delta_r^\theta}[s^2(x,y)]$. To decrease the computing time, it is then possible to adapt the fast algorithm described in Section 5.2.2 which transforms a 2-D summation over a region Δ_r^θ into a 1-D summation along its contour δ_r^θ. For this purpose, one has to specify how a region is deduced from its contour or more generally, how the function $\mathbf{w}^\theta$ is defined from a given state of the polygonal grid.

We can adopt a convention analogous to the one used in Section 5.2.2.1 to define the polygonal snake. The region Δ_r^θ corresponding to a contour δ_r^θ is defined as the strict interior of the contour translated by $(1/2, 1/4)$ (see Figure 6.16). With this convention, the different regions of the grid do not overlap and the equivalence between 2-D and 1-D summations can then be written:

$$T(\chi_r^{\boldsymbol{\theta}}) = \sum_{(x,y)\in\Delta_r^{\boldsymbol{\theta}}} \beta[s(x,y)] = \sum_{(x,y)\in\delta_r^{\boldsymbol{\theta}}} c(x,y)F(x,y) \qquad (6.16)$$

where $c(x,y)$ is a contour encoding associated with the convention, whose values are given in Table 5.1. In Eq. 6.16, $F(x,y)$ is a preprocessed version of the scene which is computed once for all, before the iterative process:

$$F(x,y) = \sum_{z=0}^{x} \beta\left[s(z,y)\right] \qquad (6.17)$$

Furthermore, the computation of the new value of $J(\mathbf{s},\boldsymbol{\theta})$ after a node has moved does not actually require one to sum $c(x,y)F(x,y)$ along the totality of each contour $\delta_r^{\boldsymbol{\theta}}$. Indeed, the move of the node only affects the segments to which it belongs and one can limit the summation to only these segments.

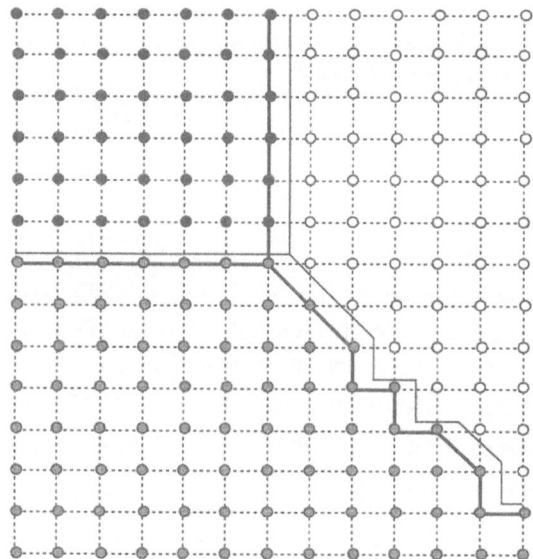

Figure 6.16. Convention adopted to deduce the partition $\mathbf{w}^{\boldsymbol{\theta}}$ from the polygonal grid (linked pixels). The three regions (pixels in white, gray, and black) are defined as the strict interiors of the contours (thick lines) translated by $(1/2, 1/4)$ (represented by the thin lines). ©2001 Elsevier Science.

6.3.3   Some segmentation examples

In this section, we present segmentation results with approximative computing times to illustrate the performance of the active grid (see Table 6.1). These

	No. of regions	Final no. of nodes	Image size	Computing time
Aircraft	3	34	307×210	350 ms
Ship	3	24	256×256	400 ms
SAR	7	16	64×96	200 ms

Table 6.1.   Computing times (obtained on a 700-MHz Pentium III PC) for the segmentation of the three presented images.

results were obtained on a PC under Linux (Mandrake 7.0) with a 700-MHZ Pentium III processor and a 256 Mo RAM. The same optimization scheme (see Section 6.3.1.2) is applied to all images. The number of nodes is progressively increased in a three-step convergence: $d = 20$ at the beginning of the second step and $d = 15$ at the beginning of the third one.

In Figure 6.17, an aircraft appears on a background constituted of two regions: the sky and the ground. A two-region statistical snake is not able to segment the object in this case. Using the Gaussian criterion $J_{gau}(\mathbf{s}, \boldsymbol{\theta})$, one obtains a satisfactory result with a three-region active grid. A similar result is presented in Figure 6.18 with a binary image. The Bernoulli criterion is used here to segment the ship appearing on a sky and sea background. Figure 6.19 illustrates another application: the segmentation of agricultural scenes acquired with SAR [77]. Since these images are corrupted by speckle noise, the Gamma criterion is well suited in this case. The active grid is used here as a refinement stage to locate precisely the boundaries of the fields. It is initialized with a rough segmentation obtained by edge detection and watershed extraction [73].

6.3.4   Conclusion

The automatic partition of an image into several homogeneous regions is one of the most studied problems in image processing. The first approaches based on modeling the images as a "patchwork" of statistically homogeneous regions have used random Markov fields to simultaneously recover the contours and the region graylevels [118, 119]. More recently, several different solutions using active contours have been proposed. Leclerc [120] and Zhu and Yuille [96] have designed segmentation methods using the MDL principle. Paragios and Deriche [121] and Vese and Chan [122] have proposed multiregion segmentation methods with region-based active contours implemented with level sets.

In this context, the statistical active grid described in this section is quite fast when an accurate initialization is available. Recently, Galland *et al.* [123] have proposed a generalization of the statistical active grid which no longer assumes that the number and the approximate frontiers of the regions are known. The

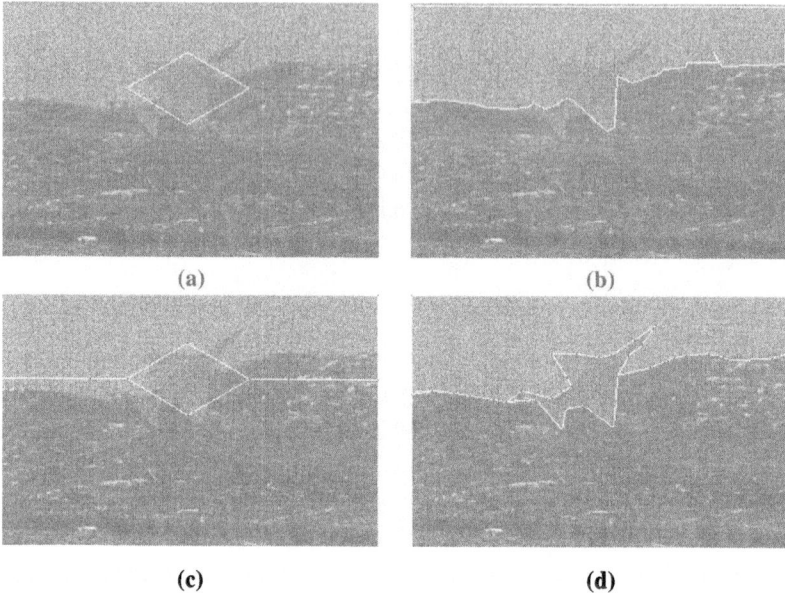

Figure 6.17.   **(a)** Image of an aircraft and initialization of the two-region snake. **(b)** The two-region snake fails in segmenting the aircraft because the background is composed of two distinct regions. **(c)** Initialization of the active grid with three regions. **(d)** Active grid after convergence. ©2001 Elsevier Science.

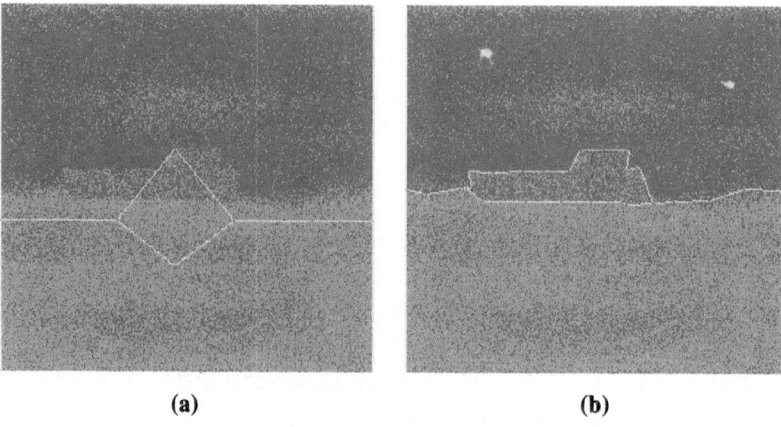

Figure 6.18.   **(a)** Binary image of a ship and initialization of the active grid with three regions. **(b)** Active grid after convergence. ©2001 Elsevier Science.

number of regions and of nodes is estimated with an MDL approach similar to that introduced in Section 6.2.

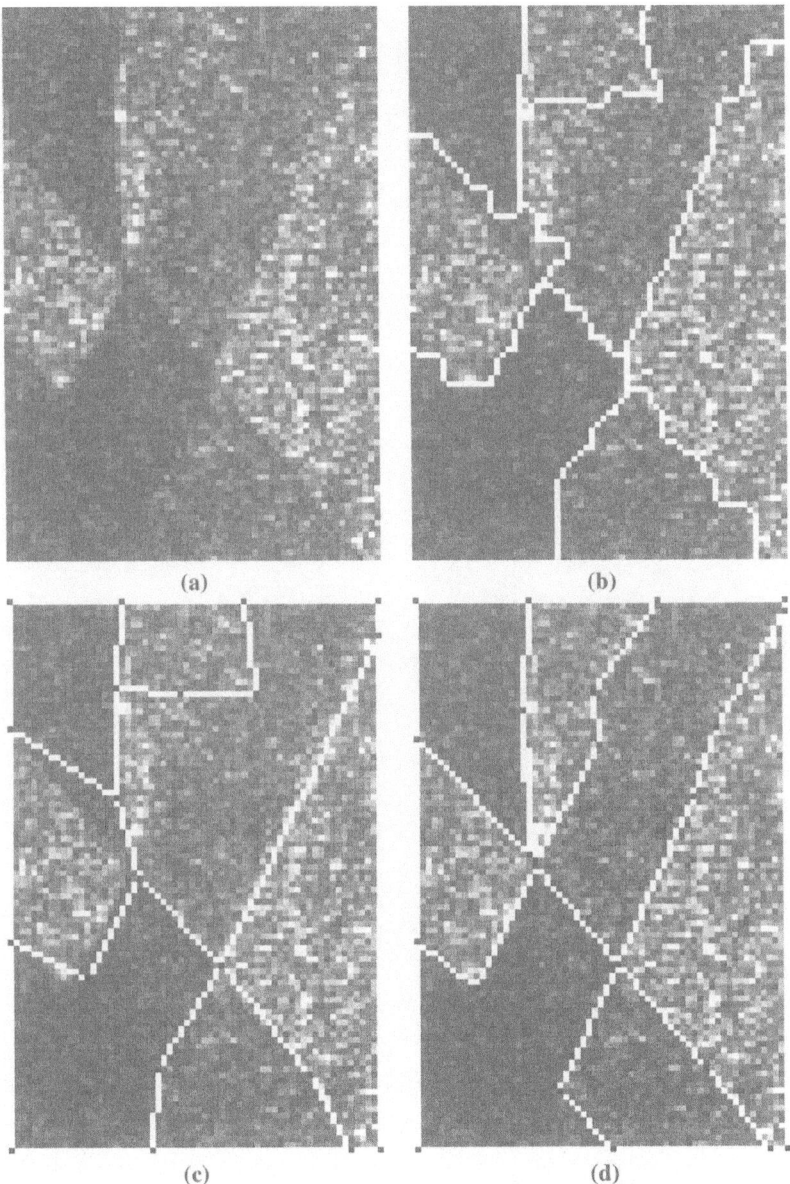

Figure 6.19.   (a) Extract of an ERS-1 three-look SAR image, provided by the CNES and delivered by ESA. (b) Segmentation obtained by edge detection and watershed edge extraction. (c) Initialization of the active grid obtained by sampling the presegmentation presented in (b). The nodes are in black. (d) Active grid after convergence. ©2001 Elsevier Science.

Chapter 7

AN EXAMPLE OF APPLICATION: PROCESSING OF COHERENT POLARIMETRIC IMAGES

In this chapter, we will illustrate the statistical image processing techniques described throughout this book on a nonconventional imaging modality: active polarimetric imagery. This will provide us with the opportunity of introducing important new aspects of the design of algorithms based on decision and estimation theory. In particular, we will analyze how to design computationally efficient detection and segmentation algorithms when the noise pdf does not belong to the exponential family. We will also present a general method for defining the contrast between two image regions in scalar or vectorial images with arbitrary statistics.

In Section 7.1, we introduce the basics of active polarimetric imaging with coherent light and the statistical description of speckle in such images. We address in Section 7.2 the case of images of degree of polarization (DOP) and illustrate some of their advantages over intensity images. DOP images are single channel and represent the degree of polarization of the light reflected or transmitted by each point of a scene. We characterize their graylevel statistics in Section 7.3 and present some applications to target detection and object segmentation in Section 7.4. In Section 7.5, we consider Stokes images, which are vectorial images that represent the complete polarimetric state of light. We first address the problem of defining a contrast parameter in these vectorial images and show that this contrast can be defined by two parameters which have a clear physical meaning. We then present in Section 7.6 some image processing applications to edge detection and segmentation in coherent Stokes images.

7.1. Basics of polarimetric imaging

The polarization state of light contains important information about a scene, which are complementary to information provided by light intensity and color.

197

Forming an image of the polarization properties of the light reflected or emitted by a scene is thus useful in such applications as scene analysis, robotics, and automatic target recognition [124-126]. This can reveal contrasts between two zones of the scene which have the same intensity reflectivity (and thus no contrast appears in the intensity image) but different polarimetric properties. It can thus improve the detectability of small, low-contrast objects in images [127].

Polarimetric imaging systems can be classified into two categories: *passive* and *active*. In passive systems, it is the natural reflection of ambient light, or the natural emission for infrared light, which is used for forming the image. The polarimetric camera measures the polarization state of the incoming light, usually in the form of its Stokes parameters. By analyzing the direction of polarization and the degree of partial polarization of the reflected light, important information can be obtained about the presence and the orientation of occluding edges [126, 128] or the surface orientation of the observed objects [126,129-131]. It has also been shown that polarization imagery could help in distinguishing a real object with its reflected image [126] or detecting the reflection of an outdoor scene through a semireflector (e.g., a window pane) [132]. These cues are useful in image understanding applications [130, 131] and are complementary with those provided by intensity and color imaging. Another important application of polarimetric imagery is the determination of the nature of the materials present in a scene from their polarimetric properties [133]. It can help for example discriminating between grass, dirt, and rocks in a natural scene [131], which can be difficult in intensity or even color images. Polarimetric imagery has also been found useful in detecting rust during ship hulls inspection [131] and in discriminating between dielectric and metallic areas on circuit boards [126, 130].

Active polarimetric imaging systems use artificial illumination of the scene. The image is formed from polarimetric analysis of the reflected (or transmitted) fraction of this light. Active images are useful for the same type of applications as passive ones, but their main advantage is that the polarization state of the illuminating light can be controlled, which makes it easier to interpret the reflected polarization state and thus the underlying physical properties of the scene. Active polarimetric imagery can be used for enhancing visibility through water or fog [127] and a number of active polarimetric imaging systems have been developed for outdoor operations and remote sensing [134-137]. Active polarimetric imaging is also used in medical imaging, for example in order to form images of the superficial layers of the skin [138-141] or of the retina in ophthalmologic inspection [142].

In active systems, scene illumination is often performed with laser light. Lasers are preferred for their beam directivity, which makes it possible to image at long distances. Another advantage of laser light is its small spectral bandwidth, which can improve optical detection. However, lasers produce coherent light and the images are thus perturbed with speckle noise in the same way as microwave SAR images (see Section 4.4). This makes it necessary to develop specific processing algorithms.

In the following of this section, we briefly introduce the notions of polarimetry used throughout this chapter. We then describe the basic principle of active polarimetric imaging systems and finally give a few definitions about the statistical distribution of the electric field in coherent polarimetric images.

7.1.1   The representation of polarized light

The evolution of the electric vector $\mathcal{E} = (\mathcal{E}_1, \mathcal{E}_2)^T$ of a purely polarized monochromatic wave of pulsation ω as a function of time is classically represented by its analytic signal, which is a complex vector $\mathbf{E}$ whose real part is $\mathcal{E}$ and which is defined as:

$$\begin{cases} E_1 &= A \cos \alpha \, e^{j\omega t} \\ E_2 &= A \sin \alpha \, e^{j(\omega t - \varphi)} \end{cases} \tag{7.1}$$

As time varies, the extremity of the real part of vector $\mathbf{E}$ describes an ellipse whose characteristics (angle of the main axis and ellipticity) are related to the angles α and φ. The vector $\mathbf{E}$ can be written as $A \, \mathbf{J} \, e^{j\omega t}$, where $\mathbf{J} = [\cos \alpha, \sin \alpha \, e^{-\varphi}]^T$. $\mathbf{J}$ is usually called the *Jones vector* and defines the polarization state of purely polarized light with unit intensity.

In practice, light is often partially polarized. Partially polarized light is classically represented by a 2-D circular Gaussian random vector [143], that is, with a pdf equal to:

$$P_\Gamma(\mathbf{E}) = \frac{1}{\pi^2 \det \Gamma} \exp\left[-\mathbf{E}^\dagger \, \Gamma^{-1} \, \mathbf{E} \right] \tag{7.2}$$

Γ is the covariance matrix, which we denote as follows:

$$\Gamma = \begin{bmatrix} \langle |E_1|^2 \rangle & \langle E_1 E_2^* \rangle \\ \langle E_2 E_1^* \rangle & \langle |E_2|^2 \rangle \end{bmatrix} = \begin{bmatrix} \mu_1 & \delta \\ \delta^* & \mu_2 \end{bmatrix} \tag{7.3}$$

where the symbol $< . >$ denotes ensemble averaging and $\det \Gamma$ is the determinant of Γ. This covariance matrix Γ defines the polarization state of the vector. In the optics community, it is frequently called *coherency matrix* [143-146] . Γ is Hermitian and thus completely defined by four real-valued parameters: the average intensities μ_1, μ_2 and the complex correlation coefficient δ. Moreover, it can be diagonalized into the following form:

$$\Gamma = U \, D \, U^\dagger \tag{7.4}$$

where U is a unitary matrix and D a diagonal matrix whose eigenvalues d_1 and d_2 are positive [78]. We will assume in the following that the eigenvalues are ordered so that $d_1 \geq d_2$. The matrix U being unitary, it is uniquely defined by the eigenvector $\mathbf{u}$ associated with d_1. $\mathbf{u}$ is the Jones vector of a purely polarized state termed *principal polarization state* of the light. The degree to which a given state of light is purely polarized can be quantified by the degree of polarization (DOP), defined as follows [143]:

$$\mathcal{P} = \frac{d_1 - d_2}{d_1 + d_2} \tag{7.5}$$

When $\mathcal{P} = 0$, Γ is proportional to the unit matrix: there is no privileged polarization state, the light is then termed totally depolarized. On the other hand, $\mathcal{P} = 1$ means that $d_2 = 0$, or, in other words, Γ is rank-deficient. In this case, the light is purely polarized in the state defined by the Jones vector $\mathbf{u}$.

It can be noted that the information contained in the coherency matrix is totally equivalent to that contained in the well-known Stokes vector $\mathbf{S} = [S_0, S_1, S_2, S_3]^T$ which is also frequently used to represent the polarimetric properties of light [147, 148] since:

$$\begin{cases} S_0 &= \mu_1 + \mu_2 \\ S_1 &= \mu_1 - \mu_2 \\ S_2 &= 2\,\mathrm{Re}(\delta) \\ S_3 &= 2\,\mathrm{Im}(\delta) \end{cases} \tag{7.6}$$

The Stokes vector has interesting properties in terms of the representation of polarization. Since the component S_0 represents the total intensity of the light, $\mathbf{S}$ can be normalized so as to define a 3-D subvector $\mathbf{s} = [S_1/S_0, S_2/S_0, S_3/S_0]^T$. $\mathbf{s}$ can be rewritten as:

$$\begin{cases} s_1 &= \mathcal{P}\,\cos 2\psi\,\cos 2\chi \\ s_2 &= \mathcal{P}\,\sin 2\psi\,\cos 2\chi \\ s_3 &= \mathcal{P}\,\sin 2\chi \end{cases} \tag{7.7}$$

The norm of $\mathbf{s}$ is equal to the degree of polarization $\mathcal{P}$ and its direction can be described by the polar angles 2ψ and 2χ [147]. It can be shown that the angle ψ represents the direction of the polarization ellipse of the principal polarization state $\mathbf{u}$ and χ represents its ellipticity (see Figure 7.1). These three parameters completely describe a polarization state of unit intensity: They constitute the Poincaré representation of the polarization state [147]. For example, purely polarized states, which correspond to $\mathcal{P} = 1$, lie on the sphere of unit radius (called Poincaré sphere). Linearly polarized states lie on the equator of this sphere, since they correspond to a null ellipticity. On the other hand, circularly polarized states lie on the north and south poles depending on the sense of rotation of the polarization vector. Partially polarized states with DOP $\mathcal{P}$ lie on a sphere of radius $\mathcal{P}$ and the position of the representative point on this sphere depends on the characteristics of the principal polarization state. The Poincaré representation of polarized light will be used for defining the polarimetric contrast in Section 7.5.2.

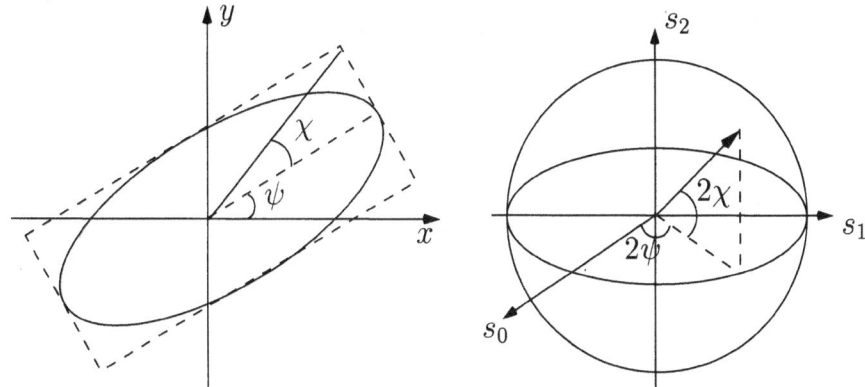

Figure 7.1. Left: Representation of the polarization ellipse of a pure polarization state. Right: Poincaré representation of the same polarization state.

It is often necessary to describe how the state of polarization of the light is modified when interacting with a material. Let S_{in} denote the Stokes vector of the light incoming on a given material and S_{out} the Stokes vector of the light reflected or transmitted by this material. It is classically assumed that there is a linear relation between the incident Stokes vector S_{in} and the outgoing Stokes vector S_{out}, which is described by a 4×4 matrix M, so that $S_{out} = M \, S_{in}$ [149]. The matrix M is classically termed the *Mueller matrix*. Provided the hypothesis of linearity is valid, this matrix completely describes the properties of the material with respect to interaction with polarized light. The Mueller matrix is thus an intrinsic property of materials, whereas the Stokes vector is a property of polarized light.

7.1.2 Active polarimetric imaging systems

An example of active optical polarimetric imaging system is sketched in Figure 7.2 [136]. The scene is illuminated with collimated light from a laser. The polarization state of the emitted light is controlled by a Polarization State Generator (PSG). This device can generate any purely polarized state on the Poincaré sphere. The light backscattered by the scene is then collected by a telescope to form an image on a CCD camera. However, before reaching the CCD, the light is analyzed by a Polarization State Analyzer (PSA) which can analyze the backscattered light in any polarization state [134, 136]. At least four intensity measurements are necessary to measure the Stokes vector of the backscattered light, although more measurements can improve the accuracy of the polarimetric parameter estimation.

With an active system, it is in principle possible to measure the whole Mueller matrix by illuminating the scene with four different polarization states and measuring the polarization of the backscattered light for each of these states [135,136,150]. However, this requires the acquisition of at least 16 images.

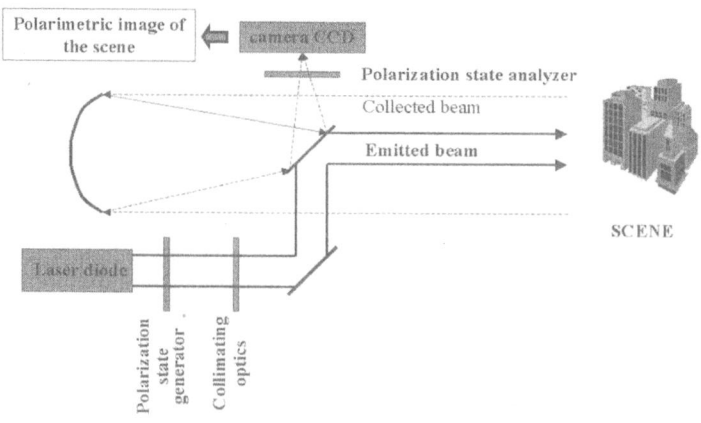

Figure 7.2.   Principle of an active polarimetric imaging system.

7.1.3   Model of coherent polarimetric images

The main specificity of coherent polarimetric images is the presence of speckle noise. This noise is very disturbing since its standard deviation has the same order of magnitude as the average value of the signal. The speckle phenomenon is caused by the backscattering of coherent light on rough surfaces whose average roughness is larger than the wavelength of the illuminating light. The electric vector $\mathbf{E} = (E_1, E_2)$ of the light backscattered by any point of such a surface is classically represented by a 2-D circular Gaussian random vector with zero mean distributed with the pdf described in Eq. 7.2. Its polarization state is thus characterized by its covariance matrix Γ [143]. One can note that in coherent polarimetric images, each speckle can be purely polarized, since it corresponds to a given realization of the random coherent electric field. However, this polarization state may vary spatially from one speckle to another, so that on a spatial averaging point of view, the light may be totally or partially depolarized.

7.2. Processing degree of polarization (DOP) images

In this section, we consider a simple but useful configuration of polarimetric imaging, which consists in imaging the DOP of the light reflected or transmitted from the scene. This case is simple since the DOP images are single-channel images. However, DOP images can reveal contrasts that do not appear in intensity images [127,136,151,152]. For example, consider a metallic object appearing against a natural background and whose intensity reflectivity is similar to that of the background. The metallic object will in general depolarize the incident light less than the natural background and thus a contrast will appear in the DOP image. Using the DOP image can thus improve the detectability of targets with low intensity contrast.

Our goal with this particular example is to illustrate how one can generalize the previously described techniques of detection, estimation, and segmentation, which have been developed for pdf which belong to the exponential family. Indeed, we will see that the pdf of the DOP does not belong to this family but possesses specific properties which make it possible to accurately approximate its statistical distribution with Gaussian laws provided an appropriate transformation of the data is performed.

7.2.1 Principle of DOP imaging

In its simplest configuration, the polarimetric imaging system described in Section 7.1.2 illuminates the scene with a single elliptical polarization state and analyzes the backscattered light in the polarization states parallel and orthogonal to the incident one [134, 136]. Let $\mathbf{u}_1$ denote the incident polarization state and $\mathbf{u}_2$ the polarization state which is orthogonal. At each pixel (i, j) within the image, the electric field is $\mathbf{E}(i, j) = (E_1(i, j), E_2(i, j))^T$. According to the model described in Section 7.1.3, it is a random vector with zero mean and coherency matrix $\Gamma(i, j)$. When the light passes through an analyzer parallel to the incident state $\mathbf{u}_1$, one measures an intensity $s_1(i, j) = |E_1(i, j)|^2$, which is a random variable whose mean is equal to $< s_1(i, j) > = \mu_1$ (see Eq. 7.3). In the same way, when the light passes through an analyzer parallel to the orthogonal state $\mathbf{u}_2$, one measures an intensity $s_2(i, j) = |E_2(i, j)|^2$ whose mean is equal to $< s_2(i, j) > = \mu_2$.

In the following, we will denote the ensemble of two images $\mathbf{s}_1$ and $\mathbf{s}_2$ as the two "channels" which form a Two-Channel Image (TCI). It is obvious that the sum of these images corresponds to the intensity backscattered by the scene. One can also compute the Orthogonal State Contrast Image (OSCI) as follows:

$$\rho(i, j) = \frac{s_1(i, j) - s_2(i, j)}{s_1(i, j) + s_2(i, j)} \tag{7.8}$$

This image can be considered as an estimation of $\frac{\mu_1-\mu_2}{\mu_1+\mu_2}$ at each pixel. Let us now assume that the materials which compose the scene only modify the degree of polarization of incident light without modifying its principal polarization state. This kind of material is called a *pure depolarizer*. By definition of the notion of pure depolarizer, the coherency matrix of the backscattered light is diagonal and the average intensities μ_1 and μ_2 are equal to the eigenvalues d_1 and d_2. In this case, one can see that the OSCI is an estimate of the DOP of the light ($\mathcal{P} = \frac{d_1-d_2}{d_1+d_2}$, see Eq. 7.5) at pixel (i,j).

Since the incident light is purely polarized, the OSCI is linked to the depolarization capacity of the observed materials. If $\rho(i,j)$ is close to 1, the degree of polarization of the backscattered light is large, which means that the depolarization capacity of the corresponding material is low. On the other hand, if $\rho(i,j)$ is close to 0, the corresponding zone in the scene is strongly depolarizing. Experimental studies [136] have shown that when observed in backscattered light, most materials encountered in real outdoor scenes behave as pure depolarizers. In this section, we will assume that this is true and we will consider images in which the contrast between the different objects comes from their different depolarization capacities. However, let us note that even if this hypothesis is not perfectly fulfilled, the OSCI, which does not exactly represent the DOP in this case, can still reveal interesting contrasts in images.

Polarimetric images built in a similar way as the OSCI have been used in the literature [124, 125, 127, 141]. It has early been noticed [124] that images of this type are independent of the spatial nonuniformity of the illumination since they are normalized by the local intensity. Indeed, the spatial nonuniformity of the illumination is a standard problem in active imaging. It is particularly crucial for long-range systems which use high-power sources that often produce nonuniform wave fronts. Moreover, even if the beam is uniform at the output of the source, it can be deformed by atmospheric perturbations. This nonuniformity is thus difficult to eliminate or even to compensate (the atmospheric perturbations being difficult to exactly predict). It constitutes a problem for detection algorithms, since it can create false target-like patterns.

The *nonuniformity cancellation* property of the OSCI is thus an important advantage in terms of signal processing. We show in Figure 7.3 real-world polarimetric images obtained with the system presented in Section 7.1.2. The observed scene is composed of a piece of paper (in the center of the image) appearing on a background made with a different type of paper. The images in the parallel and orthogonal channels (respectively s_1 and s_2) and the corresponding OSCI are represented. We can notice a spatially varying pattern covering the channels s_1 and s_2. The details of this pattern can be easily taken for low-contrast small targets and will thus create false alarms. This pattern is not due to the observed scene, in which the background is homogeneous: it is due to the nonuniformity of the illumination beam. We can observe in Figure 7.3 that this illumination pattern has been canceled in the OSCI. In this image, the object of interest appears on a more homogeneous background and will be easier to detect than in the images s_1 or s_2. We will show in the following that this feature makes the OSCI preferable for small target detection and segmentation when using statistical processing algorithms based on the SIR model.

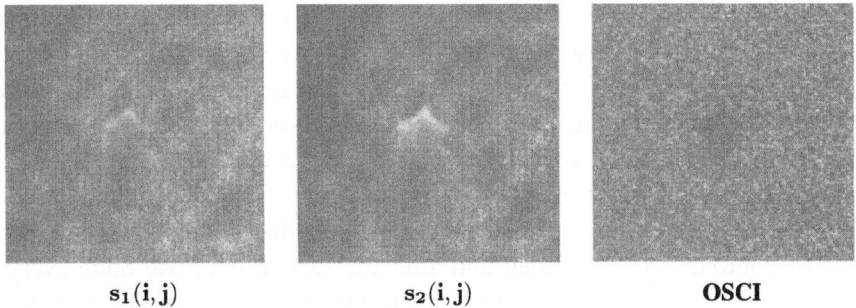

| $s_1(i,j)$ | $s_2(i,j)$ | **OSCI** |

Figure 7.3.   Example of real polarimetric image, with parallel channel s_1, orthogonal channel s_2, and OSCI. ©2002 OSA.

7.2.2   Influence of illumination nonuniformity on segmentation performance

In this section, we intend to evaluate the gain in processing performance that is obtained by using the OSCI instead of intensity images in the presence of illumination nonuniformities [153]. In order to do so, we consider the task of segmenting a uniform object on a homogeneous background by using the polygonal statistical snake introduced in Chapter 5.

The TCI delivered by a polarimatric imaging system of the type defined in Section 7.1.2 is a vectorial image whose two channels are perturbed with Gamma noise. This is exactly the case analyzed in Section 6.1, and we have seen that in this case, the snake energy has the following expression:

$$J_{tci}(\mathbf{s}, \boldsymbol{\theta}) = N_a^{\boldsymbol{\theta}} \sum_{k=1}^{2} \log \left[\widehat{\mu_a^{(k)}}(\boldsymbol{\theta}) \right] + N_b^{\boldsymbol{\theta}} \sum_{k=1}^{2} \log \left[\widehat{\mu_b^{(k)}}(\boldsymbol{\theta}) \right] \qquad (7.9)$$

where $N_a^{\boldsymbol{\theta}}$ and $N_b^{\boldsymbol{\theta}}$ are the numbers of pixels in regions $\Delta_a^{\boldsymbol{\theta}}$ and $\Delta_b^{\boldsymbol{\theta}}$, $\widehat{\mu_a^{(k)}}(\boldsymbol{\theta})$ and $\widehat{\mu_b^{(k)}}(\boldsymbol{\theta})$ the average pixel intensities for the channel k estimated in these two regions. The active contour procedure maximizing the function in Eq. 7.9 will be referenced as the *TCI-snake*.

Let us now consider segmenting the target on the OSCI. If the images s_1 and s_2 are Gamma distributed, we will show in Section 7.3 that the pdf of the OSCI only depends on the parameter $\gamma = \mu_1/\mu_2$, where μ_1 is the average graylevel in image s_1 and μ_2 the average graylevel in s_2. The exact pdf of the OSCI will be determined later (in Section 7.3), but for the sake of simplicity, we will apply to the OSCI the active contour procedure which is optimal for Gaussian data with unknown mean and variances $J_{gau}(s, \boldsymbol{\theta})$ (see Eq. 5.20). This segmentation procedure will be called in the following *OSCI-snake*. Let us compare the results of segmentation of real images acquired with the active polarimetric imager described in Ref. 136. We show in Figure 7.4 two examples of two-channel images acquired with this device. The results of the segmentation of the images with the TCI-snake and the OSCI-snake are displayed in the last two columns of this figure. The initial contour is a small square with four nodes centered on the target, as shown in the first column of Figure 7.4. For segmentation, a three-step multiresolution approach is used. After the first iteration, one node every 10 contour pixels is added and after the second one, one node every 5 contour pixels. In this last step, we have introduced a regularization term as in Eq. 5.38 with $\lambda = 0.2$, which tends to smooth the contours.

Let us consider the first row of Figure 7.4. We can see that in images s_1 and s_2, the upper part of the target has a good (negative) contrast with the background, but the lower part is perturbed by a bright illumination patch. The TCI-snake only segments the upper part of the target. On the other hand, in the OSCI, the effect of the illumination has been canceled and the whole target appears with a good contrast. The OSCI-snake is thus able to segment the whole target. In the second row of Figure 7.4, it can be seen that the illumination pattern strongly perturbs the visibility of the target. The TCI-snake is unable to segment it and produces very poor segmentations similar to those which have been characterized in the previous section. On the other hand, the OSCI-snake is able to correctly segment the target.

In conclusion, these experimental results show that in the presence of nonuniform illumination, the OSCI-snake leads to much better performance than the TCI-snake, provided there is a sufficient contrast of depolarization capabilities between the object of interest and the background. In the next section, we will characterize more precisely the statistical properties of noise in the OSCI.

7.3.   The statistics of the OSCI and its natural representation

In order to perform optimal processing of the OSCI, it is necessary to determine the pdf of the fluctuations of the graylevels in this image. This is a nontrivial task since the OSCI is a nonlinear combination of two speckled images. Indeed, when the pdf of s_1 and s_2 are assumed Gamma distributed, we will see that the pdf of the OSCI does not belong to the exponential family. However, we will show that, under quite general assumptions, the pdf of the OSCI

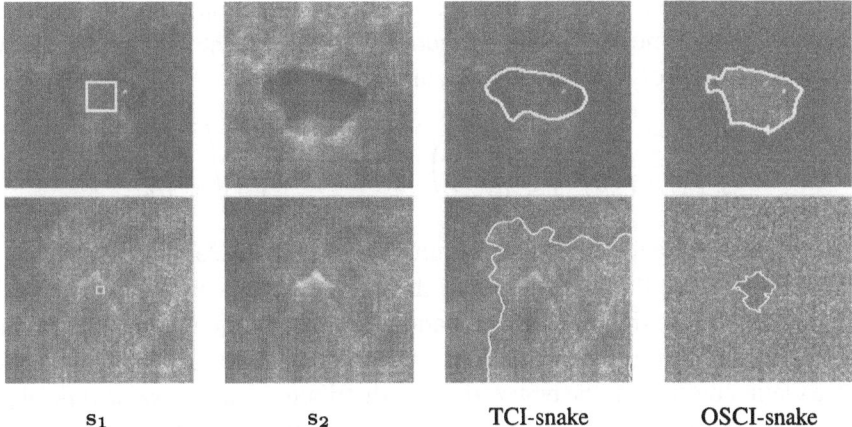

| s_1 | s_2 | TCI-snake | OSCI-snake |

Figure 7.4. Two examples of segmentation results on real polarimetric images. First column: s_1 with initial contour. Second column: s_2. Third column: s_1 with the result of TCI-snake.Fourth column: OSCI with the result of OSCI-snake. ©2002 OSA.

possesses interesting properties thanks to which it can be easily approximated by a Gaussian pdf [154].

In Section 7.3.1, we will determine the general properties of the OSCI pdf for a wide class of different speckle statistics. We then determine in Section 7.3.2 the general expression of the pdf of the OSCI for this type of speckle statistics, and comment on some of its properties. Finally, we introduce in Section 7.3.3 a new representation of the OSCI in which the pdf of the noise is easier to handle.

7.3.1 Speckle and multiplicative noise

As seen in the previous section, the value measured by the polarimetric imager at each pixel of the image $s_k(i,j)$, $k = 1, 2$, fluctuates around the coefficient μ_k of the coherency matrix. These fluctuations are called "speckle." As seen above, the usual model for speckle perturbations is Gamma law. However, the Gamma model is not always the best one to describe real-world speckle. For example, phenomenological studies performed mainly on radar images have shown that in some cases, speckle fluctuations are better represented by a Weibull law [155]:

$$P(x) = \frac{b}{\mu_k} \left(\frac{x}{\mu_k} \right)^{b-1} \exp\left[-\left(\frac{x}{\mu_k} \right)^b \right] \tag{7.10}$$

where b is the shape parameter of the law.

For some coarse materials, it may be possible that the reflectivity μ_k varies spatially inside a zone composed of the same material. If one assumes that this value μ_k is a random variable which follows a Gamma law of order ν and of mean $< \mu_k >$, and if one models the speckle phenomenon with a Gamma law of order L and of mean equal to a random variable μ_k, the observed graylevel follows a "K law" of parameters L, ν, and $< \mu_k >$ [77]:

$$P(x) = \frac{2}{x\, \Gamma(L)\, \Gamma(\nu)} \left(\frac{L\nu x}{<\mu_k>}\right)^{\frac{(L+\nu)}{2}} K_{\nu-L}\left(2\sqrt{\frac{L\nu x}{<\mu_k>}}\right) \quad (7.11)$$

where $K_n(x)$ is the modified Bessel function of second class of order n.

We can thus see that depending on the experimental conditions and on the observed materials, the speckle fluctuations are represented with different pdf. These pdf are multiplicative, in the sense that if the random variable is multiplied with a scalar coefficient, the new variable is distributed with the same type of pdf as the initial variable, but with a parameter μ_k multiplied with this coefficient. The general model we will consider for joint pdf of $s_1(i,j)$ and $s_2(i,j)$ has the following form:

$$P_{s_1,s_2}(x,y) = \frac{1}{\mu_1 \mu_2} U\left(\frac{x}{\mu_1}, \frac{y}{\mu_2}\right) \quad (7.12)$$

where μ_1 and μ_2 are the parameters in the two channels. The function $U(x,y)$ represents the pdf of channels s_1 and s_2 when $\mu_1 = \mu_2 = 1$. This model includes the Gamma, Weibull, K distributions, and many others. It does not take into account the correlations of electric field that may appear with arbitrary coherency matrices but does not assume that the intensities in both channels are independent.

7.3.2   The probability density function of the OSCI

Having chosen a model for the fluctuations in images $s_1(i,j)$ and $s_2(i,j)$, we analyze in this section the fluctuations in the OSCI. Let us first define the random variables $\tilde{s}_1$ and $\tilde{s}_2$ which are equal to the values of images $s_1(i,j)$ and $s_2(i,j)$ at a given location (i,j) when $\mu_1 = \mu_2 = 1$, and the random variable $\eta = (\tilde{s}_1 - \tilde{s}_2)/(\tilde{s}_1 + \tilde{s}_2)$. η represents the OSCI ρ when light is totally depolarized. It is easy to show that the random variable ρ, which represents the OSCI for any values μ_1 and μ_2, deduces from η in the following way:

$$\rho = \frac{s_1 - s_2}{s_1 + s_2} = \frac{\mu_1 \tilde{s}_1 - \mu_2 \tilde{s}_2}{\mu_1 \tilde{s}_1 + \mu_2 \tilde{s}_2} = \frac{\eta + u}{1 + u\,\eta} \quad (7.13)$$

where

$$u = \frac{\mu_1 - \mu_2}{\mu_1 + \mu_2} \quad (7.14)$$

In particular, one can check that the OSCI depends on parameters μ_1 and μ_2 only through parameter u. When the materials are purely depolarizing, the parameter u is equal to the DOP (see Eq. 7.5) since in this case $\mu_1 = d_1$ and $\mu_2 = d_2$ where d_1 and d_2 are the eigenvalues of the coherency matrix.

On the other hand, one can note that the parameter $\gamma = \frac{\mu_1}{\mu_2}$ is a bijective function of u. Indeed, one can write:

$$u = \frac{\gamma - 1}{\gamma + 1} \quad \text{and} \quad \gamma = \frac{1 + u}{1 - u} \tag{7.15}$$

Consequently, the parameters u and γ can be used interchangeably to specify the properties of the random variable ρ.

Let us now consider the pdf of η, which will be denoted $\Psi(x)$. From Eq. 7.13, one can determine the pdf $P_u^{(\rho)}(x)$ of ρ as a function of $\Psi(x)$ by using the classical relation $P_u^{(\rho)}(\rho)d\rho = \Psi(\eta)d\eta$. One obtains (see Appendix 7.A):

$$P_u^{(\rho)}(\rho) = \frac{1 - u^2}{(1 - u\rho)^2} \Psi\left(\frac{\rho - u}{1 - u\rho}\right) \tag{7.16}$$

It is interesting to notice that the parameter u, which represents the true degree of polarization, is not equal to the average of the random variable ρ. Indeed, we show in Appendix 7.A, by using a series development for low values of u, that the ensemble average of ρ has the following approximate expression:

$$<\rho> -u \simeq (u^3 - u)\sigma_\eta^2 \tag{7.17}$$

where σ_η^2 is the variance of η. On the other hand, we show in Appendix 7.A that u is equal to the median of ρ

As an example, let us consider that the channels s_1 and s_2 are independent and Gamma distributed. Using Eq. 7.A.4 of Appendix 7.A, it can be shown that the pdf of the OSCI has the following expression:

$$P_u^{(\rho)}(\rho) = \frac{(2L - 1)!}{2^{2L-1}[(L - 1)!]^2}(1 - u^2)^L \frac{(1 - \rho^2)^{L-1}}{(1 - u\rho)^{2L}} \tag{7.18}$$

We have represented in Figure 7.5 the functions $P_u^{(\rho)}(\rho)$ for three different values of $u = 0, 0.5, 0.8$ and three values of $L = 1, 5, 50$. Recall that by definition, $P_0^{(\rho)}(x)$ corresponds to the pdf $\Psi(x)$ which characterizes totally unpolarized light. $\Psi(x)$ is bell-shaped except when $L = 1$. One can also observe how $P_u^{(\rho)}(\rho)$ "deforms" and becomes asymmetric when u takes nonnull values.

We have plotted in Figure 7.6 the average, the standard deviation, and the mode[1] of ρ as a function of u for different orders. The average and the standard deviation have been estimated by numerical integration of the analytical form of the pdf in Eq. 7.18. In Figure 7.6, it can be checked that the mean is not equal

---

[1] The mode is the position of the maximum of the pdf.

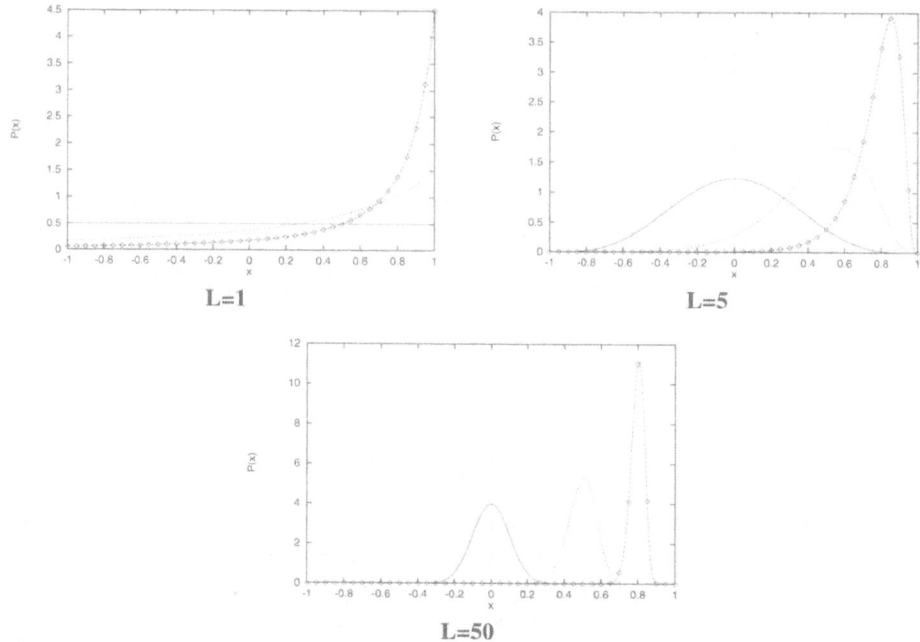

Figure 7.5. pdf of OSCI when the graylevels in both channels are distributed according to Gamma pdf of different orders L and with values of u equal to 0 (continuous line), 0.5 (long dotted line), and 0.8 (short dotted line with $\diamond$). ©2001 OSA.

to u except when u is zero or tends to one, as indicated by Eq. 7.17. On the other hand, the standard deviation decreases when u increases. Finally, when $L = 1$, the mode, i.e., the position of the maximum of the pdf, is equal to 1 if $u > 0$, -1 if $u < 0$, and is not defined for $u = 0$, since the pdf is constant in this case (see Figure 7.5). On the other hand, when L is large, the mode is close to u.

7.3.3   The natural representation of the OSCI

We have seen in the previous section that as u varies, the pdf of ρ undergoes a nonlinear transformation (see Eq. 7.16) which modifies its shape and makes it asymmetric. However, the transformation defined in Eq. 7.13 belongs to a subgroup of the Moebius group [156]. Furthermore, this subgroup is isomorphic to the additive group. In other words, there exists a nonlinear transform of the random variable ρ so that in this new representation, a variation of u only produces a translation of the pdf. In this new representation, the perturbations which affect the OSCI are transformed into an additive noise. In order to show this property, we can notice that there exists a bijective relation between the

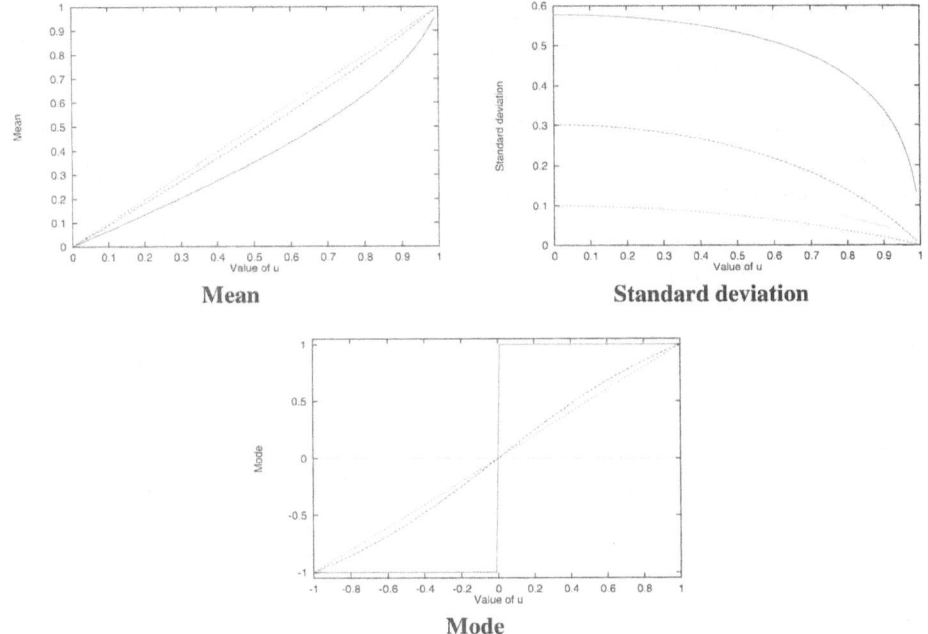

Mean
Standard deviation
Mode

Figure 7.6. Mean, standard deviation, and mode of OSCI when the channels s_1 and s_2 are Gamma distributed, as a function of parameter u for different values of L: $L = 1$ (continuous line), $L = 5$ (long dotted line), and $L = 50$ (short dotted line). ©2001 OSA.

OSCI ρ and the random variable $r = s_1/s_2$. If one takes the logarithm of r, one obtains the random variable β defined as follows:

$$\beta = \log\left[\frac{s_1}{s_2}\right] = \log\left[\frac{\tilde{s}_1}{\tilde{s}_2}\right] + \log\gamma \qquad (7.19)$$

where $\tilde{s}_1$ and $\tilde{s}_2$ are defined in Section 7.3.2, and $\gamma = \mu_1/\mu_2$. The random variable $z = \log\left[\frac{\tilde{s}_1}{\tilde{s}_2}\right]$ only depends on the function $U(x,y)$ and not on the parameter u. We will call $\Phi(x)$ the pdf of the random variable z. It is easy to show that it is even. Indeed, one has $\log\left[\frac{\tilde{s}_1}{\tilde{s}_2}\right] = \log\tilde{s}_1 - \log\tilde{s}_2$, which is the difference between two identically distributed random variables and thus corresponds to a symmetric pdf. This property of symmetry of the pdf is the essential difference between the present approach and the homomorphic treatment of speckled signals, which consists in transforming a multiplicative noise into an additive one by taking its logarithm [157]. Indeed, for intensity speckle signal, the pdf of the resulting noise is not symmetric in general.

From Eq. 7.19, it is clear that the pdf of β is equal to:

$$P_\gamma^{(\beta)}(x) = \Phi(x - \log\gamma) \tag{7.20}$$

As $\dot{\Phi}(x)$ is even, one has $<\beta>= \log\gamma$. Consequently, a modification of the degree of polarization produces a variation of the mean of the image β equal to $\log\gamma$, without modifying the other properties of the distribution. We will call β the *natural representation* of the OSCI, since it corresponds to the representation of the data with the simplest group representation, i.e., the additive group. The bijective relation which exists between the representations ρ and β is simply $\beta = 2\tanh^{-1}(\rho)$.

In the case where the variables s_1 and s_2 are distributed with Gamma pdf, it is an easy matter to determine the pdf $\Phi(x)$. It suffices to notice that $\beta = \log(1 + \rho) - \log(1 - \rho)$ and to apply the classical formula for variable changes (see Ref. 82) to the pdf of ρ (see Eq. 7.18). One obtains the following result:

$$\Phi(x) = \left[\frac{1}{2}\right]^{2L} \frac{(2L - 1)!}{[(L - 1)!]^2} \left[\frac{1}{\cosh(x/2)}\right]^{2L} \tag{7.21}$$

where $\cosh(x)$ denotes hyperbolic cosine. We show in Figure 7.7 the aspect of $\Phi(x)$ for three different values of speckle order L. Let us recall that for a given value of u, and thus of γ, the pdf of image β is simply the function $\Phi(x)$ translated of the value $\log\gamma$. We have also plotted in the same graphs Gaussians of the same variance as $\Phi(x)$. We can see that $\Phi(x)$ is close to a Gaussian and that the similarity increases with the speckle order.

The concept of natural representation remains valid whatever the nature of the speckle in images s_1 and s_2. Let us consider speckles distributed with Weibull and K laws. It is difficult in these cases to determine the expression of $\Phi(x)$, but we have represented in Figure 7.8 estimations of functions $\Phi(x)$ for these laws obtained by determining the histogram of a sufficiently large sample of realizations. One can notice that the functions are similar to those obtained in the case of Gamma laws: They are even and bell-shaped. We have also represented in the same figures the Gaussian pdf with the same variance as $\Phi(x)$. We can see that as in the case of Gamma pdf, $\Phi(x)$ is close to a Gaussian.

7.4.   Applications to image processing of the OSCI

With the help of the statistical analysis performed in the previous section, we will study how to efficiently perform detection of contrasted regions (small targets or edges) and object segmentation on the OSCI. It is clear that on such noisy images as OSCI, it is necessary to take into account the characteristics of the noise in order to design efficient algorithms. We will thus make use of the decision and estimation theory tools that have been introduced in the previous chapters. The originality of the present problem comes from the fact that, as

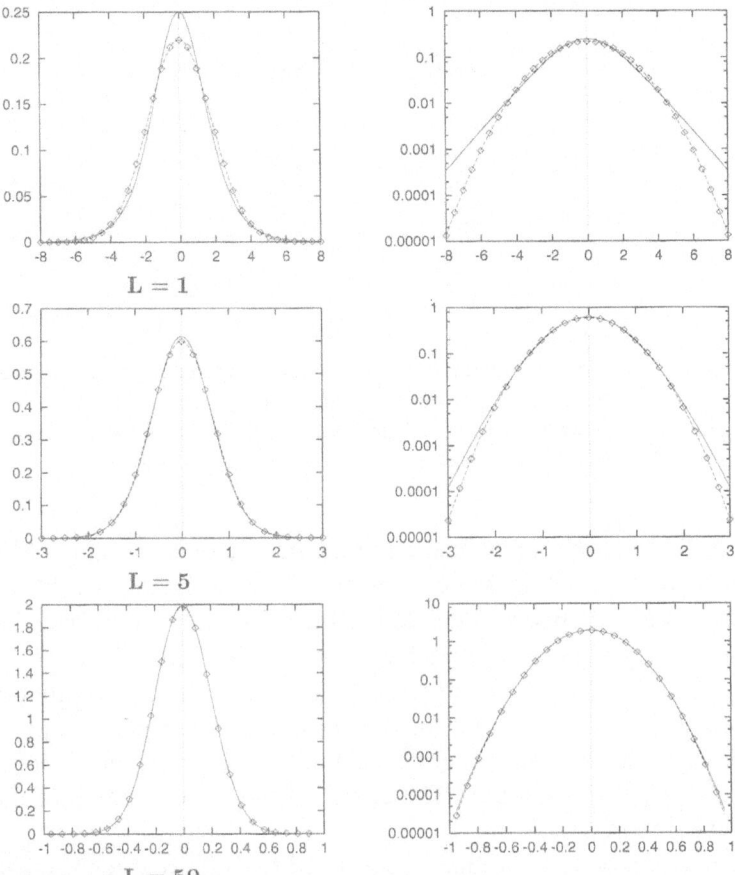

Figure 7.7.   pdf $\Phi(x)$ of graylevels in image β (see Eq. 7.21) for channels whose graylevels are distributed according to Gamma laws of order L (continuous line) and Gaussian pdf with same variance (dotted line with $\diamond$). Left column: pdf. Right column: logarithm of the pdf. ©2001 OSA.

shown in the previous section, the noise pdf does not belong to the exponential family, but a good compromise between signal processing performance and computational efficiency can be obtained by using the natural representation of the OSCI.

7.4.1   Target and edge detection

In this section, we deal with target and edge detection in OSCI with the GLRT approach (see Section 4.2.2). If the graylevel pdf of the OSCI is known, it is possible to determine the exact expression of the GLRT. However, comput-

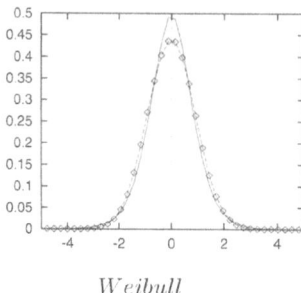 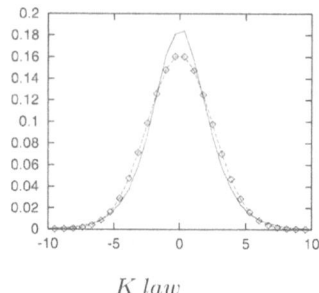

Weibull   *K law*

Figure 7.8. pdf $\Phi(x)$ of the graylevels of the natural representation β for channels whose graylevels are distributed according to different laws. The pdf have been estimated from 10^5 value samples. Left: Weibull law (see Eq. 7.10) with $b = 2$. Right: K law (see Eq. 7.11) with parameters $\nu = 0.7$ and $L = 5$. Dotted lines: pdf of Gaussians with the same variance.

ing the GLRT requires the estimation of the pdf parameter u in the ML sense. No explicit solution can be found in general to this estimation problem, which makes the GLRT relatively computation intensive. Thus for designing algorithms as simple and as fast as possible, a simple way consists in approximating the true pdf of the OSCI by a pdf whose parameters are more easily estimated in the ML sense – such as a pdf belonging to the exponential family. In this case, it is clear that the final performance will depend on the match between the true and the approximated pdf. The choice of the representation of the OSCI will thus be important.

In the previous section, we have considered two different representations of the OSCI: ρ (see Eq. 7.13) and β (see Eq. 7.19). These two representations can be deduced from each other by a bijective transformation. Consequently, they describe exactly the same physical information. However, if we want to apply to the OSCI the GLRT adapted to a simpler pdf – for rapid computation time – the performance will depend on the representation of the OSCI which is used. On this point of view, the natural representation β has important advantages over the standard representation ρ.

First of all, as can be seen by comparing Figures 7.5 and 7.7, the shape of the pdf in representation β, which is symmetrical whatever the value of u, is closer to a Gaussian than that of ρ. Moreover, in representation β, only the mean depends on u. Thus two regions having different values of u will have pdf with the same shape and in particular the same variance. This corresponds to a configuration where the noise is additive. One can thus apply to this image the GLRT adapted to Gaussian laws with identical variances and different means. We show in Appendix 7.B that its expression is:

$$r_{gau_1}(\boldsymbol{\tau}) = A\,[\widehat{m}_a(\boldsymbol{\tau}) - \widehat{m}_b(\boldsymbol{\tau})]^2 \qquad (7.22)$$

where $A = \frac{1}{2\sigma^2} \frac{N_a N_b}{N_F}$, $\sigma = \sigma_a = \sigma_b$ is the common variance of the regions a and b, and $\widehat{m}_u(\tau)$ is the average graylevel in region $u = a$ or b. One can notice that this GLRT is equivalent to a linear filtering and is thus potentially very fast.

In the natural representation β, the pdf of the target and background graylevels differ only by their means $\log \gamma_a$ and $\log \gamma_b$. Thus the performance of the GLRT $r_{gau_1}(\tau)$ only depends on the difference of these averages. Consequently, one can define the contrast coefficient between two regions of respective DOP u_a and u_b in the following way (see Figure 7.9):

$$C = |\log \gamma_a - \log \gamma_b| = \left| \log \left[\frac{\mu_1^a \mu_2^b}{\mu_2^a \mu_1^b} \right] \right| = \left| \log \left[\frac{(1+u_a)(1-u_b)}{(1-u_a)(1+u_b)} \right] \right| \quad (7.23)$$

where one has used the relation between γ and u defined in Eq. 7.15. One can also define the value of the contrast in decibels in the following way:

$$C_{dB} = 10 \frac{C}{\log 10} = 10 \left| \log_{10} \left[\frac{(1+u_a)(1-u_b)}{(1-u_a)(1+u_b)} \right] \right| \quad (7.24)$$

One can notice that this contrast parameter depends in a nontrivial way on the values of u in the two regions.

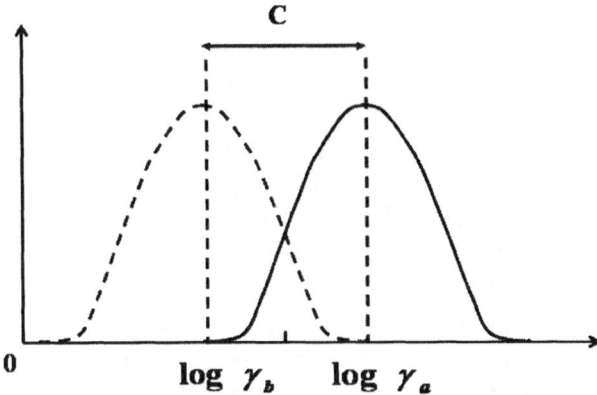

Figure 7.9. Definition of the contrast parameter C on natural representation β. Continuous line: pdf of target graylevels. Dotted line: pdf of background graylevels.

In the representation β, the performance of the Gaussian-based GLRT only depends on the contrast C, but this is not the case in the standard representation ρ. Indeed, in this representation, the pdf on the target and on the background have different shapes, which vary with u_a and u_b. In particular, as u varies, the mean and the variance of the pdf are modified. Consequently, contrary to what happens in representation β, the fluctuations can no longer be considered as additive. In order to perform detection on images ρ, one can use the GLRT adapted to Gaussian pdf with different means and variances in regions a and b, which is $r_{gau}(\tau)$ defined in Eq. 4.41. Moreover, the contrast parameter C no longer completely defines the processing performance of image ρ with $r_{gau}(\tau)$. We will in the following compare the performance of the following two detection approaches:

- $GLRT_{\beta}$, which consists in applying the detector $r_{gau1}(\tau)$ defined in Eq. 7.22 to the natural representation β.

- $GLRT_{\rho}$, which consists in applying the detector $r_{gau}(\tau)$ defined in Eq. 4.41 to the standard representation ρ.

In order to compare these two approaches, we plot the ROCs (see Section 3.2.3) obtained for different parameters u_a and u_b. These curves were estimated on 10^6 noise realizations. The number of pixels in the target (N_a) and in the background (N_b) is chosen equal to 50 and four different combinations of parameters u_a and u_b which all correspond to the same value of C_{dB} are considered. We can see in Figure 7.10 that the performance of the $GLRT_{\rho}$ is very dependent on u_a and u_b. In particular, this performance is low when u_b is high, that is, when the target depolarizes more than the background. On the other hand, the performance of $GLRT_{\beta}$ only depends on C_{dB}.

These two detection algorithms have been tested on real images acquired with a polarimetric imager of the type described in Section 7.1.2 [136]. Consider the scene represented in Figure 7.11. The images are 512×512 pixels and the target is approximately 85 pixels. It is located in the upper right part of the images and its C_{dB} is approximately 0.9 dB. For detection, the mask $\mathbf{w}$ has been chosen to match approximately the shape of the real target. The scanning window $\mathbf{F}$ has been obtained by morphologically dilating $\mathbf{w}$ 5 times with a 3×3 pixel square-shaped structuring element.

Figure 7.12.a displays the results obtained by applying to channel $\mathbf{s_1}$ the GLRT adapted to Gamma pdf (see Eq. 4.42). This algorithm does not compensate enough the variations of the illumination, and the results are not satisfying, since there are too many false alarms generated by the illumination pattern.

Figure 7.12.b represents the results of $GLRT_{\rho}$. We can see that there is a clear peak at the position of the target and it is possible to detect the target without false alarms if a proper threshold is selected. This illustrates the superiority of using the OSCI rather than the intensity images for target detection in real polarimetric images. Finally, Figure 7.12.c represents the results obtained with $GLRT_{\beta}$. We can see that compared to Figure 7.12.b, the peak is sharper over the background noise.

We will now illustrate in Figure 7.13 the performance of these two detection approaches on an edge detection application. In this case, the mask $\mathbf{w}$ and its complementary $\bar{\mathbf{w}}$ are contiguous rectangles (see Figure 4.6.b). We have

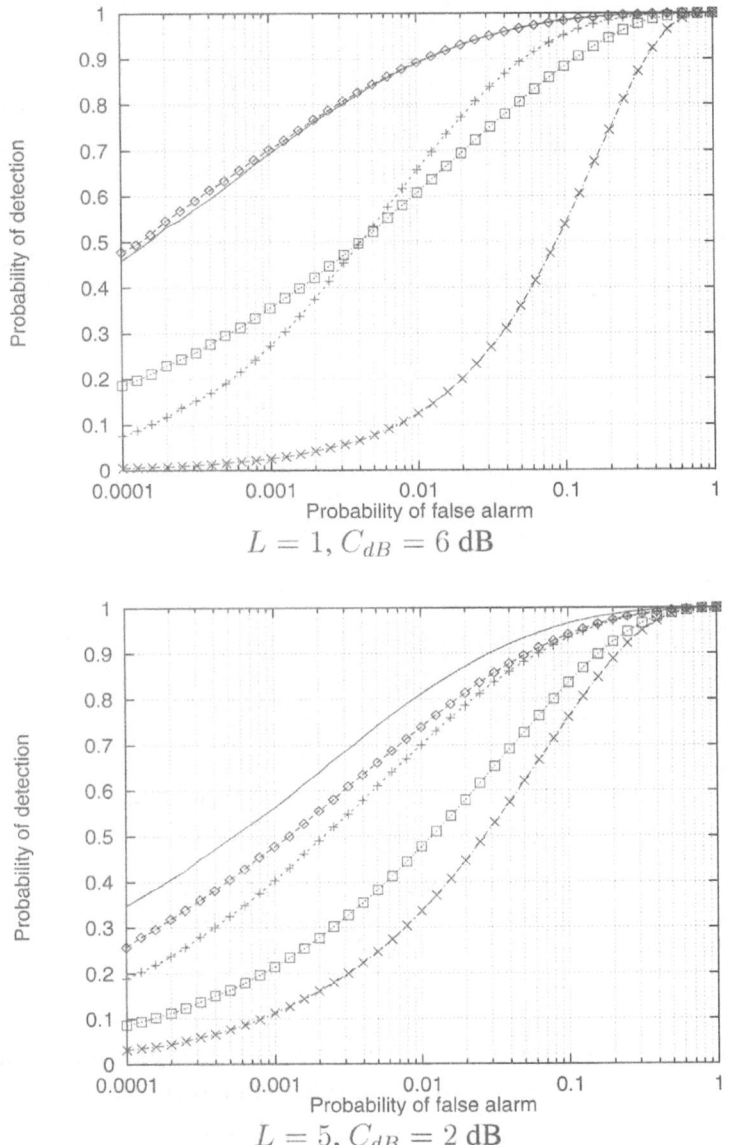

$$L = 1, C_{dB} = 6 \text{ dB}$$

$$L = 5, C_{dB} = 2 \text{ dB}$$

Figure 7.10. Performance of detectors $GLRT_\beta$ and $GLRT_\rho$ in the presence of Gamma distributed speckle noise of orders $L = 1$ and $L = 5$. On both graphs, the continuous line represents the ROC of $GLRT_\beta$, which only depends on C_{dB}. The other curves represent the ROC of $GLRT_\rho$, for the same value of C_{dB} but different combinations of values of (u_a, u_b). Upper graph: ($\diamond$): $(0.6, 0)$; ($\square$): $(0.9, 0.67)$; ($+$): $(0, 0.6)$; ($\times$): $(0.67, 0.9)$. Lower graph: ($\diamond$): $(0.23, 0)$; ($\square$): $(0.78, 0.67)$; ($+$): $(0, 0.23)$; ($\times$): $(0.67, 0.78)$. In all cases, $N_a = 50$ pixels and $N_F = 100$ pixels. The ROC have been estimated on 10^6 realizations.

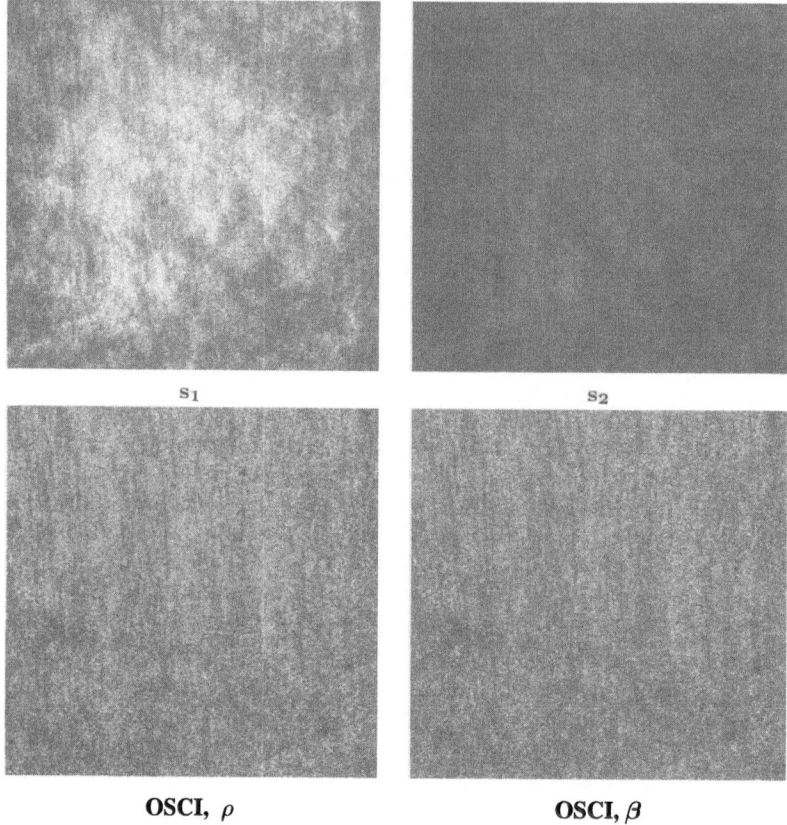

$$s_1 \qquad\qquad s_2$$

OSCI, ρ   **OSCI, β**

Figure 7.11. Example of real polarimetric image, with parallel channel s_1, orthogonal channel s_2, and OSCI in standard representation ρ and in natural representation β. The target, which is small, is in the upper right part of the scene. ©2001 OSA.

generated images having two zones with different values of u for a speckle of order 1 and applied the GLRT with rectangular windows of size 5×10 pixels. Comparing the results of $GLRT_\rho$ and $GLRT_\beta$ detectors, we can see that the output fluctuations of $GLRT_\rho$ do not have the same amplitude on the dark and on the bright parts of the image. It is an illustration of the fact that the detector does not possess the CFAR property, contrary to $GLRT_\beta$. The thresholded images also show that better results are obtained with $GLRT_\beta$.

Finally, Figure 7.14 represents some examples of edge detection on a real polarimetric image using $GLRT_\beta$. In order to be able to detect edges with different orientations, we have used the same approach as in Section 4.4.1 by combining the results of filtering with four windows oriented at different angles $\{0^\circ, 45^\circ, 90^\circ, 135^\circ\}$.

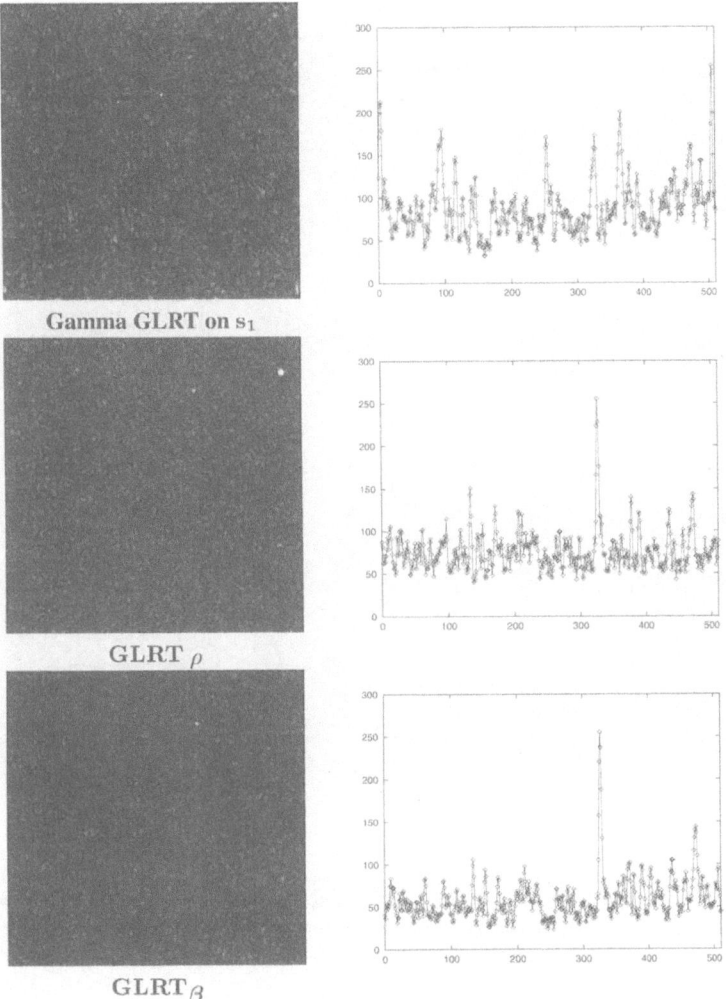

Figure 7.12. Results of different detection algorithms applied to images in Figure 7.11. The images in the left column represent the GLRT $r(\tau)$ at each point τ of the scene. In the right column are represented the maximum values of each column of these planes. ©2001 OSA.

7.4.2   Statistical snake segmentation of OSCI

Let us now address the problem of segmenting the OSCI with the statistical snake approach described in Chapter 5. Here again, it would be possible to design the snake to maximize the likelihood corresponding to the true pdf of the OSCI. However, we have shown in Section 5.2.2 that the implementation of the statistical snake can lead to a fast algorithm when the pseudo-likelihood is a

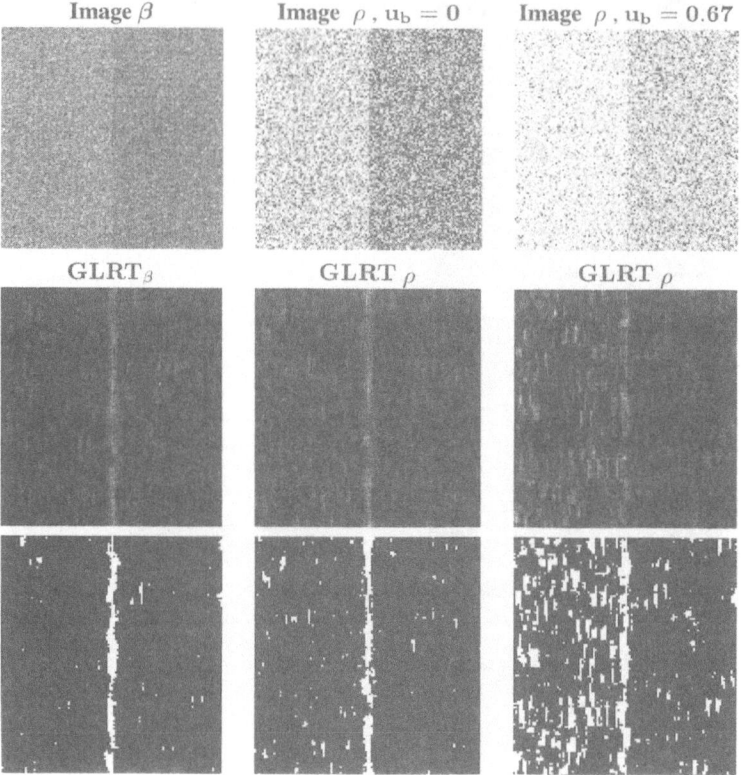

Figure 7.13. First row: Natural representation β and standard representation ρ for two different values of u_b (parameter of the darker part of the image). In all cases, $L = 1$ and $C_{dB} = 6$ dB. Second row: result of the application of the GLRT adapted to the three images. Third row: binarized versions of the previous results.

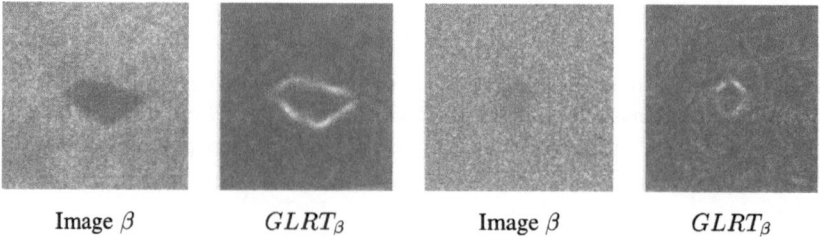

Figure 7.14. Two examples of edge detection on real-world polarimetric images. Successively, image β and result of $GLRT_\beta$. The edge detector has been designed with four windows oriented at 0, 45, 90, and 135 degrees.

function of a simple sufficient statistics. This is for example the case for some pdf of the exponential family. Otherwise, in the general case, the statistical snake can be 100 times slower. Thus for segmentation even more than for detection, it can be important to be able to approximate the true noise pdf with a pdf which leads to a fast algorithm, such as a Gaussian pdf.

As in the previous section, we will compare two approaches for processing the images:

- J_β, which consists in using the external energy function $J_{gau1}(\mathbf{s}, \boldsymbol{\theta})$ defined as follows:

$$J_{gau1}(\mathbf{s}, \boldsymbol{\theta}) = -N_a(\boldsymbol{\theta})\, \widehat{m}_a^2(\boldsymbol{\theta}) - N_b(\boldsymbol{\theta})\, \widehat{m}_b^2(\boldsymbol{\theta}) \qquad (7.25)$$

on the natural representation β. The criterion $J_{gau1}(\mathbf{s}, \boldsymbol{\theta})$ is adapted to the segmentation of images with Gaussian distributed graylevels with the same variance and different means (see Appendix 7.B).

- J_ρ, which consists in using the external energy function $J_{gau}(\mathbf{s}, \boldsymbol{\theta})$ defined in Eq. 5.20 on the standard representation ρ.

In order to compare the two approaches, different combinations of values u_a and u_b corresponding to a fixed value of C_{dB} have been considered. From these values, we have generated images with the polygonal shape represented in Figure 7.15. These images have been segmented with the MDL approach described in Section 6.2, which allows one to automatically estimate the number of nodes necessary to perform the segmentation. In order to quantify the segmentation quality, the Averaged Number of Misclassified Pixels (ANMP) has been estimated over 1000 realizations. The values of the ANMP for each combination of parameters u_a and u_b appear in Table 7.1. One can notice that the performance of J_β depends on u_a and u_b only through C_{dB}, which is not the case with J_ρ. Moreover, the performance of J_β is in all cases better than that of J_ρ. Finally, we can see in Figure 7.16 examples of segmentations of real-world OSCI using J_β.

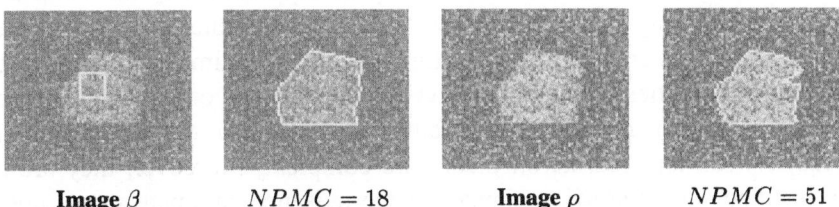

Image β $\qquad$ $NPMC = 18$ $\qquad$ **Image** ρ $\qquad$ $NPMC = 51$

Figure 7.15. From left to right: Image β, $L = 5$, $C_{dB} = 4$ dB (the white square represents the initial contour); result of the segmentation with the ANMP estimated on the segmented images; image ρ, $L = 5$, $C_{dB} = 4$ dB, $u_a = 0.43$, $u_b = 0$; segmentation result.

$$L = 1 \, , \, C_{dB} = 6 \text{ dB}$$

u_a	u_b	ANMP - β	ANMP - ρ
0.6	0	104	114
0.9	0.67	105	330
0	0.6	103	104
0.67	0.9	101	192

$$L = 5 \, , \, C_{dB} = 4 \text{ dB}$$

u_a	u_b	ANMP - β	ANMP - ρ
0.43	0	36	37
0.85	0.67	35	56
0	0.43	35	38
0.67	0.85	35	59

Table 7.1.　ANMP obtained by segmentation with J_β and J_ρ, for fixed value of C_{dB} but different values of u_a and u_b . Upper table: $L = 1$, $C_{dB} = 6$ dB. Bottom table: $L = 5$, $C_{dB} = 4$ dB. Segmented images are similar to those represented in Figure 7.15. The ANMP has been estimated on 1000 realizations.

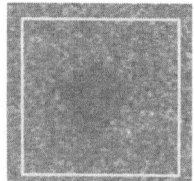 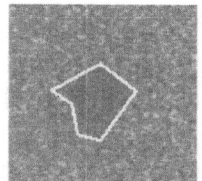 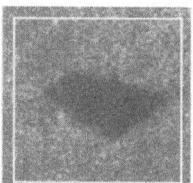 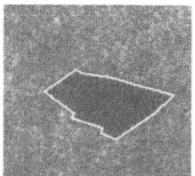

Initial snake　　**Segmentation result**　　**Initial snake**　　**Segmentation result**

Figure 7.16.　Two examples of segmentation of real-world OSCI with the snake J_β. The MDL approach has been used to estimate the number of nodes.

To conclude this section, let us sum up the approach that has been used to design processing algorithms on the OSCI. A possible strategy in image processing is to determine the algorithms which are optimally adapted to the prbabilistic properties of the data. However, in the case of the OSCI, the pdf does not belong to the exponential family and the algorithms based on statistical decision and estimation theory are complex. Moreover, they are not fully justified since the ML estimates of the nuisance parameters may not be minimum variance even if they are unbiased. Then, another approach can be implemented, which consists in finding a representation of the data in which the pdf of the noise can be approximated by a simple pdf which belongs to the exponential family. This is the case of the natural representation of the OSCI on which the algorithms optimal for additive Gaussian noise lead to satisfactory results. Such tasks as target detection and object segmentation are then performed rapidly and efficiently.

7.5. Defining a contrast in Stokes images

In this section, we address the problem of processing Stokes images, which represent the full polarimetric state of the light coming from each point of a scene. Contrary to the OSCI, these images are multichannel, since the Stokes vector has four components (see Eq. 7.6). Our goal with this particular example is to illustrate how the previously described methods for designing statistical processing algorithms can be efficiently applied to vectorial images. In particular, vectorial images are defined by a large number of parameters, which can make the characterization of the performance of processing algorithms very tedious. However, we will see that, thanks to statistical invariance properties, the problem can be largely simplified [158]. We will then apply the algorithm design techniques based on decision and estimation theory to these vectorial and correlated data in Section 7.6.

7.5.1   Position of the problem

Let us consider the simple example of detection in a classical intensity image of a region with known shape and mean intensity I_a on a homogeneous background of mean intensity I_b. To simplify the notation, we will consider one-dimensional images (see Figure 7.17.a). A simple definition of the contrast can be $C = (I_a - I_b)/(I_a + I_b)$. A motivation for defining such a contrast is to characterize the image, which was *a priori* defined by the two parameters I_a and I_b, with the single value C. C will be a good measure of the contrast if, for different values of I_a and I_b, the difficulty to detect region a is the same as soon as the images have the same value of C. In other words, the definition of a contrast should enable one to reduce the number of parameters necessary to characterize an image with respect to the task which has to be performed with it.

It is clear that a valid definition of the contrast depends on the statistical properties of the image. In order to address this point, let us first give a few definitions. Suppose that the image $I(i)|i \in [1, N]$ is composed of N pixels. Region a is composed of N_a pixels whose graylevels are distributed with a pdf $P_a(x)$. For the sake of simplicity, we will assume that the graylevels of each pixel are statistically independent. Similarly, region b is composed of N_b pixels whose graylevels are distributed with pdf $P_b(x)$. Let us assume that the image is perturbed by additive Gaussian noise with standard deviation σ (see Figure 7.17.b). Then $P_a(x)$ is a Gaussian distribution with mean I_a and standard deviation σ. Similarly, $P_b(x)$ is also a Gaussian distribution with mean I_b and

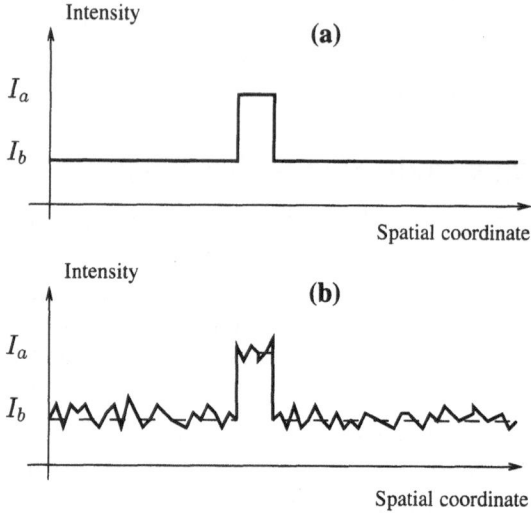

Figure 7.17.   (**a**) Ideal image model, composed of a target zone of intensity I_a and a background zone of intensity I_b. (**b**) Image (a) perturbed with additive Gaussian noise. ©2002 OSA.

standard deviation σ. Thus the performance of any processing task depends *a priori* on three parameters I_a, I_b, and σ. We show in the following that in fact, the performance depends on a single parameter.

One can apply to the image the affine transform $\mathcal{T}(.)$ defined as $\mathcal{T}[I(i)] = [I(i) - I_b]/\sigma$. This transform is bijective and leaves the statistics Gaussian. In the transformed image, the background pdf is thus Gaussian with zero mean and unit variance. The pdf of region a has also unit variance, and its average is equal to $C = (I_a - I_b)/\sigma$. Let us now consider a processing task on image $I(i)$, such as detection or segmentation of region a. Any optimal processing algorithm operating on $I(i)$ cannot lead to a worse performance than the optimal algorithm adapted to the transformed image $\mathcal{T}[I(i)]$. Indeed, if this were not the case, the algorithm which consists of applying the transformation $\mathcal{T}(.)$ to $I(i)$ and then the optimal algorithm adapted to the transformed image would lead to better performance than the optimal algorithm adapted to $I(i)$, which is a contradiction. Since $\mathcal{T}(.)$ is an invertible transform, one can show in the same manner that the optimal algorithm for $I(i)$ cannot lead to better performance than the optimal algorithm adapted to $\mathcal{T}[I(i)]$. Thus the images $I(i)$ and $\mathcal{T}[I(i)]$ lead to the same processing performance when optimal algorithms are applied.

This result means that images corrupted with additive Gaussian noise and with different statistical parameters (σ^2, I_a, I_b) but with the same value of $C = |I_a - I_b|^2/\sigma^2$ will correspond to the same difficulty (or easiness) of processing. From this point of view, C is a good definition of the contrast of the target on a homogeneous background. This is not the case of the contrast definition

$(I_a - I_b)/(I_a + I_b)$. Indeed, there exist different images corrupted with additive Gaussian noise and with different statistical parameters (σ^2, I_a, I_b) but with the same value of $(I_a - I_b)/(I_a + I_b)$ that will not lead to the same signal processing performance.

Different interesting conclusions can be drawn from this simple example. First, the expression of valid contrast parameters depends on the statistics of the fluctuations in the images. We have sketched a practical method for determining the expression of such contrast parameters. It will be shown that it is very general, since the obtained contrast parameters are valid for pure detection tasks, target location, boundary localization, or segmentation of different regions.

However, there is no proof that this method can always reduce all the parameters that define the signal down to a single scalar parameter. One can easily imagine physical situations for which the set of images which lead to the same detection performance is characterized by two, or more, scalar functions of the image statistical parameters. We will encounter such a situation in Section 7.5.2, where we will find that in the case of coherent polarimetric images, two contrast parameters are necessary.

7.5.2 Contrast parameters for coherent polarimetric signals

We determine the polarimetric contrast parameters between two states of partially polarized coherent light. For that purpose, we apply the method discussed in the previous section.

As discussed in Section 7.1.3, in coherent polarimetric imaging, the electric field can be represented by a circular Gaussian random vector $\mathbf{E}$ whose pdf is defined in Eq. 7.2. The polarimetric properties of the random vector $\mathbf{E}$ are defined by the coherency matrix Γ (see Eq. 7.3). The problem at stake is thus defined by the pdf's of the two polarization states $P_{\Gamma_a}(\mathbf{E})$ and $P_{\Gamma_b}(\mathbf{E})$ and thus by their coherency matrices Γ_a and Γ_b representative of the target and of the background. As seen in Section 7.1.1, a coherency matrix is defined by four real-valued parameters. Consequently, any processing task on an image composed of two regions is *a priori* defined by eight parameters.

We have seen in the previous section that contrast parameters for an image processing problem can be obtained by applying to the image some transforms which preserve the pdf family of the noise. Let us now define the class of transforms that will be used to solve the present problem. Let $\mathbf{E}_a$ denote the electric field – that is, a random vector – arriving at a pixel in region a and $\mathbf{E}_b$ the electric field arriving at a pixel in region b. Let us consider the group $\mathcal{G}$ of mappings $\mathbf{E}_b = g(\mathbf{E}_a)$ consisting in all linear nonsingular transformations such as $\mathbf{B} = T\mathbf{A}$ where T is a nonsingular 2×2 matrix. All the transforms belonging in this group transform Gaussian 2-D complex vectors in other Gaussian complex vectors. Note that we will use a Gaussian model for the optical field because it is a classical representation, and for the sake of simplicity. In fact, the Gaussian assumption is not necessary for defining a pdf family invariant with respect to the group of nonsingular 2×2 matrices. Indeed, whatever the pdf of the random variable (RV) representative of unpolarized light, it is possible to construct an invariant family by applying to this RV all the possible 2×2 nonsingular matrices. The physical assumption underlying this construction is the following: for given experimental conditions, if an RV is representative of the optical field, then this must also be the case for the RV resulting from passing through any Jones matrix, and then the pdf of this RV must belong to the same family as the initial RV.

In order to proceed, let us notice that since the matrices Γ_a and Γ_b are Hermitian and positive, they can be decomposed as follows:

$$\Gamma_a = U M_a U^\dagger \tag{7.26}$$
$$\Gamma_b = V M_b V^\dagger \tag{7.27}$$

where U and V are unitary matrices which can be defined as follows:

$$U = \begin{bmatrix} u_1 & -u_2^* \\ u_2 & u_1^* \end{bmatrix} \quad \text{and} \quad V = \begin{bmatrix} v_1 & -v_2^* \\ v_2 & v_1^* \end{bmatrix} \tag{7.28}$$

and M_a and M_b are diagonal matrices:

$$M_a = \begin{bmatrix} d_1^a & 0 \\ 0 & d_2^a \end{bmatrix} \quad \text{and} \quad M_b = \begin{bmatrix} d_1^b & 0 \\ 0 & d_2^b \end{bmatrix} \tag{7.29}$$

where $d_1^a \geq d_2^a \geq 0$ and $d_1^b \geq d_2^b \geq 0$. In the following, we will assume that the matrices Γ_a and Γ_b are nonsingular, that is, $d_2^a > 0$ and $d_2^b > 0$.

Let us now define the "whitening" transform which "decorrelates" the data $\mathbf{E}_a$ in region a. This transform is:

$$\Gamma_a^{-1/2} = U M_a^{-1/2} U^\dagger \tag{7.30}$$

Indeed, if $\mathbf{E}_a' = \Gamma_a^{-1/2} \mathbf{E}_a$, it is easily seen that $\mathbf{E}_a'$ is Gaussian with covariance matrix $\left\langle \mathbf{E}_a' \mathbf{E}_a'^\dagger \right\rangle = Id$, where Id is the identity matrix. Under the same transformation, the covariance matrix of the vector $\mathbf{E}_b$ in region b becomes:

$$\begin{aligned} \Gamma_b' &= \Gamma_a^{-1/2} \Gamma_b \Gamma_a^{-1/2} \\ &= U M_a^{-1/2} U^\dagger V M_b V^\dagger U M_a^{-1/2} U^\dagger \\ &= U M_a^{-1/2} T M_b T^\dagger M_a^{-1/2} U^\dagger \end{aligned} \tag{7.31}$$

with $T = U^\dagger V$. T is a unitary matrix and can thus be written:

$$T = \begin{bmatrix} t_1 & -t_2^* \\ t_2 & t_1^* \end{bmatrix} \tag{7.32}$$

Physically speaking, this transform amounts to converting the vector $\mathbf{E}_a$ into a totally depolarized state (see Figure 7.18). But since such state is invariant under any unitary transformation, one can apply to the data the transform which diagonalizes the matrix Γ_b'. Indeed, this matrix is also Hermitian and can thus be decomposed as:

$$\Gamma_b' = X \begin{bmatrix} \lambda_1 & 0 \\ 0 & \lambda_2 \end{bmatrix} X^\dagger \tag{7.33}$$

with $\lambda_1 \geq \lambda_2 > 0$ and X unitary. Thus, if we apply to the Gaussian vectors in both regions a and b the transformation:

$$Z = X^\dagger \, \Gamma_a^{-1/2} \tag{7.34}$$

the covariance matrices of the resulting vectors are respectively the identity in region a and a diagonal matrix in region b. The result of applying Z to the data is illustrated in Figure 7.18.

In conclusion, we have determined a transform Z in Eq. 7.34 which converts the initial problem defined by matrices Γ_a and Γ_b to an equivalent problem defined by the eigenvalues λ_1 and λ_2 of the matrix Γ_b' in Eq. 7.31. According to what was said in Section 7.5.1, the couple (λ_1, λ_2) thus constitutes a set of reduced parameters for the problem at hand. Moreover, it is clear that any bijective function of these two values is also a set of reduced parameters. For example, we can choose the determinant $\det[\Gamma_b'] = \lambda_1 \lambda_2$ and the trace $\text{tr}[\Gamma_b'] = \lambda_1 + \lambda_2$. We thus obtain that the two following values

$$\det \left[\Gamma_a^{-1/2} \, \Gamma_b \, \Gamma_a^{-1/2} \right] \tag{7.35}$$

$$\text{tr} \left[\Gamma_a^{-1/2} \, \Gamma_b \, \Gamma_a^{-1/2} \right] \tag{7.36}$$

are polarimetric contrast parameters for coherent polarimetric images.

In order to obtain tractable computations and physically meaningful results, we will rather choose the following set of reduced parameters:

$$\alpha = \det \left[\Gamma_a^{-1/2} \, \Gamma_b \, \Gamma_a^{-1/2} \right] = \lambda_1 \lambda_2 \,, \tag{7.37}$$

$$\beta = \frac{\text{tr} \left[\Gamma_a^{-1/2} \, \Gamma_b \, \Gamma_a^{-1/2} \right]}{2\sqrt{\det \left[\Gamma_a^{-1/2} \, \Gamma_b \, \Gamma_a^{-1/2} \right]}} = \frac{\lambda_1 + \lambda_2}{2\sqrt{\lambda_1 \lambda_2}} \tag{7.38}$$

We show in Appendix 7.C that the parameters α and β are physically meaningful, and we give some interpretations of their values in special cases.

7.6. Detection and segmentation in Stokes images

After having determined the expression of the contrast between two regions in a polarimetric image of Stokes vectors, we now address the problem of

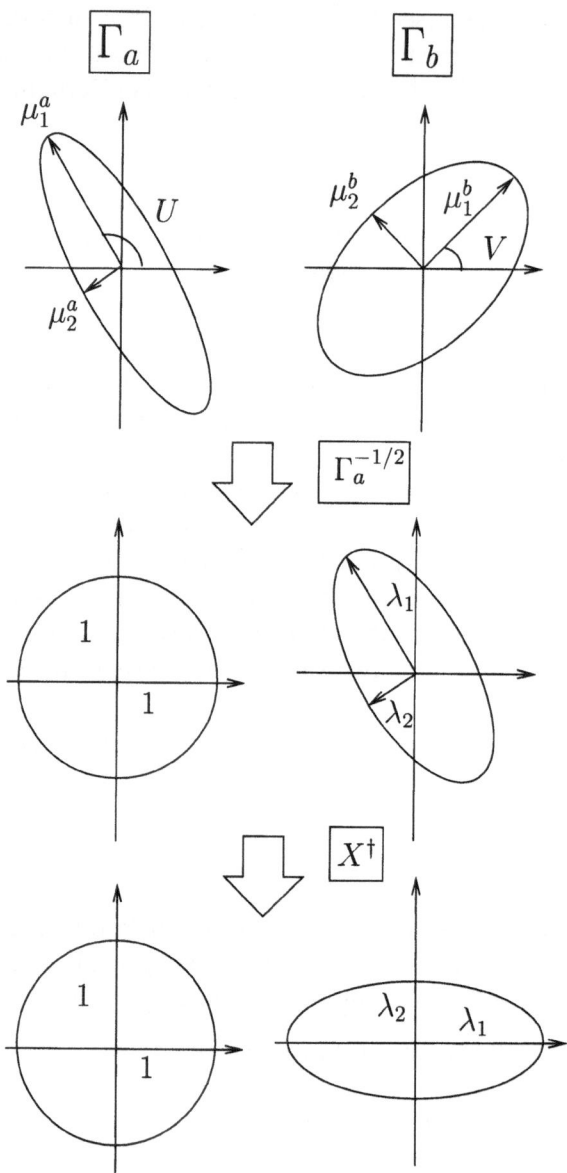

Figure 7.18.   Principle of the derivation of the polarimetric contrast parameters. On the first line, the two polarimetric states Γ_a and Γ_b are represented. The lengths of the axes of the ellipses are the eigenvalues of the matrices and the angles illustrate the principal polarization states U and V. The second line represents the results of applying the whitening transform $\Gamma_a^{-1/2}$ to both states. The last line represents the result of applying the matrix $X^\dagger$, with the result that the matrix $X^\dagger \Gamma_a^{-1/2} \Gamma_b \Gamma_a^{-1/2} X$ is diagonal. ©2002 OSA.

designing efficient processing algorithms on these images using the principles of statistical image processing that have been considered throughout this book. In particular, we will use the SIR image model to design GLRT-based detection algorithms and statistical snake-based segmentation techniques.

Solving these tasks will require the estimation of the likelihood of a zone composed of N pixels with homogeneous polarimetric properties defined by a coherency matrix Γ. We thus consider a set of N random vectors $\{\mathbf{r}_i \mid i \in [1, N]\}$ representing the electromagnetic field backscattered by the pixels of this region. According to the Gaussian circular model, the pdf of each of these vectors is of the type defined in Eq. 7.2. The loglikelihood of this sample has thus the following expression:

$$l(\Gamma) \;\; = \;\; -N \log \pi^2 - N \, \log[\det \Gamma] - \sum_{i=1}^{N} \mathbf{r}_i^\dagger \, \Gamma^{-1} \, \mathbf{r}_i \qquad (7.39)$$

In most applications, the coherency matrix Γ is unknown and will thus be estimated in the maximum-likelihood sense. A nontrivial but classic calculus [159] shows that the ML estimate of the coherency matrix is simply the empirical covariance matrix of the sample:

$$\widehat{\Gamma} = \frac{1}{N} \sum_{i=1}^{N} \mathbf{r}_i \, \mathbf{r}_i^\dagger \qquad (7.40)$$

This estimate of the covariance matrix gathers all the information about the sample which is useful to estimate the polarization state. Injecting this estimate of Γ in Eq. 7.39, one obtains the following pseudo-likelihood:

$$\ell \;\; = \;\; -N \, \log[\det \widehat{\Gamma}] - 2N(1 + \log \pi) \qquad (7.41)$$

Only the first member of Eq. 7.41 is of interest, the second member being a constant. It is important to note that what is needed to compute the pseudo-likelihood is the determinant of the estimate of the coherency matrix of the sample. This value can be determined by a Stokes imager as described in Section 7.1.2. Indeed, let us assume that the imaging system measures at each pixel i the Stokes vector $\mathbf{S}_i$. The relation between $\mathbf{S}_i$ and the covariance matrix estimate at pixel i, $\Gamma_i = \mathbf{r}_i \mathbf{r}_i^\dagger$ is given by Eq. 7.6. If we denote $\widehat{\mathbf{S}} = \frac{1}{N} \sum_{i=1}^{N} \mathbf{S}_i$ the average Stokes vector associated with the sample, one obtains by inverting Eq. 7.6:

$$\begin{cases} \widehat{\Gamma}_{11} & = & 1/2(\widehat{S}_0 + \widehat{S}_1) \\ \widehat{\Gamma}_{22} & = & 1/2(\widehat{S}_0 - \widehat{S}_1) \\ \mathrm{Re}[\widehat{\Gamma}_{12}] & = & \widehat{S}_2/2 \\ \mathrm{Im}[\widehat{\Gamma}_{12}] & = & \widehat{S}_3/2 \end{cases} \qquad (7.42)$$

Using this relation, the determinant of $\widehat{\Gamma}$ can be expressed as a function of the estimated Stokes vector:

$$\det \widehat{\Gamma} = \frac{1}{4} \left[(\widehat{S}_0)_i^2 - (\widehat{S}_2)_i^2 - (\widehat{S}_2)_i^2 - (\widehat{S}_3)_i^2 \right] \tag{7.43}$$

One can remark that we have started our analysis with the model defined in Eq. 7.39, in which the electric field r_i at each pixel was assumed to be known. However, the pseudo-likelihood of Eq. 7.41 depends only on the determinant of the coherency matrix or, in other words, on the components of the Stokes vector. This is a very interesting property since the Stokes vector is easy to measure whereas it is very difficult to obtain the values of the electric field.

We now turn to define the ML-based algorithms adapted to several image processing applications.

7.6.1   Target detection/localization

In order to perform target detection in Stokes images, we will use the GLRT principle (see Section 4.2.2). Let us denote the observed scene $r(x, y)$. According to Eq.7.41, the pseudo-loglikelihoods of hypotheses $H_{0,\tau}$ and $H_{1,\tau}$ are:

$$\ell_0(\tau) = -N_F \log[\det[\widehat{\Gamma}_F(\tau)]] - 2N_F(1 + \log \pi) \tag{7.44}$$

$$\ell_1(\tau) = -N_a \log[\det[\widehat{\Gamma}_a(\tau)]] - N_b \log[\det[\widehat{\Gamma}_b(\tau)]] \\ -2N_F(1 + \log \pi) \tag{7.45}$$

where N_a (N_b) is the number of pixels in region a (b), $N_F = N_a + N_b$, and:

$$\widehat{\Gamma}_a(\tau) = \frac{1}{N_a} \sum_{(x,y) \in \Delta_a^{\mathcal{T}}} r(x, y) r^\dagger(x, y) \tag{7.46}$$

$$\widehat{\Gamma}_b(\tau) = \frac{1}{N_b} \sum_{(x,y) \in \Delta_b^{\mathcal{T}}} r(x, y) r^\dagger(x, y) \tag{7.47}$$

and $\widehat{\Gamma}_F = \frac{1}{N_F} \left(N_a \widehat{\Gamma}_a(\tau) + N_b \widehat{\Gamma}_b(\tau) \right)$ is the estimate of the coherency matrix in region $\Delta_F^{\mathcal{T}}$. The GLRT has thus the following expression:

$$r(\tau) = -N_a \log[\det[\widehat{\Gamma}_a(\tau)]] - N_b \log[\det[\widehat{\Gamma}_b(\tau)]] \\ +N_F \log[\det[\widehat{\Gamma}_F(\tau)]] \tag{7.48}$$

We will illustrate this approach with an edge detection application. Our goal is to detect the presence of an edge at each point of a coherent polarimetric image. To deal with the different possible orientations of edges, one can use the approach described in Section 4.4.1 which consists in fusing the results obtained with several edge masks at different orientations. This technique has been applied to the polarimetric image represented in Figure 7.19. Such an image has four components, which can be represented in several ways. The components of the coherency matrix are shown in the first row of Figure 7.19 and in the second row, the Poincaré representation: intensity I, degree of polarization $\mathcal{P}$, and the two polar angles ψ and χ. It is interesting to note that the image of DOP $\mathcal{P}$ is uniformly equal to 1. Indeed, each speckle is purely polarized so that its DOP is unity. In other words, the determinants of matrices $\mathbf{r}(x, y)\mathbf{r}^\dagger(x, y)$ are zero since these matrices are of rank 1. This is not the case of the true covariance matrix Γ_a (Γ_b) nor of the matrices estimated over a whole region $\widehat{\Gamma}_a(\tau)$ ($\widehat{\Gamma}_b(\tau)$), and this is why there may exist a contrast of DOP between two regions.

The image in Figure 7.19 has been constructed so that the contrast between the object and the background is zero in the intensity and DOP channels. The contrast is thus only created by a difference in the angles ψ and χ, that is, in the orientation and the ellipticity of the polarization ellipse. The values of the contrast parameters are $\alpha = 1$ and $\beta = 1.8$. The result of edge detection on this image, that is, the map $r_f(\tau)$ (see Eq. 4.71), is represented in the middle row of Figure 7.20.

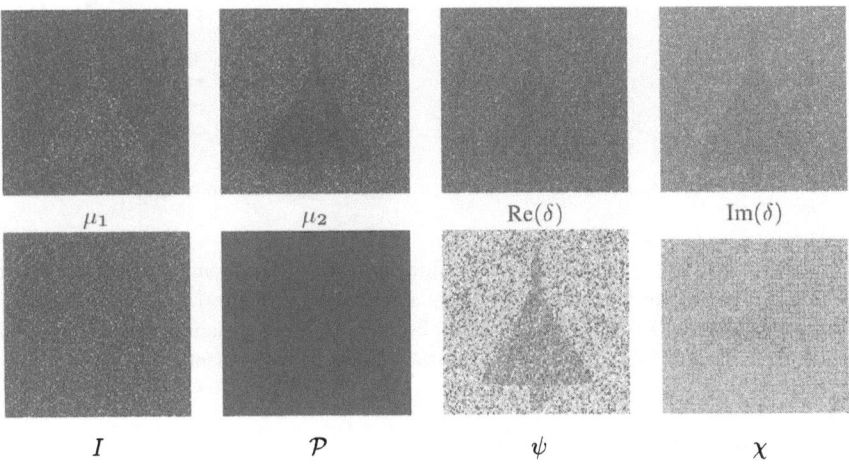

Figure 7.19. Simulated polarimetric image of an airplane. First row: the four components of the coherency matrix representation. Second row: the components of the Poincaré representation. In this image, $\alpha = 1$ and $\beta = 1.8$.

7.6.2 Statistical snake-based segmentation

According to the expression of the pseudo-likelihood in Eq. 7.41, performing statistical snake-based segmentation on Stokes images consists in optimizing

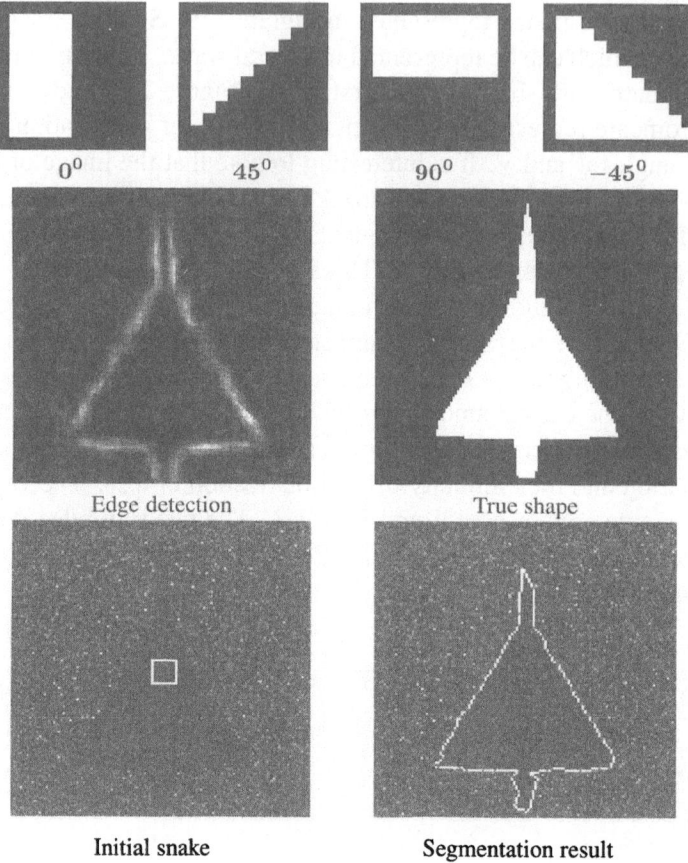

$$0^0 \qquad 45^0 \qquad 90^0 \qquad -45^0$$

Edge detection   True shape

Initial snake   Segmentation result

Figure 7.20. Top row: pattern used for omnidirectional edge detection. Middle row: Result of the omnidirectional edge detection on the image represented in Figure 7.19. Bottom row: result of the segmentation of the image represented in Figure 7.19 with the statistical snake. The initial snake and the result of the segmentation are displayed on the component $\mathrm{Re}(\delta)$.

the following external energy criterion:

$$J(\mathbf{s}, \boldsymbol{\theta}) = N_a(\boldsymbol{\theta}) \, \log[\det[\widehat{\Gamma}_a(\boldsymbol{\theta})]] + N_b(\boldsymbol{\theta}) \, \log[\det[\widehat{\Gamma}_b(\boldsymbol{\theta})]] \qquad (7.49)$$

Let us illustrate this segmentation technique on the polarimetric image represented in Figure 7.19. Segmentation is performed in three steps, beginning with a four-node polygon and increasing the number of nodes at each step, as described in Section 5.3.2. The results appear in the bottom row of Figure 7.20. The image on the left represents the initial snake and that on the right the result of the segmentation. One can note that this approach can be generalized to the partition of an image in several regions using the same approach as in

Section 6.3. Indeed, in this case, the criterion to optimize would be:

$$J(\mathbf{s}, \boldsymbol{\theta}) = \sum_{r=1}^{R} N_r(\boldsymbol{\theta}) \, \log[\det[\widehat{\Gamma}_r(\boldsymbol{\theta})]]$$

(7.50)

where $\widehat{\Gamma}_r(\boldsymbol{\theta})$ is the estimate of the covariance matrix in each of the R regions.

7.6.3 Contrast parameters and detection performance

Let us now illustrate on a numerical simulation the property of the contrast parameters α and β to represent the processing performance on polarimetric images, whereas other heuristic parameters can fail. Let us consider an application of target detection where the target $\mathbf{w}$ is of size 20 pixels and the total size of the considered image subwindow is 40 pixels. This can be representative of the search of an edge at a given location in an image, since the size of the object region is half the size of the subwindow.

In order to define some polarization state configurations for regions a and b, we use the representation of the contrast parameters α and β in the Poincaré sphere (see Appendix 7.C):

$$\alpha = \left(\frac{I_b}{I_a}\right)^2 \frac{1 - \mathcal{P}_b^2}{1 - \mathcal{P}_a^2}$$

(7.51)

$$\beta = \frac{1 - \mathcal{P}_a \mathcal{P}_b \cos \Omega}{\sqrt{1 - \mathcal{P}_a^2} \sqrt{1 - \mathcal{P}_b^2}}$$

(7.52)

where I_a (I_b) denotes the intensities in regions a and b, $\mathcal{P}_a$ ($\mathcal{P}_b$) the degrees of polarization, and Ω is the angle between the representative vectors of the principal polarization states in the Poincaré sphere (see Figure 7.C.1).

Three couples of values (α, β) are considered: $(1.3, 1.3)$, $(1.5, 1.5)$, and $(1.8, 1.8)$. For each couple, we set $I_a = I_b = 1$, we choose different values of the degree of polarization $\mathcal{P}_b$, and, from the different values of α, β, and $\mathcal{P}_b$, we set the polarimetric properties of regions a and b as follows:

$$\mathcal{P}_a = \sqrt{1 - \frac{1 - \mathcal{P}_b^2}{\alpha}}$$

$$\psi_a = \chi_a = \chi_b = 0$$

$$\psi_b = \frac{1}{2} \, \mathrm{acos} \left[\frac{1 - \beta \frac{1 - \mathcal{P}_b^2}{\sqrt{\alpha}}}{\mathcal{P}_a \mathcal{P}_b} \right]$$

Finally, we modify the so-defined polarization states by applying to each of them the same transformation, which consists in rotating the principal polarization states in the Poincaré sphere of an angle 45° around an axis oriented along

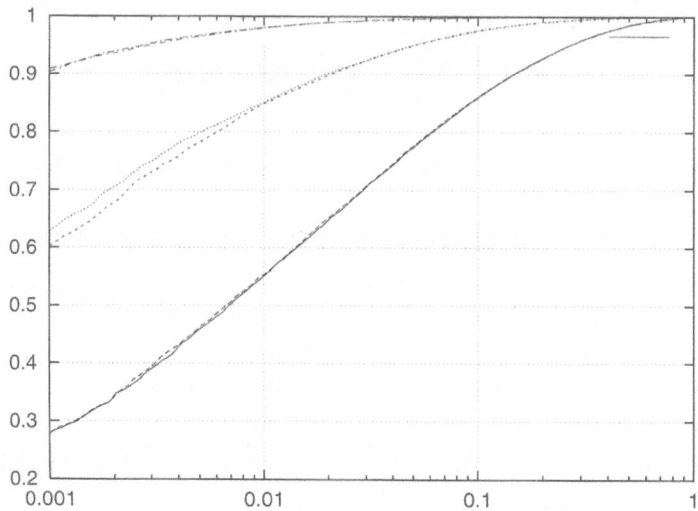

Figure 7.21.   ROC for detection by GLRT on polarimetric images, estimated on 10^5 realizations. $N_F = 40, N_v = 20$. Upper curves: $(\alpha, \beta) = (1.8, 1.8)$, $\mathcal{P}_b = 0.95$ and 0.84. Middle curves: $(\alpha, \beta) = (1.5, 1.5)$, $\mathcal{P}_b = 0.95$ and 0.84. Lower curves: $(\alpha, \beta) = (1.3, 1.3)$, $\mathcal{P}_b = 0.95$ and 0.84.

the vector $(1, 1, 1)$. At the end of this process, we obtain three couples of polarization states for the two regions. Detection is performed on the image using the GLRT algorithm defined in Eq. 4.33. The Receiver Operating Characteristics (ROC) are estimated from 10^5 realizations of the signal. The result appears in Figure 7.21 and we can see that, as expected, the curves corresponding to identical values of α and β match each other, whatever the value of $\mathcal{P}_b$.

Parameters α and β are the rigorous reduced parameters that define the processing performance of a given image processing task. Other, heuristic parameters could be used, but they do not possess this property. For example, since the polarization state is defined by the coherency matrix, one could use the Frobenius distance between the coherency matrices of regions a and b as a measure of the contrast. We have represented in Table 7.2 the value of the Frobenius distances between Γ_a and Γ_b, that is:

$$\mathcal{F} = \sum_{i=1}^{i=2} \sum_{j=1}^{j=2} [(\Gamma_a)_{ij} - (\Gamma_b)_{ij}]^2 \qquad (7.53)$$

for the different combinations of parameters α, β, and $\mathcal{P}_b$ used to plot the ROC in Figure 7.21. It can be seen that this norm does not represent correctly the

α	β	$\mathcal{P}_b$	Frobenius distance
1.3	*1.3*	*0.95*	*0.16*
1.3	1.3	0.84	0.28
1.5	*1.5*	*0.95*	*0.20*
1.5	1.5	0.84	0.34
1.8	*1.8*	*0.95*	*0.24*
1.8	1.8	0.84	0.41

Table 7.2. Frobenius distance between Γ_a and Γ_b for different combinations of values of α, β and $\mathcal{P}_b$.

contrast between the two regions, since for the same value of (α, β), different values of $\mathcal{P}_b$ lead to different values of $\mathcal{F}$ although the ROC are identical.

7.7. Conclusion

In this chapter, we have illustrated how statistical decision algorithms can be applied to design image processing techniques on polarimetric images formed in coherent light. Two examples have been considered, which both present interesting properties in terms of noise statistics. These two types of noisy image have allowed us to illustrate different aspects of the techniques described in the previous chapters and to emphasize two important points: We have presented a way of adapting the statistical processing techniques to pdf which do not belong to the exponential family and a method of determination of contrast parameters.

APPENDIX 7.A: Statistical properties of the OSCI

In this appendix, we demonstrate the results about the statistical properties of the OSCI used in Section 7.3.2.

Expression of $\Psi(x)$ in the general case. Let (s_1, s_2) be a two-dimensional random vector with joint pdf:

$$P_{s_1, s_2}(x, y) = \frac{1}{\mu_1 \mu_2} U\left(\frac{x}{\mu_1}, \frac{y}{\mu_2}\right) \qquad (7.A.1)$$

We define the vector $(\tilde{s}_1, \tilde{s}_2)$ as (s_1, s_2) when the parameter values are $\mu_1 = \mu_2 = 1$. The pdf of the vector $(\tilde{s}_1, \tilde{s}_2)$ is thus the function $U(x, y)$. Let us also define the following auxiliary random variable:

$$\eta = \frac{\tilde{s}_1 - \tilde{s}_2}{\tilde{s}_1 + \tilde{s}_2} \qquad (7.A.2)$$

Our goal is to determine the pdf $\Psi(\eta)$ of η. In order to do so, we use the classical technique defined in Ref. 82, p. 124. Let us consider the following variable change, as well as its inverse:

$$\begin{cases} \xi &= \tilde{s}_1 \\ \eta &= \frac{\tilde{s}_1 - \tilde{s}_2}{\tilde{s}_1 + \tilde{s}_2} \end{cases} \quad \text{and} \quad \begin{cases} \tilde{s}_1 &= \xi \\ \tilde{s}_2 &= \xi \frac{1-\eta}{1+\eta} \end{cases} \qquad (7.A.3)$$

The pdf of the vector (ξ, η) equals:

$$P_{\xi,\eta}(\xi, \eta) = \frac{1}{|J|} U(\tilde{s}_1, \tilde{s}_2)$$

where $|J|$ is the absolute value of the determinant of the transform defined in Eq. 7.A.3. One can show that:

$$|J| = \begin{vmatrix} \frac{\partial \xi}{\partial \tilde{s}_1} & \frac{\partial \xi}{\partial \tilde{s}_2} \\ \frac{\partial \eta}{\partial \tilde{s}_1} & \frac{\partial \eta}{\partial \tilde{s}_2} \end{vmatrix} = \frac{2\tilde{s}_1}{(\tilde{s}_1 + \tilde{s}_2)^2} = \frac{(1 + \eta)^2}{2\xi}$$

Let us remark that the variable ξ is strictly positive. One obtains:

$$P_{\xi,\eta}(\xi, \eta) = \frac{2\xi}{(1 + \eta)^2} U\left(\xi, \xi \frac{1 - \eta}{1 + \eta}\right)$$

and

$$\Psi(\eta) = P_\eta(\eta) = \int_0^{+\infty} P_{\xi,\eta}(\xi, \eta)d\xi = \frac{2}{(1 + \eta)^2} \int_0^{+\infty} \xi U\left(\xi, \xi \frac{1 - \eta}{1 + \eta}\right) d\xi$$

Let us consider the variable change $\nu = \xi/(1 + \eta)$. The previous equation becomes:

$$\Psi(\eta) = \int_0^{+\infty} 2\nu \, U\left[(1 + \eta)\nu, (1 - \eta)\nu\right] d\nu \qquad (7.A.4)$$

Approximate expression of $< \rho >$. Equation 7.13 provides a relation $\rho = g(\eta)$ between the random variables ρ and η. The mean of ρ is equal to $< \rho >= \int \rho P_u^{(\rho)} d\rho$ and since $P_u^{(\rho)} d\rho = \Psi(\eta)d\eta$, one has:

$$< g(\eta) >= \int \frac{\eta + u}{1 + u\eta} \Psi(\eta)d\eta$$

One easily deduces from this equation the following relation:

$$< \rho > -u = (1 - u^2) \int \frac{\eta}{1 + u\eta} \Psi(\eta)d\eta$$

Since $|u| < 1$ and $|\eta| < 1$, one can use the relation:

$$\frac{\eta}{1 + u\eta} = \sum_{n=0}^{+\infty}(-u)^n \, \eta^{n+1}$$

One thus obtains:

$$< \rho > -u = (1 - u^2) \sum_{n=0}^{+\infty}(-u)^n \, \mathcal{M}_{n+1}$$

with $\mathcal{M}_{n+1} = \int \eta^{n+1}\Psi(\eta)d\eta$. For small values of u, one can consider a second-order expansion and since $\mathcal{M}_1 = 0$, one obtains:

$$< \rho >= u + (u^3 - u)\sigma_\eta^2$$

where $\sigma_\eta^2 = \int \eta^2 \Psi(\eta)d\eta$.

Properties of the median of ρ. The median m of the random variable ρ is defined as:

$$\int_{-1}^{m} P_u^{(\rho)} d\rho = \frac{1}{2}$$

Let us consider the bijective transformation $\rho = g(\eta)$ defined by Eq. 7.13. Since $\Psi(x)$ is the pdf of η, one has:

$$\int_{-1}^{m} P_u^{(\rho)} d\rho = \int_{g^{-1}(-1)}^{g^{-1}(m)} \Psi(x) dx = \frac{1}{2} \tag{7.A.5}$$

Moreover, according to Eq. 7.13, $g^{-1}(x) = (x - u)/(1 - ux)$. Since $g^{-1}(-1) = -1$ and η takes values between -1 and 1, Eq. 7.A.5 means that $g^{-1}(m)$ is equal to the median of $\Psi(x)$, that is, 0, since $\Psi(x)$ is even. Consequently,

$$m = g(0) = u$$

which shows that u is the median of ρ.

APPENDIX 7.B: GLRT and statistical snake for Gaussian noise with common variance

In Section 4.2.2 and in Section 5.2.1, we have determined the expressions of the GLRT and of the statistical snake in the presence of Gaussian noise (see Eqs. 4.41 and 5.20). In that case, we have assumed that the means and the variances of the Gaussian pdf describing the target and the background graylevel fluctuations were different and unknown, and thus had to be estimated in the ML sense. If one assumes that the variances on the two regions are equal, say, to σ^2, and that σ^2 is known, then the expressions of the GLRT and of the statistical snake energy are different. We determine these expressions in this appendix.

Let us first consider the GLRT in the SIR image model (see Eq. 3.70). If we consider that the target and the background regions have Gaussian pdf with variance σ^2 known, the loglikelihoods of the two hypotheses are:

$$\mathcal{L}_0 = -N_F \log \sqrt{2\pi} - N \log \sigma - \frac{1}{2\sigma^2} \sum_{(x,y) \in \Delta_F^T} [s(x,y) - m_F]^2 \tag{7.B.1}$$

$$\mathcal{L}_1 = -N_a \log \sqrt{2\pi} - N_a \log \sigma - \frac{1}{2\sigma^2} \sum_{(x,y) \in \Delta_a^T} [s(x,y) - m_a]^2$$

$$-N_b \log \sqrt{2\pi} - N_b \log \sigma - \frac{1}{2\sigma^2} \sum_{(x,y) \in \Delta_b^T} [s(x,y) - m_b]^2 \tag{7.B.2}$$

The only parameters that have to be estimated are the means m_a, m_b, and m_F under the two hypotheses. It is easily seen that the corresponding ML estimates $\hat{m}_a$, $\hat{m}_b$, and $\hat{m}_F$ on the different regions have the following expressions:

$$\hat{m}_v(\tau) = \frac{1}{N_v} \sum_{(x,y) \in \Delta_v^T} s(x,y) \tag{7.B.3}$$

and $v = a, b$, or F. Injecting these estimates in Eqs. 7.B.1 and 7.B.2 and forming the log-GLRT, one obtains:

$$r(\tau) = \frac{1}{2\sigma^2} \left[N_a [\hat{m}_a(\tau)]^2 + N_b [\hat{m}_b(\tau)]^2 - N_F [\hat{m}_F(\tau)]^2 \right] \tag{7.B.4}$$

However, one has the following relation: $N_F \widehat{m}_F(\tau) = N_a \widehat{m}_a(\tau) + N_b \widehat{m}_b(\tau)$. Replacing $\widehat{m}_F(\tau)$ with this expression in Eq. 7.B.4, and after a few simple manipulations, one obtains:

$$r(\tau) = \frac{1}{2\sigma^2} \frac{N_a N_b}{N_a + N_b} \left[\widehat{m}_a(\tau) - \widehat{m}_b(\tau)\right]^2 \qquad (7.B.5)$$

Similarly, the expression of the external energy of the statistical snake corresponding to this model is simply the opposite of the sum of the pseudo-loglikelihoods of regions a and b, which has the following expression:

$$\ell(\mathbf{s}, \boldsymbol{\theta}) = -N_a \log \sqrt{2\pi} - N_a \log \sigma - \frac{1}{2\sigma^2} \sum_{i \in \mathbf{M}} x_i^2 +$$
$$\frac{1}{2\sigma^2} \left[N_a [\widehat{m}_a(\boldsymbol{\theta})]^2 + N_b [\widehat{m}_b(\boldsymbol{\theta})]^2\right] \qquad (7.B.6)$$

where M represents the whole image. Leaving the terms that do not depend on θ and taking the opposite, one obtains the external energy term of the statistical snake:

$$J(\mathbf{s}, \boldsymbol{\theta}) = -N_a [\widehat{m}_a(\boldsymbol{\theta})]^2 - N_b [\widehat{m}_b(\boldsymbol{\theta})]^2 \qquad (7.B.7)$$

APPENDIX 7.C: Interpretation of the contrast parameters

We have first to determine the expression of α and β as functions of the polarimetric states defined by the coherency matrices Γ_a and Γ_b. Besides the coherency matrix, there are many other ways of representing partially polarized light [147]. For example, the polarization state can be defined by the two eigenvalues μ_1^a and μ_2^a of the matrix Γ_a, plus the eigenvector $\mathbf{u} = (u_1, u_2)$ which defines the matrix U in Eq. 7.28. However, the expressions of α and β are particularly simple and physically meaningful if the Poincaré representation of the polarization states is used (see Eq. 7.7). Indeed, using this representation, it can be shown [158] that α and β have the following expression:

$$\alpha = \left(\frac{I_b}{I_a}\right)^2 \frac{1 - \mathcal{P}_b^2}{1 - \mathcal{P}_a^2} \qquad (7.C.1)$$

$$\beta = \frac{1 - \mathcal{P}_a \mathcal{P}_b \cos \Omega}{\sqrt{1 - \mathcal{P}_a^2} \sqrt{1 - \mathcal{P}_b^2}} \qquad (7.C.2)$$

where I_a (I_b) denotes the intensities in regions a and b, $\mathcal{P}_a$ ($\mathcal{P}_b$) the degrees of polarization, and Ω is the angle between the representative vectors of the principal polarization states in the Poincaré sphere (see Figure 7.C.1).

We can first notice that when the polarization states in both regions a and b are strictly identical, the two parameters α and β are equal to 1. Furthermore, if only one of the polarization states is fully polarized, then $\mathcal{P}_a$ or $\mathcal{P}_b$ is equal to 1 and the two parameters α and β diverge. This is because in our model, the only source of fluctuations is the random nature of the partially polarized light. In other words, in case of a totally polarized state, it is possible to design an analyzer orthogonal to this state. The intensity measured for a signal in this state will always be 0, which constitutes a zero-failure decision rule. Of course, this is no longer true if measurement noise is introduced.

It is easy to show that β increases when Ω varies so that the polarization goes from parallel (if $\Omega = 0$) to orthogonal states (if $\Omega = \pi$). One can also note that when $\mathcal{P}_a = \mathcal{P}_b = 0$, α is similar to the contrast parameter in the presence of speckle (that is, I_a/I_b), and that it can be either greater or smaller than 1.

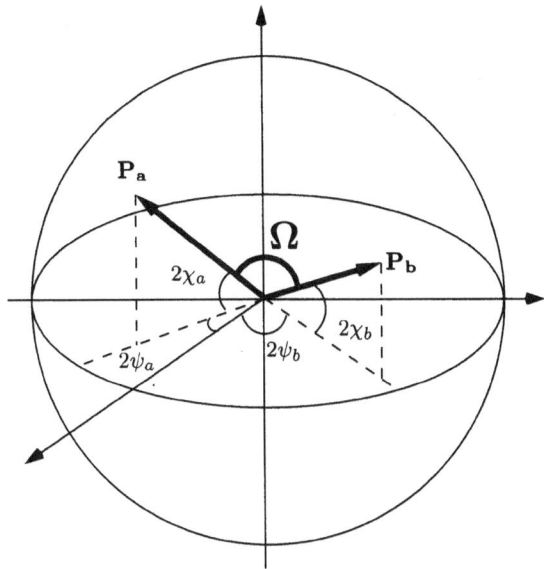

Figure 7.C.1.   Definition of the angle Ω between two polarization states in the Poincaré sphere. ©2002 OSA.

Let us now consider special cases when only one type of polarimetric parameter is different in the two regions.

■ If only the intensities differ, the sufficient parameter is I_a/I_b. This is normal since in this case, the noise reduces to simple intensity speckle, for which it is well known that the expression of the contrast is the ratio of the average intensities (see Section 7.5.1). On the other hand, if only the polarization degrees differ, one can notice that the performance depends on both $\mathcal{P}_a$ and $\mathcal{P}_b$.

■ Now, if only the principal polarization states vary, that is, $\Omega \neq 0$, then $\alpha = 1$ and β is maximal when $\Omega = \pi$, which corresponds to two orthogonal states [147]. This conclusion is physically statisfying.

■ Finally, if at least one of the states Γ_a or Γ_b is fully depolarized, that is, if $\mathcal{P}_a = 0$ or $\mathcal{P}_b = 0$, then β does not depend on Ω which is, in fact, not defined.

Credits

We thank the following scientific societies and publishing companies for authorizing us to reproduce in this book figures that had been previously published in some of their journals:

- The International Society for Optical Engineering (SPIE), for Refs. 20, 30, 46, 69.
- The Optical Society of America (OSA), for Refs. 63,113,153,154,158.
- Elsevier Science, for Ref. 101.
- The IEEE, for Refs. 62, 80, 103

- Figure 2.8, p. 32; Figure 2.9, p. 35; Figure 2.10, p. 35; Figure 2.11, p. 36; Figure 2.12, p. 37; Figure 2.13, p. 37: reprinted from *Euro-American Workshop on Pattern Recognition*, Bahram Javidi, Philippe Réfrégier, eds.; Philippe Réfrégier and Vincent Laude, "Critical analysis of filtering techniques for optical pattern recognition: are the solutions of this inverse problem stable ?", p. 58-84. Copyright 1994, with permission of the International Society for Optical Engineering (SPIE).

- Figure 2.14, p. 38; Figure 2.15, p. 41; Figure 2.16, p. 42 : reprinted from *Optical Engineering*, vol. 33 no. 6, Jean Figue and Philippe Réfrégier, "Angle determination of airplanes by multicorrelation techniques with optimal trade-off synthetic discriminant filters", p. 1821-1828. Copyright 1994, with permission of the International Society for Optical Engineering (SPIE).

- Figure 3.1, p. 51; Figure 3.2, p. 51; Figure 3.3, p. 53; Figure 3.7, p. 72; Figure 3.8, p. 73: reprinted from *Optoelectronic Information Processing*, Bahram Javidi, Philippe Réfrégier, eds.; Philippe Réfrégier and François Goudail, "Decision theory applied to object location and nonlinear joint-transform correlation", p. 137-165. Copyright 1997, with permission of the International Society for Optical Engineering (SPIE).

- Figure 4.2, p. 102; Figure 4.3, p. 103; Figure 4.4, p. 104 : reprinted from *Optoelectronic Information Processing*, Ph. Réfrégier, B. Javidi, eds.; Philippe Réfrégier, François Goudail and Christophe Chesnaud, "Statistically independent region models applied to correlation and segmentation techniques ", p. 193-224. Copyright 1999, with permission of the International Society for Optical Engineering (SPIE).

- Figure 4.7, p. 110; Figure 4.8, p. 110; Figure 4.9, p. 112: reprinted from *Journal of the Optical Society of America A*, vol. 15 no. 12, Henrik Sjöberg, François Goudail and Philippe

References

[1] H. J. Caulfield and W. T. Maloney, "Improved discrimination in optical character recognition," Appl. Opt. **8**, 2354–2355 (1969).

[2] D. Casasent and D. Psaltis, "Position, rotation and scale invariant optical correlation," Appl. Opt. **15**, 1795–1799 (1976).

[3] J. L. Horner, "Light utilization in optical correlators," Appl. Opt. **21**, 4511–4514 (1982).

[4] Y. N. Hsu, H. H. Arsenault, and G. April, "Rotation-invariant digital pattern recognition using circular harmonic expansion," Appl. Opt. **21**, 4012–4015 (1982).

[5] Y. N. Hsu and H. H. Arsenault, "Optical pattern recognition using circular harmonic expansion," Appl. Opt. **21**, 4016–4019 (1982).

[6] B. V. K. Vijaya Kumar, "Efficient approach for designing linear combination filters," Appl. Opt. **22**, 1445–1448 (1983).

[7] J. L. Horner and P. D. Gianino, "Phase-only matched filtering," Appl. Opt. **23**, 812–816 (1984).

[8] D. Casasent, "Unified synthetic discriminant function computation formulation," Appl. Opt. **23**, 1620–1627 (1984).

[9] F. M. Dickey and L. A. Romero, "Normalized correlation for pattern recognition," Opt. Lett. **16**, 1186–1188 (1991).

[10] F. M. Dickey and K. T. Stalker, "Binary phase only filters: Implications of bandwidth and uniqueness on performance," J. Opt. Soc. Am. A **4**, 69 (1987).

[11] F. M. Dickey, T. K. Stalker, and J. J. Mason, "Bandwidth considerations for binary phase-only filter," Appl. Opt. **27**, 3811–3818 (1988).

[12] Z. Zouhir Bahri and B. V. K. Vijaya Kumar, "Fast algorithms for designing optical phase-only-filters (POFs) and binary phase-only-filters (BPOFs)," Appl. Opt. **29**, 2992–2996 (1990).

[13] F. M. Dickey, B. V. K. Vijaya Kumar, L. A. Romero, and J. M. Connelly, "Complex ternary matched filters yielding high signal-to-noise ratios," Opt. Eng. **29**, 994–1001 (1990).

[14] A. A. S. Awwal, M. A. Karim, and S. R. Jahan, "Improved correlation discrimination using an amplitude modulated phase-only-filter," Appl. Opt. **29**, 233–236 (1990).

[15] B. V. K. Vijaya Kumar, "Minimum variance synthetic discriminant functions," J. Opt. Soc. Am. A **3**, 1579–1584 (1986).

[16] A. Mahalanobis, B. V. K. Vijaya Kumar, and D. Casasent, "Minimun average correlation energy filters," Appl. Opt. **26**, 3633–3640 (1987).

[17] Ph. Réfrégier, "Filter design for optical pattern recognition: Multi-criteria optimization approach," Opt. Lett. **15**, 854–856 (1990).

[18] J. Figue and Ph. Réfrégier, "Influence of the noise model on correlation filters: peak sharpness and noise robustness," Opt. Lett. **17**, 1476–1478 (1992).

[19] Ph. Réfrégier, "Application of the stabilizing functional approach to pattern recognition filters," J. Opt. Soc. Am. A **11**, 1243–1251 (1994).

[20] Ph. Réfrégier and V. Laude, "Critical analysis of filtering techniques for optical pattern recognition: Are the solutions of this inverse problem stable ?," in *Euro-American Workshop on Pattern Recognition*, Ph. Réfrégier and B. Javidi, eds., pages 58–84 (Proc. Soc. Photo-Opt. Instrum. Eng., Bellingham, Washington, 1994).

[21] C. S. Weaver and J. W. Goodman, "Technique for optically convolving two functions," Appl. Opt. **5**, 1248–1249 (1966).

[22] B. Javidi, "Nonlinear joint power spectrum based optical correlation," Appl. Opt. **28**, 2358–2367 (1989).

[23] L. Pichon and J.-P. Huignard, "Dynamic joint-Fourier transform correlator by Bragg diffraction in photorefractive BSO crystal," Opt. Commun. **36**, 277–280 (1981).

[24] F. Turon, E. Ahouzi, J. Campos, K. Chalasinska-Macukow, and M. J. Yzuel, "Nonlinearity effects in the pure phase correlation method in multiobject scenes," Appl. Opt. **33**, 2188–2191 (1994).

[25] R. D. Juday, "Optimal realizable filters and the minimum Euclidean distance principle," Appl. Opt. **32**, 5100–5111 (1993).

[26] B. V. K. Vijaya Kumar, C. Hendrix, and D. W. Carlson, "Tradeoffs in the design of correlation filters," in *Optical Pattern Recognition*, J. L. Horner and B. Javidi, eds., pages 191–215, SPIE Optical Engineering Press, (1992).

[27] G. Ravichandran and D. P. Casasent, "Minimum noise and correlation energy optical correlation filter," Appl. Opt. **31**, 1823–1833 (1992).

[28] B. V. K. Vijaya Kumar and L. Hassebrook, "Performance measures for correlation filters," Appl. Opt. **29**, 2997–3006 (1990).

[29] H. L. Van Trees, *Detection, Estimation and Modulation Theory. Part I : Detection, Estimation and Linear Modulation Theory* (John Wiley and Sons, Inc., New York, 1968).

[30] B. Javidi and J. L. Horner, "Single spatial light modulator joint transform correlator," Appl. Opt. **28**, 1027–1032 (1989).

[31] L. P. Yaroslavsky, "The theory of optimal methods for localization of objects in pictures," in *Progress in Optics, XXXII*, E. Wolf, ed., pages 145–201 (Elsevier Science Publishers, Amsterdam, 1993).

[32] Ph. Réfrégier, B. Javidi, and V. Laude, "Nonlinear joint-transform correlation: an optimal solution for adaptive image discrimination and input noise robustness," Opt. Lett. **19**, 405–407 (1994).

[33] Ph. Réfrégier, "Optimal introduction of optical efficiency for pattern recognition filters," in *Optical Information Processing Systems and Architectures IV*, B. Javidi, ed., pages 104–115, Proc. SPIE, (1992).

[34] Ph. Refregier, "Filter design for optical pattern recognition: multicriteria optimization approach," Opt. Lett. **15**, 854–856 (1990).

[35] V. Laude and Ph. Réfrégier, "Multicriteria characterization of coding domains with optimal Fourier spatial light modulator filters," Appl. Opt. **33**, 4465–4471 (1994).

[36] B. R. Frieden, "Restoring with maximum likelihood and maximum entropy," J. Opt. Soc. Am. A **26**, 511–518 (1972).

[37] S. J. Wernecke and L. R. d'Addario, "Maximum entropy image reconstruction," IEE Trans. Comput. **C-26**, 351–364 (1977).

[38] M. Fleisher, U. Mahalab, and J. Shamir, "Entropy optimized filter for pattern recognition," Appl. Opt. **29**, 2091–2098 (1990).

[39] A. N. Tikhonov and V. Y. Arsenin, *Solutions of Ill-Posed Problems* (John Wiley and Sons, Inc., New York, 1977).

[40] J. Figue and Ph. Réfrégier, "Angle determination of airplanes by multicorrelation techniques with optimal trade-off synthetic discriminant filters," Opt. Eng. **33**, 1821–1828 (1994).

[41] J. Figue, *Study of optimal correlation methods including learning capabilities for pattern recognition, and applications to planes attitude estimation* (in French) (Ph.D. Thesis, Université de Paris VI, 1993).

[42] J. Figue and Ph. Réfrégier, "Evaluation of optimal synthetic discriminant filters on an application to angle determination," in *Photonics for Processors, Neural Networks, and Memories*, J.L. Horner, B. Javidi, S. T. Kowel, and W. J. Miceli, eds., Proc. SPIE **3707**, 58–69 (1993).

[43] P. M. Woodward, *Probability and Information Theory* (McGraw-Hill, New York, 1953).

[44] B. Javidi and J. Wang, "Limitation of the classic definition of the correlation signal-to-noise ratio in optical pattern recognition with disjoint signal and scene noise," Appl. Opt. **31**, 6826–6829 (1992).

[45] F. Goudail, V. Laude, and Ph. Réfrégier, "Influence of non-overlapping noise on regularized linear filters for pattern recognition," Opt. Lett. **20**, 2237–2239 (1995).

[46] Ph. Réfrégier and F. Goudail, "Decision theory applied to object location and nonlinear joint-transform correlation," in *Optoelectronic Information Processing*, Ph. Réfrégier and B. Javidi, eds., pages 137–165 (Proc. Soc. Photo-Opt. Instrum. Eng., Bellingham, Washington, 1997).

[47] R. O. Duda and P. E. Hart, *Pattern Classification and Scene Analysis* (John Wiley and Sons, Inc., New York, 1973).

[48] H. V. Poor, *An Introduction to Signal Detection and Estimation* (Springer-Verlag, New York, 1994).

[49] S. M. Kay, "Statistical decision theory II," in *Fundamentals of Statistical Signal Processing. Volume II: Detection Theory*, pages 186–247 (Prentice-Hall, Upper Saddle River, New Jersey, 1998).

[50] A. H. Jazwinski, *Stochastic Processes and Filtering Theory* (Academic Press, New York, 1970).

[51] C. P. Robert, *The Bayesian Choice – A Decision-Theoretic Motivation* (Springer-Verlag, New York, 1996).

[52] Ph. Réfrégier and F. Goudail, "A decision theoretical approach to nonlinear joint-transform correlation," J. Opt. Soc. Am. A **15**, 61–67 (1998).

[53] Ph. Réfrégier, "Bayesian theory for target location in noise with unknown spectral density," J. Opt. Soc. Am. A **16**, 276–283 (1999).

[54] B. Javidi, Ph. Réfrégier, and P. Willet, "Optimum receiver design for pattern recognition with nonoverlapping target and scene noise," Opt. Lett. **18**, 1660–1662 (1993).

[55] F. Guérault and Ph. Réfrégier, "Unified statistically independent region processor for deterministic and fluctuating target in non-overlapping background," Opt. Lett. **23**, 412–414 (1998).

[56] F. Goudail and Ph. Réfrégier, "Optimal and suboptimal detection of a target with random gray levels imbedded in non-overlapping noise," Opt. Commun. **125**, 211–216 (1996).

[57] F. Guérault, L. Signac, F. Goudail, and Ph. Réfrégier, "Location of target with random gray levels in correlated background with optimal processors and preprocessings," Opt. Eng. **36**, 2660–2670 (1997).

[58] T. S. Ferguson, "Exponential families of distributions," in *Mathematical Statistics, a Decision Theoretic Approach*, pages 125–132 (Academic Press, New York, 1967).

[59] M. Figueiredo, J. Leitão, and A. K. Jain, "Unsupervised contour representation and estimation using B-splines and a minimum description length criterion," IEEE Trans. Image Process. **9**, 1075–1087 (2000).

[60] A. Chakraborty, L. H. Staib, and J. S. Duncan, "Deformable boundary finding in medical images by integrating gradient and region information," IEEE Trans. Med. Imaging **15**, 859–870 (1996).

[61] O. Germain and Ph. Réfrégier, "Optimal snake-based segmentation of a random luminance target on a spatially disjoint background," Opt. Lett. **21**, 1845–1847 (1996).

[62] C. Chesnaud, Ph. Réfrégier, and V. Boulet, "Statistical region snake-based segmentation adapted to different physical noise models," IEEE Trans. Pattern Anal. Mach. Intell. **21**, 1145–1157 (1999).

[63] H. Sjöberg, F. Goudail, and Ph. Réfrégier, "Optimal algorithms for target location in non-homogeneous binary images," J. Opt. Soc. Am. A **15**, 2976–2985 (1998).

[64] J. W. Goodman, *Statistical Optics* (John Wiley and Sons, Inc., New York, 1985).

[65] J. W. Goodman, "Laser speckle and related phenomena," in *Statistical Properties of Laser Speckle Patterns*, pages 9–75 (Springer-Verlag, New York, 1975).

[66] C. J. Oliver, D. Blacknell, and R. G. White, "Optimum edge detection in SAR," IEE Proc. Radar Sonar Navigat. **143**, 31–40 (1996).

[67] J. Dias and J. Leitão, "Wall position and thickness estimation from sequences of echocardiographic images," IEEE Trans. Med. Imaging **15**, 25–38 (1996).

[68] F. Guérault and Ph. Réfrégier, "Statistically independent region processor for target and background with random textures: whitening preprocessing approach," Opt. Commun. **142**, 197–202 (1997).

[69] Ph. Réfrégier, F. Goudail, and Ch. Chesnaud, "Statistically independent region models applied to correlation and segmentation techniques," in *Euro-American Workshop on Optoelectronic Information Processing*, Ph. Réfrégier and B. Javidi, eds., pages 193–224 (Proc. Soc. Photo-Opt. Instrum. Eng., Bellingham, Washington, 1999).

[70] Ph. Réfrégier, F. Goudail, and Th. Gaidon, "Optimal location of random targets in random background: random Markov fields modelization," Opt. Commun. **128**, 211–215 (1996).

[71] F. Goudail and Ph. Réfrégier, "Optimal location of an object with random gray levels embedded in a correlated background - A Gaussian Markov random field modelization," in *Optoelectronic Information Processing*, B. Javidi and Ph. Réfrégier, eds., pages 167–193 (SPIE Optical Engineering Press, Bellingham, Washington, 1997).

[72] R. Touzi, A. Lopès, and P. Bousquet, "A statistical and geometrical edge detector for SAR images," IEEE Trans. Geosci. Remote Sensing **26**, 764–773 (1988).

[73] R. Fjørtoft, A. Lopès, P. Marthon, and E. Cubero-Castan, "An optimum multiedge detector for SAR image segmentation," IEEE Trans. Geosci. and Remote Sensing **36**, 793–802 (1998).

[74] R. C. Gonzalez and R. E. Woods, *Digital Image Processing* (Addison-Wesley, Reading, Massachusetts, 1992).

[75] H. Sjöberg, F. Goudail, and Ph. Réfrégier, "Comparison of the maximum likelihood ratio test algorithm and linear filters for target location in binary images," Opt. Commun. **163**, 252–258 (1999).

[76] A. K. Jain, *Fundamentals of Digital Image Processing* (Prentice-Hall, Englewood Cliffs, New Jersey, 1989).

[77] C. Oliver and S. Quegan, "Fundamental properties of SAR images," in *Understanding SAR Images*, pages 84–99 (Artech House, London, 1998).

[78] J. W. Goodman, "The speckle effect in coherent imaging," in *Statistical Optics*, pages 347–356 (John Wiley and Sons, Inc., New York, 1985).

[79] V. S. Frost, K. S. Shanmugan, and J.C. Holtzman, "Edge detection for SAR and other noisy images," in *Proc. International Geoscience and Remote Sensing Symposium, Munich, Germany*, **FA2**, 4.1–4.9 (1982).

[80] O. Germain and Ph. Réfrégier, "On the bias of the Likelihood Ratio edge detector for SAR images," IEEE Trans. Geosci. Remote Sensing **38**, 1455–1458 (2000).

[81] L. Vincent and P. Soille, "Watersheds in digital spaces: an efficient algorithm based on immersion simulations," IEEE Trans. Pattern Anal. Mach. Intell. **13**, 583–598 (1991).

[82] A. Papoulis, *Probability, Random Variables and Stochastic Processes* (McGraw-Hill, New York, 1991).

[83] M. Kass, A. Witkin, and D. Terzopoulos, "Snakes: Active contour models," Int. J. Computer Vision **1**, 321–331 (1988).

[84] M. Berger, *Les contours actifs: modélisations, comportement et convergence* (Ph.D. Thesis, Institut polytechnique de Lorraine, 1991).

[85] F. Leymarie and M. D. Levine, "Tracking deformable objects in the plane using an active contour model," IEEE Trans. Pattern Anal. Mach. Intell. **15**, 617–634 (1993).

[86] R. Caselles, R. Kimmel, and G. Sapiro, "Geodesic active contours," Int. Conference on Computer Vision **1**, 694–699 (1995).

[87] S. Osher and J. A. Sethian, "Fronts propagating with curvature dependent speed: Algorithms based on Hamilton-Jacobi formulation," J. Comput. Phys. **79**, 12–49 (1988).

[88] J. A. Sethian, "Numerical algorithms for propagating interfaces: Hamilton-Jacobi equations and conservation laws," J. Diff. Geom. **31**, 131–161 (1990).

[89] R. Malladi, J. Sethian, and B. Venuri, "Shape modeling with front propagation," IEEE Trans. Pattern Anal. Mach. Intell. **17**, 158–175 (1995).

[90] L. Rudin, S. Osher, and E. Fatemi, "Nonlinear total variation based noise removal algorithms," Physica D **60**, 259–268 (1992).

[91] N. Paragios and R. Deriche, "Geodesic active contours and level sets for the detection and tracking of moving objects," IEEE Trans. Pattern Anal. Mach. Intell. **22**, 266–280 (2000).

[92] R. Ronfard, "Region-based strategies for active contour models," Int. J. Computer Vision **2**, 229–251 (1994).

[93] M. Figueiredo and J. Leitão, "Bayesian estimation of ventricular contours in angiographic images," IEEE Trans. Med. Imaging **11** (1992).

[94] G. Storvik, "A Bayesian approach to dynamic contours through stochastic sampling and simulated annealing," IEEE Trans. Pattern Anal. Mach. Intell. **16**, 976–986 (1994).

[95] A. K. Jain, Y. Zhong, and S. Lakshmanan, "Object matching using deformable template," IEEE Trans. Pattern Anal. Mach. Intell. **18**, 268–278 (1996).

[96] S. C. Zhu and A. Yuille, "Region competition: unifying snakes, region growing, and Bayes/MDL for multiband image segmentation," IEEE Trans. Pattern Anal. Mach. Intell. , 884–900 (1996).

[97] N. Paragios and R. Deriche, "Geodesic active regions and level set methods for supervised texture segmentation," Int. J. Computer Vision **46**, 223 (2002).

[98] G. J. McLachlan and T. Krishnan, *The EM Algorithm and Extensions* (John Wiley and Sons, Inc., New York, 1997).

[99] L. A. Vese and F. C. Chan, "Active contours without edges," IEEE Trans. Image Process. **10**, 266–277 (2001).

[100] S. Jehan-Besson, G. Aubert, and M. Barlaud, "A 3-step algorithm using region-based active contours for video objects detection," EURASIP J. Appl. Signal Process. **6**, 572–581 (2002).

[101] O. Germain and Ph. Réfrégier, "Statistical active grid for segmentation refinement," Pattern Recognition Lett. **22**, 1125 –1132 (2001).

[102] O. Germain, *Segmentation d'images radar: caractérisation de détecteurs de bords et apport des contours actifs.* (Ph.D. Thesis, Université Aix-Marseille III, 2001).

[103] O. Germain and Ph. Réfrégier, "Edge location in SAR images: Performance of the Likelihood Ratio filter and accuracy improvement with an active contour approach," IEEE Trans. Image Process. **10**, 72–78 (2001).

[104] D. J. Williams and M. Shah, "A fast algorithm for active contours and curvature estimation," CVGIP: Image Understanding **55**, 14–26 (1992).

[105] S. Kirkpatrick, S. D. Gelatt, and M. P. Vecchi, "Optimization by simulated annealing," IBM Research Report **RC** (1982).

[106] C. Chesnaud, V. Pagé, and Ph. Réfrégier, "Robustness improvement of the statistically independent region snake-based segmentation method," Opt. Lett. **23**, 488–490 (1998).

[107] R. Deriche, "Using Canny's criteria to derive a recursive implemented optimal edge detector," Int. J. Computer Vision **1**, 167–187 (1987).

[108] F. Guérault, *Techniques statistiques pour l'estimation de la position d'un objet dans des images bruitées.* (Ph.D. Thesis, Université Aix-Marseille III, 1999).

[109] H. Akaike, "A new look at the statitical model identification," IEEE Trans. Autom. Control **19**, 716–723 (1974).

[110] G. Schwartz, "Estimating dimension of a model," Ann. Stat. **9**, 461–464 (1978).

[111] J. Rissanen, "Modeling by shortest data description," Automatica **14**, 465–471 (1978).

[112] J. Rissanen, *Stochastic Complexity in Statistical Inquiry* (World Scientific, Singapore, 1989).

[113] O. Germain and Ph. Réfrégier, "Snake-based method for the segmentation of objects in multichannel images degraded by speckle," Opt. Lett. **24**, 814–816 (1999).

[114] O. Ruch and Ph. Réfrégier, "Minimal-complexity segmentation with a polygonal snake adapted to different optical noise models," Opt. Lett. **41**, 977–979 (2001).

[115] D.B. William, "Counting the degrees of freedom when using AIC and MDL to detect signals," IEEE Trans. Signal Process. **42**, 3282–3284 (1994).

[116] C. E. Shannon, "A mathematical theory of communication," Bell Syst. Tech. J. **27**, 379–423 , 623 – 656 (1948).

[117] T. M. Cover and J. A. Thomas, *Elements of Information Theory* (Wiley-Interscience, New York, 1991).

[118] S. Geman and D. Geman, "Stochastic relaxation, Gibbs distribution and the Bayesian restoration of images," IEEE Trans. Pattern Anal. and Mach. Intell. **6**, 721–741 (1984).

[119] F. S. Cohen and D. B. Cooper, "Simple parallel hierarchical and relaxation algorithms for segmenting noncausal Markovian random fields," IEEE Trans. Pattern Anal. Mach. Intell. **9**, 195–219 (1987).

[120] Y.G. Leclerc, "Constructing simple stable descriptions for image partitioning," Computer Vision **3**, 73–102 (1989).

[121] N. Paragios and R. Deriche, "Geodesic active regions: a new framework to deal with frame partition problems in computer vision," Journal of Visual Communication and Image Representation, Special Issue on Partial Differential Equations in Image Processing, Computer Vision and Computer Graphics **13**, 249–268 (2002).

[122] L. A. Vese and T. Chan, "A multiphase level set framework for image segmentation using the Mumford and Shah model," Int. J. Computer Vision **50**, 271–293 (2002).

[123] F. Galland, N. Bertaux, and Ph. Réfrégier, "Minimum Description Length Synthetic Aperture Radar image segmentation," IEEE Trans. Image Process. **12**, 995–1006 (2003).

[124] J. E. Solomon, "Polarization imaging," Appl. Opt. **20**, 1537–1544 (1981).

[125] W. G. Egan, W. R. Johnson, and V. S. Whitehead, "Terrestrial polarization imagery obtained from the space shuttle: characterization and interpretation," Appl. Opt. **30**, 435–442 (1991).

[126] L. B. Wolff, "Polarization camera for computer vision with a beam splitter," J. Opt. Soc. Am. A **11**, 2935–2945 (1994).

[127] J. S. Tyo, M. P. Rowe, E. N. Pugh, and N. Engheta, "Target detection in optical scattering media by polarization-difference imaging," Appl. Opt. **35**, 1855–1870 (1996).

[128] L. B. Wolff and T. E. Boult, "Constraining object features using a polarization reflectance model," IEEE Trans. Pattern Anal. Mach. Intell. **13**, 635–657 (1991).

[129] K. Koshikawa, "A polarimetric approach to shape understanding of glossy objects," in *Proc. 6th IJCAI*, 493–495 (1979).

[130] B. F. Jones and P. T. Fairney, "Recognition of shiny dielectric objects by analyzing the polarization of reflected light," Image and Vision Computing **7**, 253–258 (1989).

[131] L. B. Wolff, "Polarization vision: a new sensory approach to image understanding," Image and Vision Computing **15**, 81–93 (1997).

[132] Y. Y. Schechner, J. Shamir, and N. Kiryati, "Polarization and statistical analysis of scenes containing a semireflector," J. Opt. Soc. Am. A **17**, 276–284 (2000).

[133] L. B. Wolff, "Polarization-based material classification from specular reflection," IEEE Trans. Pattern Anal. Mach. Intell. **12**, 1059–1071 (1990).

[134] P. Clémenceau, S. Breugnot, and L. Collot, "Polarization diversity imaging," in *Laser Radar Technology and Applications III*, SPIE Proc. **3380**, 284–291 (1998).

[135] M. Floc'h, G. Le Brun, C. Kieleck, J. Cariou, and J. Lotrian, "Polarimetric considerations to optimize lidar detection of immersed targets," Pure Appl. Opt. **7**, 1327–1340 (1998).

[136] S. Breugnot and Ph. Clémenceau, "Modeling and performances of a polarization active imager at lambda=806 nm," in *Laser Radar Technology and Applications IV*, G. W. Kamerman and Ch. Werner, eds., Proc. SPIE **3707**, 449–460 (1999).

[137] A. Gleckler and A. Gelbart, "Multiple-slit streak tube imaging lidar MS-STIL applications," in *Laser Radar Technology and Applications V*, G. W. Kamerman, U. N. Singh, C. H. Werner, and V. V. Molebny, eds., SPIE Proc. **4035**, 266–278 (2000).

[138] S. G. Demos and R. R. Alfano, "Optical polarization imaging," Appl. Opt. **36**, 150–155 (1997).

[139] V. Sankaran, K. Schnenberger, J. T. Walsh, and D. J. Maitland, "Polarization discrimination of coherently propagating light in turbid media," Appl. Opt. **38**, 4252–4261 (1997).

[140] S. Jiao and L. V. Wang, "Two-dimensional depth-resolved Mueller matrix of biological tissue measured with double-beam polarization-sensitive optical coherence tomography," Opt. Lett. **27**, 101–103 (2002).

[141] S. L. Jacques, J. C. Ramella-Roman, and K. Lee, "Imaging skin pathology with polarized light," J. Biomed. Opt. **7**, 329–340 (2002).

[142] J. M. Bueno and P. Artal, "Double-pass imaging polarimetry in the human eye," Opt. Lett. **24**, 64–66 (1999).

[143] J. W. Goodman, "Some first-order properties of light waves," in *Statistical Optics*, pages 116–156 (John Wiley and Sons, Inc., New York, 1985).

[144] N. Wiener, "Generalized harmonic analysis," Acta Math. **55**, 119–260 (1930).

[145] E. Wolf, "Optics in terms of observable quantities," Nuovo Cimento **12**, 884–888 (1954).

[146] L. Mandel and E. Wolf, *Optical Coherence and Quantum Optics* (Cambridge University Press, New York, 1995).

[147] S. Huard, "Polarized optical wave," in *Polarization of Light*, pages 1–35 (Wiley, Masson, Paris, 1997).

[148] C. Brosseau, *Fundamentals of Polarized Light – A Statistical Approach* (John Wiley and Sons, Inc., New York, 1998).

[149] S. Huard, "Propagation of states of polarization in optical devices," in *Polarization of Light*, pages 86–130 (Wiley, Masson, Paris, 1997).

[150] J. L. Pezzaniti and R. A. Chipman, "Mueller matrix imaging polarimetry," Opt. Eng. **34**, 1558–1568 (1995).

[151] R. A. Chipman, "Polarization diversity active imaging," in *Image Reconstruction and Restoration II*, T. J. Schulz, ed., Proc. SPIE **3170**, 68–73 (1997).

[152] B. Johnson, R. Joseph, M. L. Nischan, A. Newbury, J. P. Kerekes, H. T. Barclay, B. C. Willard, and J. J. Zayhowski, "Compact active hyperspectral imaging system for the detection of concealed targets," in *Detection and Remediation Technologies for Mines and Minelike Targets IV*, A. C. Dubey, J. F. Harvey, J. T. Broach, and R. E. Dugan, eds., Proc. SPIE **3710**, 144–153 (1999).

[153] F. Goudail and Ph. Réfrégier, "Target segmentation in active polarimetric images by use of statistical active contours," Appl. Opt. **41**, 874–883 (2002).

[154] F. Goudail and Ph. Réfrégier, "Statistical algorithms for target detection in coherent active polarimetric images," J. Opt. Soc. Am. A **18**, 3049–3060 (2001).

[155] D. C. Schleher, "Radar detection in Weibull clutter," IEEE Trans. Aerosp. Electr. Syst. **12**, 736–743 (1976).

[156] F. Goudail and Ph. Réfrégier, "Statistical techniques for target detection in polarisation diversity images," Opt. Lett. **26**, 644–646 (2001).

[157] A. K. Jain and C. R. Christensen, "Digital processing of images in speckle noise," in *Applications of Speckle Phenomena*, SPIE Proc. **243**, 46–50 (1980).

[158] Ph. Réfrégier and F. Goudail, "Invariant polarimetric contrast parameters for coherent light," J. Opt. Soc. Am. A **19**, 1223–1233 (2002).

[159] R. J. Muirhead, *Aspects of Multivariate Statistical Theory* (John Wiley ans Sons, Inc., New York, 1982).

Index